Modern Methods of Plant Analysis

Editors
H.F. Linskens, Nijmegen/Siena/Amherst
J.F. Jackson, Adelaide

Volume 19

Springer

Berlin
Heidelberg
New York
Barcelona
Budapest
Hong Kong
London
Milan
Paris
Santa Clara
Singapore
Tokyo

Volumes Already Published in this Series:

Volume 1: Cell Components
1985, ISBN 3-540-15822-7

Volume 2: Nuclear Magnetic Resonance
1986, ISBN-3-540-15910-X

Volume 3: Gas Chromatography/Mass Spectrometry
1986, ISBN 3-540-15911-8

Volume 4: Immunology in Plant Sciences
1986, ISBN 3-540-16842-7

Volume 5: High Performance Liquid Chromatography in Plant Sciences
1986, ISBN 3-540-17243-2

Volume 6: Wine Analysis
1988, ISBN 3-540-18819-3

Volume 7: Beer Analysis
1988, ISBN 3-540-18308-6

Volume 8: Analysis of Nonalcoholic Beverages
1988, ISBN 3-540-18820-7

Volume 9: Gases in Plant and Microbial Cells
1989, ISBN 3-540-18821-5

Volume 10: Plant Fibers
1989, ISBN 3-540-18822-3

Volume 11: Physical Methods in Plant Sciences
1990, ISBN 3-540-50332-3

Volume 12: Essential Oils and Waxes
1991, ISBN 3-540-51915-7

Volume 13: Plant Toxin Analysis
1992, ISBN 3-540-52328-6

Volume 14: Seed Analysis
1992, ISBN 3-540-52737-0

Volume 15: Alkaloids
1994, ISBN 3-540-52738-9

Volume 16: Vegetables and Vegetable Products
1994, ISBN 3-540-55843-8

Volume 17: Plant Cell Wall Analysis
1996, ISBN 3-540-59406-X

Volume 18: Fruit Analysis
1995, ISBN 3-540-59118-4

Volume 19: Plant Volatile Analysis
1997, ISBN 3-540-61589-X

Volume 20: Analysis of Wastes of Plant Origin
(In preparation)

Plant Volatile Analysis

Edited by
H.F. Linskens and J.F. Jackson

Contributors

M.S. Allen G.L. Alonso Diaz-Marta R.R. Arnts
M.L. Arrebola C.M. Bignell G. Buchbauer P.J. Dunlop
G.H. Full D.B. Hibbert S. Husted H.B. Jakobsen
L. Jirovetz M. Kawakami M.J. Lacey C.P. Milner
A. Nikiforov I. Pal Singh M.R. Salinas Fernandez
J.K. Schjoerring T.D. Sharkey E.L. Singsaas
G.R. Takeoka R.D. Trengove N. Watanabe

With 85 Figures

Springer

Professor Dr. HANS FERDINAND LINSKENS
Goldberglein 7
D-91056 Erlangen, Germany

Professor Dr. JOHN F. JACKSON
Department of Viticulture, Encology and Horticulture
Waite Agricultural Research Institute
University of Adelaide
Glen Osmond, S.A. 5064
Australia

ISBN 3-540-61589-X Springer-Verlag Berlin Heidelberg New York

The Library of Cogress Card Number 87-659239 (ISSN 0937-8340)

Printed in Germany

Typesetting: Best-set Typesetter Ltd., Hong Kong
SPIN 10527402 31/3137 5 4 3 2 1 0 – Printed on acid-free papier

Introduction

Modern Methods of Plant Analysis

When the handbook *Modern Methods of Plant Analysis*, was first introduced in 1954, the considerations were:

1. the dependence of scientific progress in biology on the improvement of existing and the introduction of new methods;
2. the difficulty in finding many new analytical methods in specialized journals which are normally not accessible to experimental plant biologists;
3. the fact that in the methods sections of papers the description of methods is frequently so compact, or even sometimes so incomplete, that it is difficult to reproduce experiments.

These considerations still stand today.

The series was highly successful, seven volumes appearing between 1956 and 1964. Since there is still today a demand for the old series, the publisher has decided to resume publication of *Modern Methods of Plant Analysis.* It is hoped that the New Series will be just as acceptable to those working in plant sciences and related fields as the early volumes undoubtedly were. It is difficult to single out the major reasons for the success of any publication, but we believe that the methods published in the first series were up-to-date at the time and presented in a way that made description, as applied to plant material, complete in itself with little need to consult other publications.

Contribution authors have attempted to follow these guidelines in this New Series of volumes.

Editorial

The earlier series of *Modern Methods of Plant Analysis* was initiated by Michel V. Tracey, at that time in Rothamsted, later in Sydney, and by the late Karl Paech (1910–1955), at that time at Tübingen. The New Series will be edited by Paech's successor H.F. Linskens (Nijmegen, The Netherlands) and John F. Jackson (Adelaide, South Australia). As were the earlier editors, we are convinced "that there is a real need for a collection of reliable up-to-date methods for plant analysis in large areas of applied biology ranging from agriculture and horticultural experiment stations to pharmaceutical and technical institutes concerned with raw material of plant origin".

The recent developments in the fields of plant biotechnology and genetic engineering make it even more important for workers in the plant sciences to become acquainted with the more sophisticated methods, which sometimes come from biochemistry and biophysics, but which also have been developed in commercial firms, pharmaceutical laboratories, non-university research institutes, and medical establishments.

Concept of the Series

Many methods described in the biochemical, biophysical and medical literature cannot be applied directly to plant material because of the special cell structure, surrounded by a tough cell wall, and the general lack of knowledge of the specific behaviour of plant raw material during extraction procedures. Therefore all authors of this New Series have been chosen because of their special experience with handling plant material, resulting in the adaptation of methods to problems of plant metabolism.

Nevertheless, each particular material from a plant species may require some modification of the described methods and usual techniques. The methods are described critically, with hints as to their limitations. In general it will be possible to adapt the methods described to the specific needs of the users of this series; however, references have been made to the original papers and authors. During the planning of this New Series the editors tried to ensure that the aims and general layout of the contributions are within the general guidelines indicated above, but in addition they tried not to interfere too much with the personal style of each author.

There are several ways of classifying the methods used in modern plant analysis. The first is according to the technological and instrumental progress made over recent years. These aspects were taken into consideration in Volumes 1 to 5 and 11 of this series describing methods in a systematic way according to the basic principles of the methods.

A second classification is according to the plant material that has to undergo analysis. The specific application of the analytical method is determined by the special anatomical, physiological, and biochemical properties of the raw material and the technology used in processing. This classification was used in Volumes 6 to 8, 10, 14, 16 and 18.

A third category is according to the classes of substances present in the plant material and the subject of analytical methods. The latter was used for Volumes 9, 12, 13, 15 and 17 of the series.

Naturally, these three approaches to developments in analytical techniques for plant materials cannot exclude some small overlap and repetition; but careful selection of the authors of individual chapters, according to their expertise and experience with the specific methodological technique, the group of substances to be analyzed, or the plant material which is the subject of chemical and physical analysis, guarantees that recent developments in analytical methodology are described in an optimal way.

Volume Nineteen – Plant Volatile Analysis

Plant volatiles are of considerable interest from many points of view, some for purely aesthetic reasons, others for commercial interest. Aromas from wines and other foods and drinks fit into both categories, analysis of wine aromas being already dealt with in Volume 6 of the series. Nevertheless, we include two chapters on this topic in the present volume as it is of great commercial importance and of wide general interest. One chapter describes adsorption-thermal desorption gas chromatography as applied to the determination of wine aroma, while the second utilises stable isotope dilution GC-MS determination of "green" aroma in wine. The latter is due to methoxypyrazines, and is important to the regional style of Sauvignon blanc wines, while it is disliked in others. It can be overpowering and unpleasant at high concentrations, but at lower concentrations it has an important role in distinguishing wines of the grape variety Sauvignon blanc and Cabernet Sauvignon.

Many plants give off volatiles with rich aromas which are very noticeable to the human nose. A walk in the Australian "bush" soon presents one with the scent of essential oils, such as eucalyptus. While an entire volume in the series, Volume 12, has been devoted to these oils, we include some chapters on this topic here. One such chapter deals with descriptions of an analytical-scale supercritical fluid extraction of essential oils from *Eucalyptus* spp. The traditional method of steam distillation has been shown to cause bicyclogermacrene to undergo reaction and rearrangement, whilst supercritical fluid extraction and vacuum distillation extracts this type of compound intact. It is considered that when extraction of essential oils is for chemotaxonomic purposes, supercritical fluid extraction and vacuum distillation are preferred methods. A chapter on chemometric analysis of data from essential oils is included in this volume also therefore, in view of these improvements. Another chapter on essential oils covers those from *Thymus* species as being especially aromatic. Tea aromas and scented teas and their analysis are also to be found in this volume, both being important commercially and of great human interest.

The volatiles given off by flowers and fruits are of interest from the point of view of insect-plant interaction; this subject is covered through an in situ headspace analysis method. Finally, a method is included for the measurement of isoprene in air, as well as a chapter describing the measurement of ammonia gas emission from plants. The description here focuses on aspects which are relevant for experiments under controlled environment conditions, including atmospheric ammonia concentrations corresponding to those found naturally.

Acknowledgements. The editors would like to express their gratitude to all contributors for their efforts in keeping the production schedules, and to Dr. Dieter Czeschlik and the staff of Springer-Verlag, especially Ms. Isolde Hummel, and Mr. K.-H. Winter for their cooperation in preparing this and other volumes of the series, *Modern Methods of Plant Analysis.*

Adelaide and Nijmegen/Siena/Amherst, Winter 1996.

H.F. LINSKENS
J.F. JACKSON

Contents

List of Contributors

ALLEN M.S., Centre for Grape and Wine Research, Charles Sturt University, P.O. Box 588, Wagga Wagga, NSW 2678, Australia

ALONSO DIAZ-MARTA G.L., Departamento de Ciencia y Technología Agroforestal. Escuela Técnica superior de Ingenieros Agrónomos, Universidad de Castilla-La Mancha, E-02071 Albacete, Spain

ARNTS R.R., National Exposure Research Laboratory, United States Environmental Protection Agency, Research Triangle Park, North Carolina 27711, USA

ARREBOLA M.L., Leiden University, Center for Drug Research, Gorlaeus Laboratories, Einsteinweg 55, NL-2300 RA Leiden

BIGNELL C.M., Department of Chemistry, The University of Adelaide, South Australia, 5005, Australia

BUCHBAUER G., Institute of Pharmaceutical Chemistry, Univ. Vienna, Althanstr. 14, 1090 Vienna, Austria

DUNLOP P.J., Department of Chemistry, The University of Adelaide, South Australia, 5005, Australia

FULL G.H., Agricultural Research Service US Dept. of Agriculture 800 Buchannan Street Albany, California 94710 USA

HIBBERT D.B., Department of Analytical Chemistry, University of New South Wale, Sydney, NSW 2052, Australia

HUSTED S., Plant Nutrition Laboratory, Department of Agricultural Sciences, Royal Veterinary and Agricultural University, Thorvalsensvej 40 DK-1871 Frederiksberg C, Copenhagen, Denmark

JAKOBSEN H.B., Danish Institute of Plant and Soil Science, Department of Food Science and Technology, Kirstinebjergvej 12, DK-5792 Aarslev, Denmark

JIROVETZ L., Institute of Pharmaceutical Chemistry, Univ. Vienna, Althanstr. 14, A-1090 Vienna, Austria

KAWAKAMI M., Shion Junior College 6-11-1 Ohmika-cho, Hitachi, Ibaraki 319-12, Japan

Lacey M.J., CSIRO Division of Entomology, GPO Box 1700, Canberra, ACT 2601, Australia

Milner C.P., The University of Western Australia, Department of Chemistry, Nedlands, Perth, Western Australia 6907, Australia

Nikiforov A., Institute of Organic Chemistry, Univ. Vienna, Währingerstr. 38, A-1090 Vienna, Austria

Pal Singh I., Department of Applied Biological Chemistry, Faculty of Agriculture, Shizuoka University, 836 Ohya, Shizuoka 422, Japan

Salinas Fernandez M.R., Departamento de Ciencia y Technología Agroforestal. Escuela Técnica superior de Ingenieros Agrónomos, Universidad de Castilla-La Mancha, E-02071 Albacete, Spain

Schjoerring J.K., Plant Nutrition Laboratory, Department of Agricultural Sciences, Royal Veterinary and Agricultural University, Thorvaldsensvej 40 DK-1871 Frederiksberg C, Copenhagen, Denmark

Sharkey T.D., Department of Botany, University of Wisconsin, 430 Lincoln Drive, Madison, Wisconsin 53706-1381, USA

Singsaas E.L., Department of Botany, University of Wisconsin, 430 Lincoln Drive, Madison, Wisconsin 53706-1381, USA

Takeoka R.G., Agricultural Research Service, US Dept. of Agriculture, 800 Buchannan Street, Albany, California 94710, USA

Trengove R.D., The University of Western Australia, Department of Chemistry, Nedlands, Perth, Western Australia 6907, Australia

Watanabe N., Department of Applied Biological Chemistry, Faculty of Agriculture, Shizuoka University, 836 Ohya, Shizuoka 422, Japan

The Preisolation Phase of In Situ Headspace Analysis: Methods and Perspectives

H.B. Jakobsen

1 Introduction

Analysis of plant volatiles is a key discipline within several branches of the scientific world, including those concerned with food analysis, perfumery, ecological biochemistry, phytochemistry, and plant physiology.

The analysis of plant volatiles comprises a long array of approaches and techniques, developed to fulfil the specific needs of the sciences mentioned above. It is the aim of most studies to approximate as closely as possible the authentic composition of compounds present in either the tissue or the air surrounding the plant (the headspace). However, several studies have demonstrated that the preisolation methods used during, for example, dynamic headspace sampling of volatiles from plants, themselves influence the composition of sampled volatiles (see Sects. 2.1 and 3). Further, the methods applied in the isolation and postisolation phases may modify the authentic composition significantly (see Sect. 2.1). It is therefore essential to continuously improve the methods in order to minimize their non grata effects during the formation, accumulation, emission, and analysis of volatiles.

This chapter centers on the preisolation phase of dynamic headspace analysis of volatiles emitted in situ. This phase comprises (1) the control of environmental factors influencing the emission of volatiles, (2) noninvasive handling and mounting of the living plant material prior to and during headspace analysis, and finally (3) the intrainment and distribution of volatiles from the headspace chamber to the adsorption columns. The actual trapping and analysis during the isolation phase and the postisolation phases are common to the in situ method and other methods. These extensive subjects will therefore not be considered further in this chapter (see Maarse and Belz 1985; Parliment 1986; Bicchi and Joulain 1990; Dobson 1991; Kaiser 1991).

The in situ headspace analysis is particularly important in studies where the development in the emission profile as a function of time is investigated: Changes in the emission profile caused by detachment of the plant prior to analysis or random fluctuations in environmental factors during the sampling period is difficult to discern from quali- and quantitative effects of treatment parameters, e.g., insect herbivore attack.

In this chapter it is the aim to bring forward a brief evaluation of the approaches used for isolating plant volatiles (Sect. 2). Further, a description is given

Modern Methods of Plant Analysis, Vol. 19
Plant Volatile Analysis
Edited by H.-F. Linskens and J.F. Jackson

of a system developed to control the environmental factors during in situ headspace analysis (Sect. 3). Finally, examples from the research in two of the fields where in situ headspace analysis is used intensively, i.e., biochemical ecology and rhythmic emission of volatiles, are presented (Sect. 4).

2 Basic Approaches to the Isolation of Volatiles

A number of different approaches used in the analysis of plant volatiles have been described in the literature. The methods may by divided into two groups according to the physiological state of the plant tissue during analysis: (1) methods used to analyze volatiles released from intact tissue with the cell membranes intact, and (2) methods applied to analyze volatiles released from disrupted cells, i.e., the composition of volatiles contained in the plant tissue (Fig. 1).

The analysis of emissions from intact tissue is performed by either static or dynamic headspace analysis. Static headspace, where only a small volume of gas is taken from the headspace and injected directly on the GC, is rarely sufficiently sensitive for the analysis of in situ emissions. Therefore only the dynamic headspace technique will be described here. This method may be further subdivided into in situ and ex situ headspace analysis (Fig. 1). In situ headspace analysis is the term used in this chapter to describe the isolation of volatiles from the headspace of intact living plants. The ex situ method is performed on plant material detached from the plant, but is otherwise identical to the in situ method. For the in situ method, the part of the plant to be analyzed, e.g., the flower, is guided into a glass bulb, the headspace chamber, and the space between the petiole and

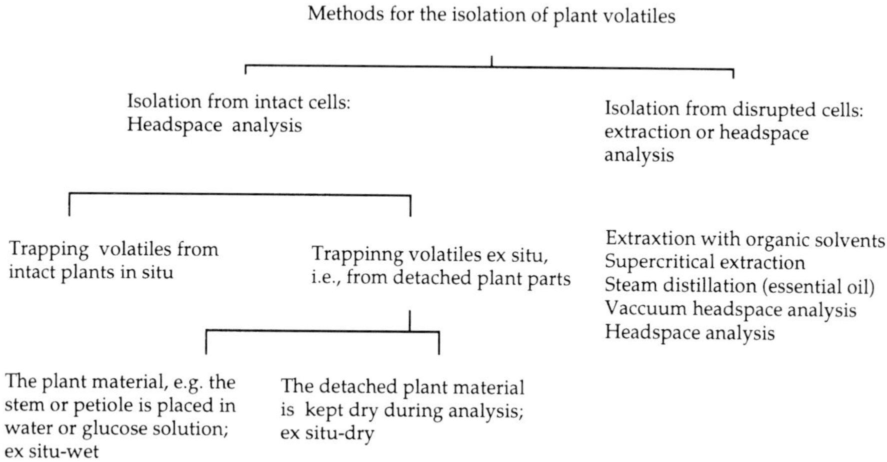

Fig. 1. An overview of approaches used for the isolation of volatiles from plants. The techniques are grouped according to the nature of the plant material to which they are applied

the bulb is sealed more or less airtight. The ex situ analysis may be performed on plant parts supplied with water or glucose solution after detachment (ex situ-wet) or without water supply (ex situ-dry) (Fig. 1). In all methods, the headspace chamber is then purged with clean air, either by suction or pressure, and volatiles are accumulating for a period of time on an adsorbent material at the outlet of the bulb (see Sects. 3.1 and 3.2 for details). Finally, the volatiles are desorbed and analyzed by, for example, gas chromatography-mass spectroscopy (GC-MS).

The methods used for isolation of volatiles from disrupted cells are based on two principles: extraction or headspace analysis. The cells may be disrupted prior to or during the isolation process. The most common methods are (1) extraction with suitable solvents, e.g., methylencloride or hexane (2) hydrodistillation, (3) supercritical extraction, (4) vacuum headspace analysis, or (5) headspace analysis of macerated material.

As shown in the next section, the methods listed above yield very different results when applied to the same plant material. The question one may ask when evaluating and selecting methods for isolating volatiles is: what do we measure using the different methods? Is it the actual content of volatiles in the tissue or headspace, or is the authentic composition to some degree modified or hidden in the effects of the analytical methods? The fact that the volatile composition is strongly influenced by the methods used for isolating the volatiles calls for a careful evaluation of the methodological processes.

2.1 Comparison of Methods: What Do We Measure?

In several investigations, two or more methods have been compared for their capacity to isolate volatiles originating from the same plant material. In many cases, the methodological impact on the volatile composition seems to depend on the plant species analyzed. Therefore, the plant species have been included in the comparison of methods in the following section.

2.1.1 The In Situ Contra the Ex Situ-Dry Methods

In a considerable percentage of the papers published on headspace analysis of plant volatiles, the ex situ-dry method is applied. This may be a suitable method within the perfumery and food sciences; however, according to the studies presented below, this method may produce results quite different from those of the in situ method. Significant qualitative and quantitative differences in the headspace composition when using the in situ contra the ex situ-dry methods have been detected in flowers of *Jasminum grandiflorum*, freesia, lilac, easter lily (Mookherjee et al. 1987), *Rosa damascena*, *Narcissus tazetta*, *Osmanthus fragrans* (Mookherjee et al. 1989), the semidried foliage of *Mentha spicata* (Mookherjee et al. 1989), the leaves of *Coriandrum sativum* (Mookherjee et al. 1990), and fruits of *Prunus persica* (peach) and *Fragaria* × *ananassa* (strawberry) (Mookherjee et al.

1987). It is important for the evaluation of these results to note that the period of collection was 6–12 h, except for *Mentha spicata*, which was dried prior to analysis. During this period, the flowers and leaves are likely to gradually dehydrate, causing serious physiological changes which may be detected in the headspace composition. However, comparing the same methods and using a sampling time of only 2–4 h, Brunke et al. (1993) detected significant differences in the composition of volatiles emitted from *Valeriana officinalis*. Kaiser (1991) noted that the ex situ-dry method may be used for collection of floral volatiles in species which display the original fragrance profile 0.5–2 h after cutting. The emission rate of volatiles appears to decrease drastically in some species even when the detached plant part receives water and glucose (see Sect. 2.1.2).

Loper and Lapioli (1971) investigated the rhythmic emission of ocimene from flowers of *Medicago sativa* by detaching flowers in 3-h intervals, and placing them in a syringe for 30 min for accumulation of volatiles, prior to injection on the GC. In this way, they demonstrated rhythmic emission of volatiles even after detachment from the plant. On the other hand, Loughrin et al. (1994) showed that rhythmic emission of volatiles was induced by herbivore damage in leaves of *Gossypium hirsutum* in situ. Although detaching a plant part when using the ex-situ headspace method is different from the damage caused by herbivore attack, this interesting study highlights the importance of checking whether rhythmic emissions observed in detached plant material are solely normal physiological phenomena, or if the mechanical damage inforced on the plant during detachment induces or influences the rhythm. The observations in this section together raise the question: to what degree is the composition of volatiles and rhythmic phenomena influenced by the ex situ-dry method? This is an important question to ask (and answer) prior to using this method as a basis for deductions in the fields of biochemical ecology and plant physiology. It is therefore suggested that the results using the in situ are compared to those of the ex situ method before applying the latter method in the experimental phase.

2.1.2 The In Situ Contra the Ex Situ-Wet Methods

Comparing the in situ method with the ex situ-wet, i.e., placing the detached petiole of the plant part to be analyzed in sugar solution or water, has received only little attention in the literature; however, some studies indicate that the effects of the ex situ-wet method vary with the plant species used in the experiment. Kaiser (1991), comparing the in situ and ex situ-wet methods, found that the flower of the hybrid tea rose continued to emit *cis*-3-hexenyl acetate, *n*-hexyl acetate, and *trans*-2-hexenyl acetate in a diurnal rhythm, for the first two periods following detachment. Further, the total amount of emitted fragrance was the same in situ and ex situ. On the other hand, in *Hoya carnosa*, cut flowers in 0.05 M glucose showed such a dramatic loss in fragrance production following detachment that the normal rhythmic nature of emission observed in flowers attached to the plant could no longer be observed (Matile and Altenburger 1988). In contrast, rhythmicity could be observed in the same study in *Odontoglossum constrictum* and *Citrus medica* for several consecutive cycles when the ex situ-wet method was applied. Relative composition of volatiles comparing cut and attached flowers was not reported. In *Trifolium repens* we have observed that the diurnal rhythm of emission in situ continues for at least 2 days in detached flowers with the petiole in deionized water (Fig. 2). The rate of emission was comparable to that of the in situ method. The inconsistent effects of this method on the emission of volatiles, calls for similar precautions as mentioned for the ex situ-dry method above (Sect. 2.1.1).

2.1.3 Headspace Analysis Contra Extraction

A series of comparisons of the headspace composition of "intact" plant material (ex- or in situ) and different extraction methods applied to the same material has been performed. The following examples in all cases showed significant differences in the composition of volatiles: ethanol extract of *Citrus medica* petals contra the headspace composition (Altenburger and Matile 1990); in situ headspace, ether extract, and the hydrodistillate of *Malus silvestris* flowers (Buchbauer et al. 1993b); in situ headspace contra hydrodistillation in *Chamomilla recutita*, *Chamaemelum nobile*, and *Santolina chamaecyparissus* (Brunke et al. 1993); in *Valeriana officinalis* comparing in- and ex situ headspace with hydrodistillation and vacuum headspace analysis (Brunke et al. 1993); in situ headspace contra hydrodistillation in *Exacum affine* (Buchbauer et al. 1994), and *Plectranthus coleoides* (Buchbauer et al. 1993a); in situ contra vacuum headspace in the flowers of *Convallaria majalis*, *Hoya carnosa*, *Phlox paniculata*, *Syringia vulgaris* hybrids, *Narcissus pseudonarcissus*, and the rose cultivar Fragrant Cloud (Surburg et al. 1992); ex situ-dry headspace contra simultaneous distillation and solvent extraction in *Actinidia chinensis* (Tatsuka et al. 1990); vacuum headspace contra the commercial essential oil of *Chamomilla recutita* (Surburg et al. 1992). Finally, significant differences were observed comparing supercritical carbon dioxide extraction with commercial

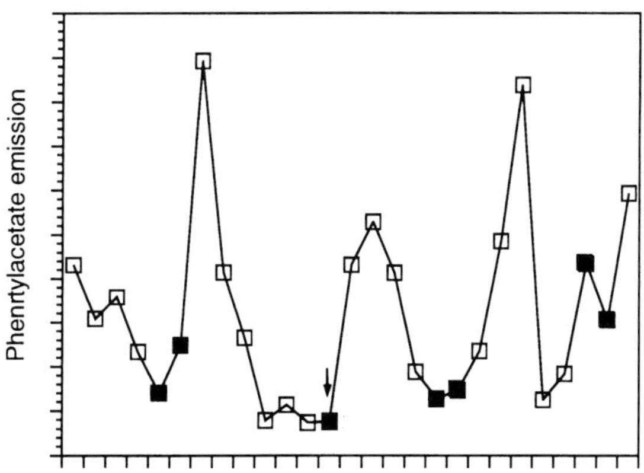

Fig. 2. Rhythmic floral emission of phenetylacetate from *Trifolium repens* prior to and after cutting the flower head from the plant, and placing it in water. *Arrow* Time of detachment. *Shaded squares* collection in darkness

essential oil and vacuum headspace in marjoram leaves (Surburg et al. 1992). The methods disrupting the tissue during analysis reflect the composition of volatiles in the tissue, which seems to be very different from that of the headspace (Altengburger and Matile 1990; Surburg et al. 1992).

It is difficult to evaluate which of the very different compositions represented by each method reflects most precisely the authentic composition of either the headspace or the extract. A complementary way to estimate the suitability of the methods is olfactory comparison of the emissions from the plant material prior to analysis with those of the isolated sample: in *Exacum affine* the in situ headspace sample was evaluated by perfumers together with the essential oil extracted by hydrodistillation (Buchbauer et al. 1994). The headspace concentrate possessed a sensoric quality closer to that of the flower compared to the hydrodistillate (Buchbauer et al. 1994). The in situ headspace, ether extract, and the hydrodistillate of *Malus silvestris* flowers were compared, and only the headspace sample of the living plant possessed the typical apple flower fragrance (Buchbauer et al. 1993b). The olfactory sensation of the product achieved from the vacuum headspace method is often described by the perfumer as "closer to nature" compared to the dynamic headspace method (Surburg et al. 1992). In comparison, the product from the dynamic headspace technique is described as comparatively flat, atypical, and less expressive. The difference may be traced to a relatively higher accumulation of lower boiling compounds in the dynamic headspace method in contrast to the higher yields of high boiling compounds in the vacuum headspace method (Surburg et al. 1992). However, the difference in olfactory sensation comparing the two products has been suggested to be caused by the differences in the very different principles of volatile emission observed from living plant tissue contra the evaporation from a smelling blotter, used by the perfumer (Surburg et al. 1992). When placed on the blotter, the evaporation of less volatile compounds is suggested to be relatively lower than the "active" emission from living plant tissue, which is thought to be supported by the emission of water vapor as a kind of aerosol effect (Surburg et al. 1992). It remains an open question whether the vacuum headspace extract comprises an overrepresentation of high boiling volatiles compensating for the relatively lower evaporation rate from the smelling blotter, thus approaching the olfactory impression of the living flower more closely than that of the dynamic headspace technique. Surburg et al. (1992) suggest that the organoleptic properties of the sample obtained by dynamic headspace should be evaluated by spraying it into the air as an aerosol rather than using a smelling blotter.

Another uncertainty using the listed methods is the formation of artifacts during sampling. The artifact may be formed on certain absorbents, for example by oxidation (Surburg et al. 1992; Brunke et al. 1994). During vacuum headspace analysis the cells are disrupted and enzymatically induced formation of artifacts has been reported (Surburg et al. 1992). In fact, all the listed methods have been reported to produce artifacts (Baigrie et al. 1985; Surburg et al. 1992; Brunke et al. 1993, 1994).

Based on the investigations and considerations listed in this section, it seems safe to conclude that cutting the flower from the plant and leaving it under dry conditions for dynamic headspace analysis may change the emission significantly. The degree of transformation seems to increase with time. The change becomes apparent, with the present level of sensitivity in GC and MS, soon after detachment in some species and somewhat later in others. Therefore the ex situ headspace method using no water supply cannot generally be recommended for use within the fields of chemical ecology and plant physiology, e.g., studies on insect-plant relationships or the phenomena of rhythmic emission of volatiles. Whether the ex situ-wet method suing water or water-glucose supply may be used seems again to be species-dependent. Considering the drastic effects on the headspace composition in some species using any of the ex situ methods, it is recommended, for the purpose of approximating the authentic emission from the plant, to use the in situ method. In order to detect artifacts formed at the adsorbent during the dynamic in situ headspace sampling, several parallel methods, e.g., vacuum headspace and a number of different absorbents (Brunke et al. 1994) should be employed. The golden rule: using a combination of parallel methods increases the possibilities for detecting errors produced by any one method is also highly relevant in the field of volatile analysis.

So it appears that the question stated in the header of this section, i.e., "what do we measure?", is a difficult one to answer. The emission of volatiles may be influenced both by detaching the plant and by environmental factors (see Sect. 3.2). Therefore, the development of an in situ headspace system for controlling the factors which may influence the emission of volatiles seems to be a precondition for creating the answer to our question. The description of such a system in detail is the subject of the next section.

3 A Modern Method for Controlling Environmental Factors and the Sequential Sampling of Plant Volatiles Emitted In Situ

The parameters which we attempt to record and control are: temperature, light intensity, photoperiod, relative air humidity (RH), air pressure, air flow rate, inlet air purity, and distribution of the headspace volatiles to a series of columns (Fig. 3). It is important to control these parameters in order to create standard conditions for the experiment. Standard conditions, inforcing a minimum of change in the growth conditions of the plant, forms a basis for detecting the effects of one parameter at a time on the changes in the response variable, for example the progressing changes in the emission of volatiles from the plant as a function of insect damage. If conditions are not kept constant during the detection of changes in the emission of volatiles in time, it will be difficult to determine which changes were induced by the insect damage and which were caused by fluctuations in the environment. Alternatively, in some cases, comparisons of damaged and undamaged plants may be used, but this calls for an exact knowledge of differences in emission composition and responses among plants. This becomes very complicated if the bias between several parameters has to be determined.

The following description refers to the method used in our lab, unless otherwise specified.

3.1 The In Situ Headspace System in Overview

The system described in this section is usually placed inside a phytotrone. The parameters controlled in the headspace collection chamber are also controlled in the phytotrone. In the following text, the numbers in brackets refer to those plotted in Fig. 3.

Basically, a vacuum pump (25) creates a vacuum in the flow path of the apparatus. The speed of the pump is continuously regulated by a mass flow meter (20). The air moves from the inlet (1) through a flow sensor (2) and a charcoal filter (3) via either the dry or wet flow line to the headspace chamber (13). This route is termed the upstream flow path. The air flow entrains the volatiles emitted in the headspace chamber to the downstream path, i.e., via the mass flow meter to the adsorption traps (22).

The information received from the various sensors, e.g., the flow sensor (20), is transferred via the transformer (26) to a PC i/o board for processing in a control program developed at this laboratory. The program, running under Windows, uses the data information for two purposes: (1) the data are compared with the setpoint of various parameters, such as the air flow rate, and the program then adjusts the speed of, in this example, the air pump. The parameters are received and adjustments take place every 0.1 s. (2) The data of all current readouts are transferred to a data file every second. In this way, the level of all parameters can be checked during extended sampling experiments, e.g., 72 h in experiments with rhythmic emission of volatiles. Further, the current setpoints and readouts of all parameters can be observed in a series of real time windows while the sampling is proceeding. Prior to sampling, all setpoints for the environmental parameters and the time intervals for each sampling column are written in the data acquisition window of the program.

3.2 The In Situ Headspace System in Detail

3.2.1 The Headspace Chamber

The chamber comprises the heart of this system. Here the plant part to be analyzed is positioned during the experiment. The function of this unit is to provide an environment similar to that prevailing for the remainder of the plant in the phytotrone outside the chamber. However, in the present setup, two factors are

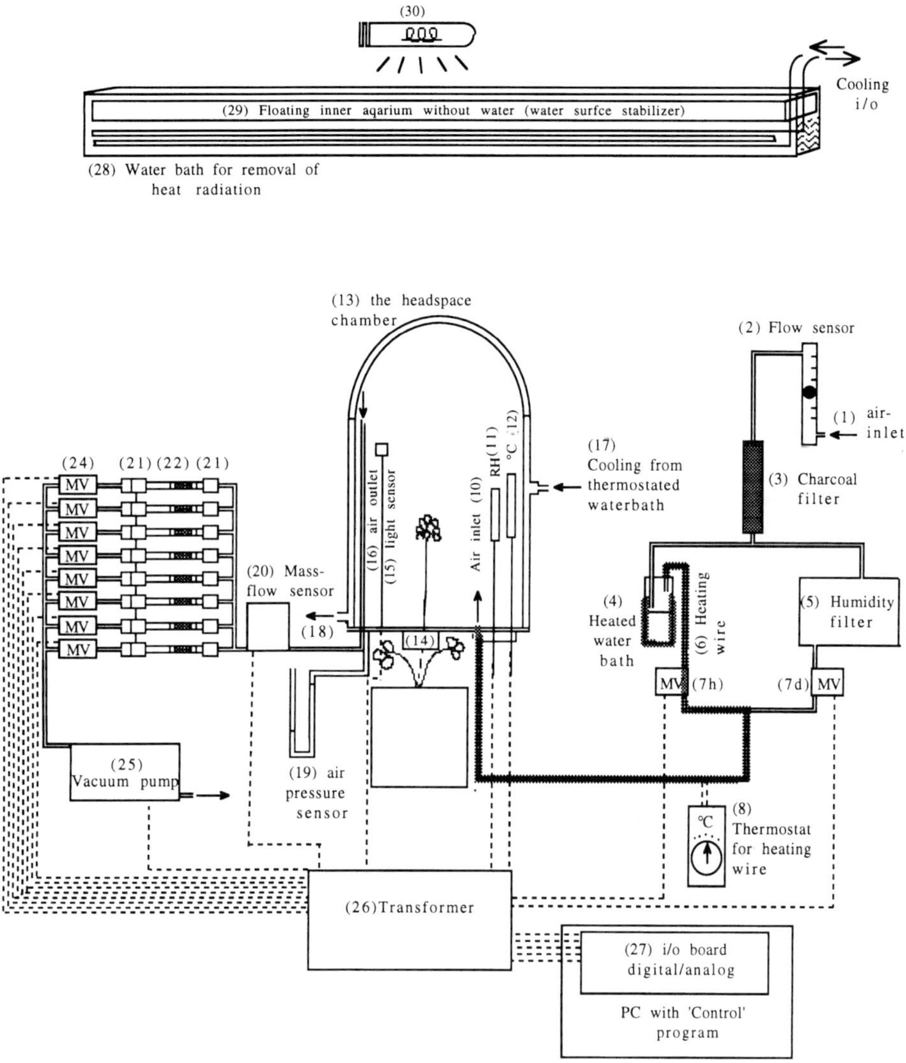

Fig. 3. Device for the collection of volatiles emitted from plants in situ. See Section 3.2 for details. *MV* Magnetic valves; (*18*) cooling water outlet; (*21*) Omnifit Teflon fittings; (*22*) adsorbtion columns; (*30*) light bulbs

different inside the chamber compared to outside: the air purging the headspace chamber has been purified for contaminating hydrocarbons, and there is a slight vacuum in the chamber, in the present setup 540 pa.

Depending on the size of the object to be placed in the chamber, we use either a double-walled 1.6-l or 5-l reaction vessel turned upside down (13). The "lid" has a central size 34/35 socket used for insertion of the plant. It is surrounded by three 24/19 sockets used for mounting sensors and flowlines. Solid Teflon stoppers are used in all sockets. Teflon tubing (3.2 mm OD) is connected to the stoppers through holes formed by precision drilling. The double walls of the vessel allow water cooling of the chamber (see details in Sect. 3.2.9).

Several different collection chambers for isolation of volatiles emitted in situ have been described in the literature. These include Plexiglas chambers with water cooling along the sides (Matile and Altenburger 1988), reaction vessels with water cooling (Jakobsen and Olsen 1994), inverted glass jars without cooling (Patt et al. 1988; Loughrin et al. 1990), and bags made of either polyvinyl flouride (Pham-Delegue et al. 1989), polyethylene (Kawano et al. 1995) or polyacetate (Tollsten and Övstedal 1994). Further, much larger chambers have been used for collection of volatiles from the entire plant including the roots (Loughrin et al. 1991; Blaakmeer et al. 1994). For the purpose of in situ studies, it is essential that the material surrounding the plant does not emit any volatiles, is inactive, and allows the light of the phytotrone to enter without much reduction in the wavelengths normally associated with plant growth. Further, the temperature of the chamber should be regulated in order to compensate for heating of the unit, which is essentially a small greenhouse. Finally, the chamber must provide sufficient space in order to keep the plant from touching the walls of the chamber. The bags of different materials mentioned above are handy for field studies. It is, however, difficult to evaluate to what degree the emission is influenced by the changes in temperature, humidity, light, and physical contact inforced by the bags.

Another often applied method described in the literature is likewise particularly suitable for field investigations. A funnel or a similar open glass container or bag is placed over the plant, and air is drawn via the headspace of the flower to the adsorbent by a vacuum pump (Bergström and Bergström 1989; Tollsten and Bergström 1989; Erhardt and Jaggi 1995). Subsequently, a blind sample is collected a little distance from the flower. This method is suitable for many experiments, because the container readily allows equalization of differences in temperature and RH, etc. with the environments. However, difficulties may occur if some compounds are emitted both by the flower to be examined and by other organisms in the area. Further, quantification of the emission, using this method, is not likely to be accurate due to the open nature of the system.

In conclusion, the headspace chamber design depends on the purpose of, and conditions for the study. In the case of long-lasting in situ experiments, it is suggested that a design which allows control of the environmental parameters is used.

3.2.2 Guiding the Plant into the Headspace Chamber

The positioning of the plant part in the vessel is a critical process. The combination of attempting to seal the space around the petiole airtight and at the same time to make sure not to damage the plant is a difficult task, particularly when handling fragile plant material. If the plant tissue is damaged during this handling, drastic changes in the emission are likely to take place (see Sects. 2.1 and 4.1). In order to allow the plant to recover from the invasive handling during the mounting described below, the flower- or leaf petiole is mounted a few days prior to initiation of the experiment. In the case of *Trifolium repens*, which is the standard plant of our laboratory, the flowerbud is guided through the socket of the lid prior to blooming in order to avoid changes in the fragrance emission induced by accidental pollination (Tollsten and Bergström 1989). The headspace chamber is mounted on the lid the day prior to sampling. The plant part to be tested is guided through the a size 34/35 socket in the lid. We make a groove in the side of a solid Teflon stopper a few millimeter larger than the diameter of the petiole or stem. The space between the groove of the Teflon stopper and the grind is sealed airtight with Vaseline (standard quality without fragrance, needless to say). The Vaseline can seal airtight only if (1) the space between the groove and the petiole allow space for the Vaseline to cover the petiole completely, and (2) the distance from the petiole to the groove does not exceed a few mm. The air tightness of the seal may be tested by placing a few mm of water on top of the Teflon stopper and then simulating a sampling routine setting the system under vacuum. A leak in the seal will be indicated by the air bubbles passing through the water layer. We have tested Vaseline together with a range of other materials at temperatures up to 25 °C, and Vaseline has been chosen for the following reasons: (1) it comprises compounds which are compatible with, and similar to, the wax or cuticle covering the plant surface; (2) it contains no compound likely to influence the metabolism of plants – and it has not shown any immediate or long-term effects on the plants we have tested; (3) it seals airtight under the circumstances used in our setup, and (4) it emits only minute amounts of a few and easily recognizable compounds to the headspace; (5) these compounds do not contribute to the olfactory quality of the sampled volatiles. This is important for the olfactory evaluation of the sample (see Sect. 2.1.3). Many ideas and a good portion of fantasy have resulted in an impressive range of interesting ways to seal the space between the plant stem or petiole and the socket or stopper more or less airtight. The materials applied include cottonwool (Brunke et al. 1992, 1993) a piece of cloth (Borg-Karlson et al. 1985; Burger et al. 1988), aluminum foil gaskets (Raguso and Pichersky 1995), filter paper (Brunke et al. 1992), parafilm (Loughrin et al. 1990), Terrostat (Matile and Altenburger 1988), low temperature melting paraffin (Jakobsen and Olsen 1994), etc. The best choice again depends on the conditions prevailing and the purpose of the study. Vaseline is not likely to be suitable under windy field conditions or at very high temperatures. Under these conditions, Terrostat (Matile and Altenburger 1988; Terosan, Heidelberg, FRG) is a very reliable sealant, although the compounds comprising this product or their effects on the formation and emission of volatiles are unknown. However, we have not observed any significant effect on the plants even after using Terrostat for extended periods of time (Jakobsen and Olsen 1994).

3.2.3 Preparing the System for Collection of Volatiles

Prior to sampling, all glassware is cleansed in hot water, then in distilled water, and finally dried in an oven. Subsequently, glassware is rinsed with distilled pentane and dried in the fume hood. In a normal in situ headspace experiment, the system is assembled without plant material in the chamber, using another lid. Collection takes place for 3 h in order to detect impurities in the system. A piece of the packing material used for sealing the petiole is placed on the stopper in the chamber during collection. The plant material is removed from the chamber after the last sampling and the test for impurities is repeated after 30 min of initial purging. The solvents, the columns with trapping material, the syringe, and the internal standard solutions are tested for impurities prior to use.

When the system is assembled with the plant in place, a test run is performed in order to adjust the temperature, RH, etc. and to detect any leaks in the system. The actual sampling is usually initiated at the same time as the photoperiod starts.

3.2.4 Flow Sensing at the Air Inlet and Air Purification

The air of the phytotrone enters through an air flow sensor (Fig. 3: 1,2). A drop in the flow rate through this sensor indicates leaks in the system between the mass flow sensor (20) and the inlet, if the former indicates that the flow is correct. If the mass flow sensor at the same time indicates a drop in the flow rate, the leak is located between the pump and the mass flow sensor. The control of the air flow level is explained below in Section 3.2.6. The incoming air is initially purified for contaminating volatiles by passing it through a cylindrical steel container with Swagelok in/outs (1) containing 250 ml of minute activated charcoal granules (ML gas puritube, Microlab, Aarhus DK). The filter is heated to 250 °C in an oven, while purged with nitrogen gas for 12 h prior to use. In most cases, this filter is sufficient, but further filters containing porous polymer adsorbent are connected in the upstream flow line when necessary (see Heath and Manukian 1992 for ways to purify inlet air).

3.2.5 Control of Relative Air Humidity in the Headspace Chamber

After having passed the charcoal filter, the air is divided via the Teflon tubes to one of two paths: either through a filter of silica gel (5) in order to remove water vapor (5), or through the humidifier path (4). The air is humidified by passing it through the headspace of a gas washer bottle with distilled water. The bottle and the tubing between the humidifier and the glass bulb are heated by a resistance heating wire (6) in order to prevent condensation before the air enters the headspace chamber. The temperature of the tube and bath is regulated by a thermostat (8) which is set between 20 and 50 °C depending on the conditions. The RH is measured in the glass bulb by an electronic humidity sensor (11) (Vaisala HMM 20D, F) mounted in a Teflon stopper. The sensor readout is passed to the i/o board (27) via the transformer (26). The control program uses the RH readout continuously to open or close the magnetic valves (MV) connected to the humid air (7h) and dry air (7d) source. Only one valve is open at a time. By this method, the RH has been controlled with an accuracy of ±2% RH in the 40 to 80% RH range.

Transpiration from the plant contributes to the water vapor content in the headspace chamber. This process is particularly significant for the RH control when leaves such as those of *Brassica napus* are analyzed. Under such circumstances, the problem is not only to regulate the RH by selecting between the dry or wet air source, but rather to remove the water vapors quickly enough to keep the RH from increasing above the setpoint of, for example, 65% at 20 °C. In such cases, the air flow rate must be raised to levels above 400–500 ml min^{-1} (see Sect. 3.2.6).

The system presented here has a “open” flow path, i.e., the air enters via the inlet and is discarded after passing the pump. Another system which has been used in several studies is based on the “closed-loop” principle (Brunke et al. 1992). Here the outlet of the pump (25) is connected to the inlet of the upstream flow path (1). One of the advantages of this principle is that contaminants are kept from continuously entering the flow path if the system is air-tight. However, in a closed system, the concentration level of compounds like carbon dioxide, oxygen, and ethylene, which show low or no retention on the adsorbents, will theoretically become different from that outside the headspace chamber, unless extensive exchange by diffusion takes place. The concentration level of these compounds strongly influences the metabolism of plants. It is suggested that these considerations should be taken into account or checked when using the closed loop method for in situ studies.

3.2.6 Mass Flow and Vacuum Control

In experiments involving a series of sequential samplings, it is inevitable to use a number of different columns. When using packed polymer or charcoal absorbent, differences in air flow resistance among columns usually appear. The resulting differences in flow rates may again cause errors when comparing the composition of smaples. It is not realistic to adjust the flow rate manually every time a new column is switched on, especially during studies of rhythmic emission of volatiles, where switching is performed between up to 24 columns in a 24-h period. We therefore inserted a mass flow sensor in the downstream line (20) between the glass bulb and the columns. The flow measured by the mass flow meter is used to adjust the speed of the air pump with an analog output signal in order to ensure a constant flow. The difference in pump speed, however, influences the size of the vacuum upstream from the pump. Therefore, the present setup has been designed to minimize the air resistance in the part of the flow path between the inlet (1) and the columns (22). This means that an increase in the speed of the pump, caused by an increased resistance going from one column to the next, causes an increase in vacuum downstream from this column, but no change in the vacuum can be observed upstream due to the low resistance in the flow path. The part of the plant placed in the chamber will be exposed to a different pressure than the remainder of the plant. This naturally has an, at least theoretical, impact on the physiology of the plant. As described in Section 3.2.10, we are presently attempting to remove the vacuum.

In most of our studies a flow rate between 200 and 400 ml min^{-1} is used; however, as mentioned in the previous section, the flow rate may in some cases be higher in order to remove water vapors emitted by the plant sufficiently fast in order to keep the RH steady.

The number of ml air purged through the system per min varies considerably in the literature, from 50 ml min^{-1} to 1000 ml min^{-1} (Raguso and Pichersky 1995) and more. The optimum flow rate depends on several parameters, including the trap temperature, retention capacity of the trapping material, and the volume of the collection chamber.

3.2.7 Trapping on Several Columns in Sequence

The columns used for trapping volatiles are mounted parallel in the flow path (21,22) using Omnifit Teflon fittings. This makes several samplings in sequence over an extended period of time possible. The control program opens one magnetic valve (24) at a time and at the same time shuts the valve opened during the previous sampling.

We use several different adsorbents, alone or in combination, in the traps. An example is Porapack Q 50–80 mesh (Waters Inc., Milford, Mass, USA) mounted between plugs of silanized glass wool in glass tubes (4 mm i.d., length 180 mm). As mentioned previously, the art of volatile trapping will not be covered in this chapter.

3.2.8 Light: Control and Effects

Light is supplied by Phillips HPI-T 400 W light bulbs (30), suiable when used as the only light source. An aquarium (28) with distilled water is placed between the light source and the collection chamber in order to remove heat radiation from the light bulbs. An empty "aquarium" (29) which fits inside the large aquarium floats on the water surface in order to avoid reflections in the water surface and at the same time limit the evaporation of water. The aquarium is cooled by a slow stream of water running through a copper pipe placed along the inside walls. A one-channel clock switches the light source on and off. The light intensity may be regulated by applying more lamps or placing thin cotton curtains below the aquarium (Jakobsen and Olsen 1994). Generally, we apply 400–700 µmol quanta m^{-2} s^{-1}.

Light intensity has been shown to affect the rate of volatile emission in situ from flowers of *Trifolium repens* (Jakobsen and Olsen 1994) and leaves of *Quercus virginiana* (Tingey et al. 1979) and *Phaseolus lunatus* (Lima bean; Takabayashi et al. 1994); see also recent review by Charron et al. (1995).

3.2.9 Air Temperature: Control and Effects

The air temperature in the glass bulb is kept constant by passing water through the space between the double walls of the headspace chamber. The cooling water is currently supplied by a thermostat-regulated water bath, which is adjusted according to the air temperature of the phytotrone. The water is continuously circulated between the water bath and the chamber (17–18). The water level between the chamber double walls is indicated at the figure (13).

Temperature has been shown to influence the in situ emission of isoprene from leaves of *Ouercus virginiana* (Tingey et al. 1979), the monoterpene in situ emission from the vegetative parts of *Pinus elliottii* (Tingey et al. 1980) and *Pinus radiata* (Juuti et al. 1990). Further, temperature influences the in situ emission of volatiles from flowers of *Trifolium repens* (Jakobsen and Olsen 1994), *Ribes nigrum* (Hansted et al. 1994), and *Hcsperis matronalis* (Nielsen et al. 1995; see also recent review by Charron et al. 1995).

3.2.10 Other Parameters and Potential Improvements

Water stress has been shown to influence the emission from leaves of *Phaseolus lunatus* (Lima bean; Takabayashi et al. 1994). Further, *Malus domestica* leaves showed significant changes in the emission as a result of drought stress, using ex situ-dry headspace analysis (Ebel et al. 1995). Plants should therefore be kept at a constant soil humidity. We water the plants every day manually, but we are attempting to develop a more accurate system based on continuous sensing of the moisture in the soil. Other improvements in the in situ system described above include programmed temperature regulation of the water cooling or heating of the headspace chamber. This would allow the introduction of lower night temperatures, thus decreasing respiration, which would be a more accurate simulation of "natural" conditions. Finally, we are working on the control of a second pump placed at the inlet (1). The idea is to equalize the pressure in the headspace chamber.

4 In Situ Headspace Analysis: Areas of Application

In the following section, examples from areas where in situ headspace analysis is fundamental are presented.

Common to both examples is the need to follow the changes taking place in the emission of volatiles through time. Under such circumstances, it seems absolutely essential to keep the plant intact in order to avoid bias between the response of detaching the plant, the effects of random flux in the environment, and the effects of the variable one is actually attempting to follow or investigate.

4.1 Allelochemicals in the Plant-Herbivore-Carnivore Interaction

Volatiles emitted from the plant affect a range of organisms in the plant sphere (Metcalf 1987). These semiochemicals have in some species been shown to include allelochemicals mediating a tritrophic interaction between the plant (first tropic level), the herbivore insect (second tropic level), and the parasitoids and predators of the herbivores (third tropic level). The emission of infochemicals attracting parasites or predators of the herbivores is termed indirect defence (see reviews by Vet and Dicke 1992; Dicke 1994). This group of volatiles, acting as synomones in the plant-carnivore relationship, is emitted after the herbivore has attacked the plant. These volatiles may also be active, as are other volatiles emitted from the plant, during mechanical damage or insect attack, in the direct defence of the plant, i.e., the defence aimed directly at the invader rather than the indirect defence, acting via the carnivores. The plant-carnivore synomones appear to be emitted systemically in several species (see review by Dicke 1994). Substantial progress in the research concerned with emission of volatiles from plants following herbivore damage has taken place during the past decade. An interesting example concerning the tritrophic system was reported by Turlings et al. (1990) studying the emission from corn seedlings during and after attack by the armyworm larva, *Spodoptera exigua*. An increase in the emission of 6-carbon alcohols, aldehydes, and acetates was recorded in the hours following the initial attack together with minor amounts of the terpenoids α-*trans*-bergamotene, (E)-β-farnesene, and (E)-nerolidol. However, the following day, a dramatic increase in the emission rate of the terpenoids could be detected. In contrast, seedlings damaged with razor blades in order to simulate caterpillars was shown to emit these compounds at far slower rates. Artificially damaged leaves treated with larval regurgitant showed emission rates comparable to those of infected leaves. Finally, undamaged leaves with regurgigant emitted only trace amounts. Females of the parasitic wasps *Cotesia marginiventris* showed a clear preference for leaves with larval damage. Further, artificially damaged leaves treated with regurgitant were preferred to those without regurgitant.

Although the C6 compounds released quickly after wounding by insects or by artificial means may attract the enemies of herbivores, this effect has been reported to decrease quickly for the parasitoid *Cotesia glomerata* in *Brassica oleracea* cv. Titurel (Brussels sprouts), unless mechanical wounding was repeated at intervals (Steinberg et al. 1993). The plants in this bioassay were tested in situ. Mattiacci et al. (1994) showed that the same parasitoid was more intensely attracted to leaves of *Brassica oleracea* var. *gemmifera* damaged by *Pieris brassicae* caterpillars than to those wounded mechanically. The latter treatment, however, attracted significantly more parasitoids the day after wounding than did undamaged leaves. The absolute release of terpenoids was unaffected whether the leaf damage was performed mechanically or by the herbivore, whereas the emission of C6 compounds increased dramatically after wounding. The leaves in this experiment were analyzed by the ex situ-wet headspace technique (Sect. 2.1.2).

Blaakmeer et al. (1994) investigated the volatile composition in the same cabbage cultivar as Mattiacci et al. (1994) and observed an emission from *P.*

brassicae-infested plants which was significantly different to that observed by Mattiacci et al. (1994). The percentage of (Z)-3-hexen-1-ol was higher and the percentage of terpenoids lower in the headspace emission. The effect of cutting the petiole is one of the potential causes for the differences observed (Mattiacci et al. 1994).

These investigations are merely examples of a long array of interesting studies of allelochemicals from the headspace of plants and herbivores (see recent review by Charron et al. 1995). It is noted that the ex situ-wet and ex situ-dry methods are used in a great proportion of the studies within this field. It should be taken into consideration that the ex situ method, in some cases (under reference to the remarks in Sects. 2.1.1 and 2.1.2), makes the conclusive phase more complicated if the effects of detaching the leaf on the emission have not been determined prior to the experiment. This, however, is an extensive challenge considering that the bias between the effect of detachment and the other variables in the experiment, e.g., the progressive development in the emission composition following wounding by herbivores, has to be checked. Further, the use of basically different approaches makes it difficult to compare the results of these studies. It is suggested that attempts be made to move towards a standard method for in situ collection of volatiles, and that this method be introduced in the fields of research where deductions concerning the interaction between the living plant and the ecosystem are made.

In the recent study by Loughrin et al. (1994), documenting that some of the volatiles induced by the beet armyworm caterpillar, *Spodoptera exigua*, was emitted from leaves of *Gossypium hirsutum* in diurnal cycles, a bridge was built to the last subject in this chapter: rhythmic phenomena in the emission of volatiles. The rhythm of emission may be correlated to the activity pattern of either the herbivores or the enemies of the herbivores. Loughrin et al. conclusively suggested that: "Perhaps the induced release of large amounts of certain odors in damaged plants is similar to the periodic release of specific odors by flowers, in that it provides a clear and reliable signal, above the background of other plant odors, that attracts natural enemies of the herbivorous attackers" (Loughrin et al. 1994).

4.2 Rhythmic Phenomena in the Emission of Volatiles

An interesting area of research, focusing on the development of volatile emissions in the course of time, is that of rhythmic phenomena in volatile emission. Several investigations have documented that emission of some, although not necessarily all, volatiles from flowers are emitted in a rhythmic manner, i.e., *Medicago sativa* analyzed by the ex situ-dry method (Loper and Lapioli 1971) *Odontoglossum constrictum*, *Citrus medica*, *Hoya carnosa*, and *Stephanotis floribunda* in situ (Altenburger and Matile 1988, 1990; Matile and Altenburger 1988), *Nicotiana sylvestris* and *N. suaveolens* in situ (Loughrin et al. 1991), *Trifolium repens* in situ (Jakobsen and Olsen 1994), *Ribes nigrum* in situ (Hansted et al. 1994), and *Hesperis matronalis* in situ (Nielsen et al. 1995). In contrast, *Nicotiana othophore* in situ and *Malus* × *domestica* (apple) analyzed ex situ-wet showed little or no rhythmicity in the emission (Loughrin et al. 1990). Further, only two floral compounds showed a distinct rhythmic nature of emission in *Brassica napus* (Jakobsen et al. 1994).

In some species, the often long array of volatiles produced by flowers are emitted in almost synchronous rhythms with a maxima of emission in either clearly the light or the dark phase of the 24-h period. However, in *Hesperis matronalis* flowers, we observed that phenetyl acetate, benzyl acetate, and their alcohols were emitted rhythmically, with maxima clearly in the dark period, whereas six monoterpenes, β-pinene, sabinene, myrcene, limonene, 1,8-cineole, and Z-β-ocimene showed maxima in the light period. Further, E-β-ocimene showed a maximum of emission slightly later in the photo-period. In contrast, α-pinene, α-terpeniol, linalool, and two linalool oxides showed no rhythmicity in the emission pattern (Fig. 4; Nielsen et al. 1995). Asynchronous emission of volatiles from different biosynthetic pathways was previously demonstrated in flowers of *Stephanotis floribunda* (Matile and Altenburger 1988) and *Nicotiana sylvestris* (Loughrin et al. 1990, 1991). In the latter, benzyl acetete, benzyl alcohol, and benzaldehyde were emitted with a nocturnal maxima of emission; however, the terpenoids linalool, caryophyllene, and α-terpineol were emitted in a nonrhythmic manner. The presence and persistence of very distinct rhythms have called for hypothesises concerning the function of these phenomena. The often strong correlation between the activity of the primary pollinators and the maxima of floral emission (Matile and Altenburger 1988; Loughrin et al. 1990, 1991; Jakobsen and Olsen 1994; Hansted et al. 1994) indicates that it is beneficial for the plant to emit the synomones mainly in the periods of pollinator activity, thus optimizing the

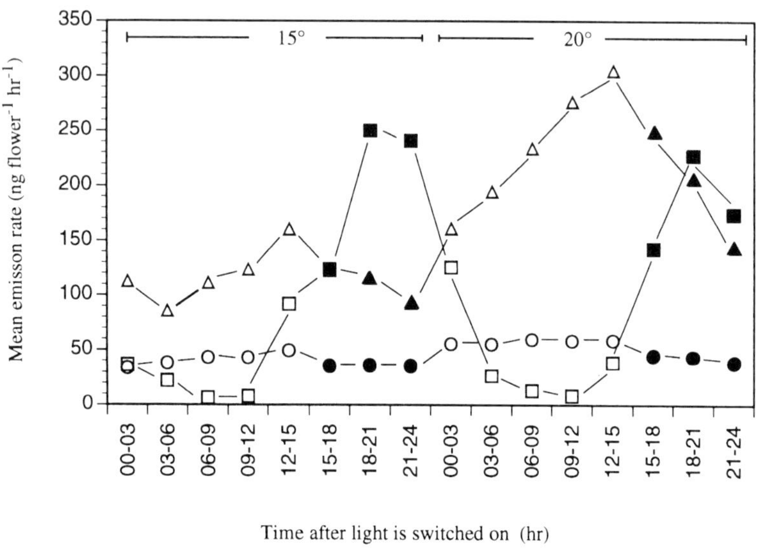

Fig. 4. Asynchronous rhythmic emission of three volatiles from flowers of *Hesperis matronalis* flowers during two 24-h periods. Symbols ○: 1,8-cineole; △: E-β-ocimene; □: benzyl acetate. *Shading of symbols* collected during the dark intervals. (Nielsen et al. 1995)

effects of the produced volatiles. However, the pronounced asynchronicity observed in some species may indicate that either several pollinating species with different activity spectra have been (or are) involved in the pollination ecology, or that the volatiles function as kariomones against herbivores with a different activity spectrum. Flower volatiles are excellent cues for insect herbivores in their attempt to locate host plants for feeding or oviposition (Metcalf 1987). On the other hand, several of the compounds isolated from flowers have been shown to serve as deterrents against herbivores (Levin 1973; Rodriguez and Levin 1976; Wood 1982). Pellmyr and Thien (1986) suggested that floral volatiles originate from compounds which initially functioned as general herbivore deterrents. Insects which could bypass this repellent barrier may use the fragrance as an orientation clue. If these insects contributed to pollination of the flower, a mutual adaptation may arise. This coadaptation could include emission of fragrance during the period where the potential pollinators are active. Indeed, recent investigations have shown that several night-pollinated and day-pollinated species emit fragrance in a rhythmic fashion which seems to correlates with the activity spectrum of the pollinators (Matile and Altenburger 1988; Loughrin et al. 1990, 1991; Hansted et al. 1994; Jakobsen and Olsen 1994). The synchrony between the emission of floral volatiles and the pollinator activity was suggested by Matile and Altenburger (1988) to be achieved via the biological clock.

The rhythmic emission from flowers may either be controlled by the endogenous clock of the plant or merely be a reflection of the light/dark cycle. The former control mechanism, which is termed circadian, has been demonstrated for a number of compounds in *Hoya carnosa* (Altenburger and Matile 1988) and *Stephanotis floribunda* (Altenburger and Matile 1990). Further, the emission of propanoid-derived volatiles in *Nicotiana sylvestris* and *N. suaveolens*occurs under the control of the biological clock, whereas volatiles from other biosynthetic pathways did not exhibit similar patterns of emission (Loughrin et al. 1991). The majority of the rhythmically emitted compounds in these species appear to have maxima in or near the nocturnal phase, although exceptions are seen, e.g., emission of 1-nitro-2-phenylethane in *S. floribunda*, which culminates at noon. In contrast, the control of emission in species tested so far, with diurnal maxima, appears to be noncircadian in nature. These include *Odontoglossum constrictum* and *Citrus midica* (Altenburger and Matile 1990), *Trifolium repens* (Jakobsen and Olsen 1994), *Ribes nigrum* (Hansted et al. 1994), and *Brassica napus* (Jakobsen et al. 1994).

5 Concluding Remarks

The research on rhythmic phenomena in volatile emission and on the significance of volatiles in the plant-insect relationship is in rapid progress. This may indeed contribute to the attempts being made to limit the use of pesticides in farming. A deeper knowledge of the defence systems acting via the accumulation and emission of volatile compounds may form a basis for improving the capacity of the plants to resist attacks by insects or pathogens. It seems that one way of increasing the speed of progress within this field would be to use a more standardized approach when analyzing the emission of volatiles. Further, if this were to be combined with the use of only a few genotypes in related research programs, the comparison and exchange of results from different research teams would be somewhat easier.

Acknowledgments. Thanks are extended to Mr. Thomas Schmitt and Mr. Peter M. Raible for developing the software controlling the in situ system, to Mr. Olivier Francois for assistance on the electronic devices, to Miss Lene Rasmussen and Mr. Brian Damm for technical assistance in carrying out the in situ headspace analysis, and to Mrs. Helle Høgild for patient assistance in the library.

References

Altenburger R, Matile P (1988) Circadian rhythmicity of fragrance emission in flowers of *Hoya carnosa* R. Brl. Planta 174:248–252

Altenburger R, Matile P (1990) Further observations on rhythmic emission of fragrance in flowers. Planta 180:194–197

Baigrie BD, Laurie WA, McHale D (1985) Artifact formation during headspace analysis. In: Adda D (ed) Progress in flavour research. Proc 4th Weurman Flavour Symp, Dourdan, 9–11 May 1984, Elsevier, Amsterdam, pp 577–582

Bergström J, Bergström G (1989) Floral scents of *Bartsia alpina* (Scrophulariaceae): chemical composition and variation between individual plants. Nord J Bot 9:363–365

Bicchi C, Joulain D (1990) Headspace-gas chromatographic analysis of medicinal and aromatic plants and flowers. Flav Fragr J 5:131–145

Blaakmeer A, Geervliet JBF, Loon JJA, Posthumus MA, van Beek TA, de Groot A (1994) Comparative headspace analysis of cabbage plants damaged by two species of *Pieris* caterpillars: consequences for in-flight host location by *Cotesia* parasitoids. Entomol Exp Appl 73:175–182

Borg-Karlson A-K, Bergström G, Groth I (1985) Chemical basis for the relationship between *Ophrys* orchids and their pollinators I. Volatile compounds of the *Ophrys lutea* and *O. fusca* as mimetic attractants/excitants. Chem Scr 25:283–294

Brunke EJ, Hammerschmidt FJ, Schmaus G (1992) The headspace analysis of flower fragrances. Dragoco Rep 39:3–31

Brunke EJ, Hammerschmidt FJ, Schmaus G (1993) Flower scent of some traditional medicinal plants. In: Teranishi R, Buttery RG, Sugisawa H (eds) Bioactive volatile compounds from plants. ASC Symp Ser 525:282–296

Brunke EJ, Hammerschmidt FJ, Schmaus G (1994) Headspace analysis of hyacinth flowers. Flav Fragr J 9:59–69

Buchbauer G, Jirovetz L, Wasicky M, Nikiforov A (1993a) Volatile constituents of the headspace and essential oil of *Plectranthus coleoides* Marginatus (Labiatae). J Ess Oil Res 5:311–313

Buchbauer G, Jirovetz L, Wasicky M, Nikiforov A (1993b) Headspace and essential oil analysis of apple flowers. J Agric Food Chem 41:116–118

Buchbauer G, Jirovetz L, Wasicky M, Nikiforov A (1994) Volatile constituents of *Exacum affine* Balf. f. flowers obtained by dynamic headspace sampling and as the essential oil. Flav Fragr J 9:55–58

Burger BV, Munro ZM, Visser JH (1988) Determination of plant volatiles 1: analysis of the insect-attracting allomone of the parasitic plant *Hydnora africana* using Grob-Habich activated charcoal traps. J High Resolut Chromatogr Chromatogr Commun 11:496–499

Charron CS, Cantliffe DJ, Heath RR (1995) Volatile emissions from plants. Hort Rev 17:43–72

Dicke M (1994) Local and systemic production of volatile herbivore-induced terpenoids: their role in plant-carnivore mutualism. J Plant Physiol 143:465–472

Dobson HEM (1991) Essential oils and waxes. In: Linskens HF, Jackson JF (eds) Modern methods of plant analysis, new ser, vol 12. Springer, Berlin Heidelberg New York, pp 231–251

Ebel RC, Mattheis JP, Buchanan DA (1995) Drought stress of apple tress alters leaf emissions of volatile compounds. Physiol Plant 93:709–712

Erhardt A, Jaggi B (1995) From pollination by Lepidoptera to selfing: the case of *Dianthus glacialis* (Caryophyllaceae). Plant Syst Evol 195:67–76

Hansted L, Jakobsen HB, Olsen CE (1994) Influence of temperature on the rhythmic emission of volatiles from *Ribes nigrum* flowers in situ. Plant Cell Environ 17:1069–1072

Heath RR, Manukian A (1992) Development and evaluation of systems to collect volatile semiochemicals from insects and plants using a charcoal-infused medium for air purification. J Chem Ecol 18:1209–1226

Jakobsen HB, Olsen CE (1994) Influence of climatic factors on emission of flower volatiles in situ. Planta 192:365–371

Jakobsen HB, Friis P, Niels JK, Olsen CE (1994) Emission of volatiles from flowers and leaves of *Brassica napus* in situ. Phytochemistry 37:695–699

Juuti S, Arey J, Atkinson R (1990) Monoterpene emission rate measurements from a Monterey pine. J Geophys Res 95:7515–7519

Kaiser R (1991) Trapping, investigation and reconstitution of flower scents. In: Müller PM, Lamparsky D (eds) Perfumes: art, science and technology. Elsevier, London, pp 213–250

Kawano S, Odaki M, Yamaoka R, Oda-Tanabe M, Takeuchi M, Kawano N (1995) Pollination biology of *Oenothera* (Onagraceae): the interplay between floral UV-absorbancy patterns and floral volatiles as signals to nocturnal insects. Plant Spec Biol 10:31–38

Levin DA (1973) The role of trichomes in plant defence. Q Rev Biol 48:3–15

Loper GM, Lapioli AM (1971) Photoperiodic effects on the emanation of volatiles from alfalfa (*Medicago sativa* L.) florets. Plant Physiol 49:729–732

Loughrin JH, Hamilton-Kemp TR, Andersen RA, Hildebrand DF (1990) Volatiles from flowers of *Nicotiana sylvestris*, *N. otophora* and *Malus* × *domestica*: headspace components and day/night changes in their relative concentrations. Phytochemistry 29:2473–2477

Loughrin JH, Hamilton-Kemp TR, Andersen RA, Hildebrand DF (1991) Circadian rhythm of volatile emission from flowers of *Nicotiana sylvestris* and *N. suaveolens*. Physiol Plant 83:492–496

Loughrin JH, Manukian A, Heath RR, Turlings TCJ, Tumlinson JH (1994) Diurnal cycle of emission of induced volatile terpenoids by herbivore-injured cotton plants. Proc Natl Acad Sci USA 91:11836–11840

Maarse H, Belz R (1985) Isolation, separation and identification of volatile compounds in aroma research. D Reidel, Dordrecht

Matile P, Altenburger R (1988) Rhythms of fragrance emission in flowers. Planta 174:242–247

Mattiacci L, Dicke M, Posthumus MA (1994) Induction of parasitoid attracting synomone in Brussels sprouts plants by feeding of *Pieris brassicae* larvae: role of mechanical damage and herbivore elicitor. J Chem Ecol 20:2229–2247

Metcalf RL (1987) Plant volatiles as insect attractants. Crit Rev Plant Sci 5:251–301

Mookherjee BD, Trenkle RW, Wilson RA, Zampino M, Sands KP, Mussinan CJ (1987) Fruits and flowers: live vs. dead – which do we want? In: Lawrence BM, Mookherjee BD, Wills BJ (eds) Flavors and fragrances: a world perspective. Proc 10th Int Congr Essential Oils, Fragrances and Flowers, Washington, DC, pp 415–424

Mookherjee BD, Trenkle RW, Wilson RA (1989) Live vs. dead part II. A comparative analysis of the headspace volatiles of some important fragrance and flavour raw materials. J Essent Oil Res 2:85–90

Mookherjee BD, Trenkle RW, Wilson RA (1990) The chemistry of flowers, fruits and spices: live vs. dead, a new dimension in fragrance research. Pure Appl Chem 62:1357–1364

Nielsen JK, Jakobsen HB, Friis P, Hansen K, Möller J, Olsen CE (1995) Asynchronous rhythms in the emission of volatiles form *Hesperis matronalis* flowers. Phytochemistry 38:847–851

Parliment TH (1986) Biogeneration of aromas. In: Parliment TH, Croteau R (eds) ACS Symp Ser 317:34–52

Patt JM, Rhoades DF, Corkill JA (1988) Analysis of the floral fragrance of *Platanthera stricta*. Phytochemistry 27:91–95

Pellmyr O, Thien BT (1986) Insect reproduction and floral fragrances: keys to the evolution of the angiosperms. Taxon 35:76–85

Pellmyr O, Thien LB, Bergström G, Groth I (1990) Pollination of New Caledonian Winterceae: opportunistic shifts of parallel radiation with their pollinators. Plant Syst Evol 173:143–157

Pham-Delegue MH, Etievant P, Guichard E, Masson C (1989) Sunflower volatiles involved in honeybee discrimination among genotypes and flowering stages. J Chem Ecol 15:329–343

Raguso RA, Pichersky E (1995) Floral volatiles from *Clarkia breweri* and *C. concinna* (Onagraceae): recent evolution of floral scent and moth pollination. Plant Syst Evol 194:55–67

Rodriguez E, Levin DA (1976) Biochemical parallelisms of repellants and attractants in higher plants and arthropods. In: Wallace JW, Mansell RL (eds) Biochemical interaction between plants and insects. Recent Adv Phylochem 10:214–270

Steinberg S, Dicke M, Vet LEM (1993) Relative importance of infochemicals from first and second trophic level in long-range host location by the larval parasitoid *Cotesia glomerata*. J Chem Ecol 19:47–59

Surburg H, Guentert M, Harder H (1992) Volatile compounds from flowers: analytical and olfactory aspects. In: Teranishi R, Buttery RG, Sugisawa H (eds) Bioactive volatile compounds from plants. ACS Symp Ser 525:168–186

Takabayashi J, Dicke M, Posthumus MA (1994) Volatile herbivore-induced terpenoids in plant mite interactions – variation caused by biotic and abiotic factors. J Chem Ecol 20:1329–1354

Tatsuka K, Suekane S, Sakai Y, Sumitani H (1990) Volatile constituents of kiwi fruit flowers: simultaneous distillation and extraction versus headspace sampling. J Agric Food Chem 38:2176–2180

Tingey DT Manning M, Grothaus LC, Burns WF (1979) The influence of light and temperature on isoprene emission from live Oak. Physiol Plant 47:112–118

Tingey DT, Manning M, Grothaus LC, Burns WF (1980) Influence of light and temperature on monoterpene emission rates from slash pine. Plant Physiol 65:797–801

Tollsten L, Bergström J (1989) Variation and post-pollination changes in floral odours released by *Platanthera bifolia* (Orchidaceae). Nord J Bot 9:359–362

Tollsten L, Övstedal DO (1994) Differentiation in floral scent chemistry among populations of *Conopodium majus* (Apiaceae). Nord J Bot 14:361–367

Turlings TC, Tumlinson JH, Lewis WJ (1990) Exploitation of herbivore-induced plant odors by host-seeking parasitic wasps. Science 250:1251–1253

Vet LEM, Dicke M (1992) Ecology of infochemical use by natural enemies in a tritrophic context. Annu Rev Entomol 37:141–172

Wood DL (1982) The role of pheromones, kairomones and allomones in the host selection and colonization behavior of bark beetles. Annu Rev Entomol 27:411–446

Analysis of Volatile Constituents of Fruit

G.R. Takeoka and G.H. Full

1 Introduction

Some of the earliest applications of capillary gas chromatography coupled with mass spectrometry (GC-MS) involved the analyses of fruit flavors (McFadden et al. 1963; Teranishi et al. 1963). Prior to the advent of this technique, it was very difficult to analyze the delicate and complex flavors of fruit. However, tremendous advances in gas chromatography and combined gas chromatography-mass spectrometry have greatly increased our knowledge of fruit flavors.

One of the primary goals in flavor research is to identify those constituents which are responsible for the typical characteristic flavor of a food. In order to study the flavor of a food, it is first necessary to isolate the volatiles from the complex of nonvolatile material. The analyst wants to obtain the maximum amount of qualitative and quantitative information about those constituents which contribute to the overall flavor. This goal is not easily achieved for the following reasons. First, the amount of volatiles in food is extremely small compared to the other constituents. In fruit, the flavor constituents are usually present in concentrations less than 50 ppm. Second, these fruit volatiles are often also extremely complex mixtures and may consist of hundreds of individual constituents. For example, there have been more than 400 volatile constituents identified in tomatoes (Petro-Turza 1986–1987). In determining the contribution of volatiles to the overall flavor one needs to consider their potency and concentration. The odor thresholds of different flavor constituents can vary tremendously, ranging from about 100 ppm for ethanol to 2 parts in 10^{14} for 1-*p*-menthene-8-thiol, a character impact compound in grapefruit. Concentrations of individual fruit constituents may vary from ppm levels to sub-ppb levels, permitting the detection by the human nose, but not identification by our most sophisticated analytical instruments.

Two subjects are discussed in this chapter. The first section details methods and techniques for the isolation and fractionation of flavor constituents from fruit. Fruit flavor constituents are a structurally diverse group of products which exhibit wide variability in their stability and chromatographic behavior. Their occurrence as minor or trace constituents has provided a major challenge to the flavor chemist. The second section discusses the use of selective detection in the analysis of chiral fruit volatiles. The advent of modified cyclodextrin capillary columns and multidimensional gas chromatography has provided great progress in the area of chirospecific analysis of fruit constituents.

Modern Methods of Plant Analysis, Vol. 19
Plant Volatile Analysis
Edited by H.-F. Linskens and J.F. Jackson

2 Sources of Error in Sample Preparation

This section discusses possible sources of error in sample preparation. First is having a non-representative sample. Obtaining a representative sample is an important part of any analysis, since any errors made in the sampling protocol cannot be corrected at any later point in the analysis. Sampling may be a problem because of the size or inhomogeneity of the sample. Other problems may occur in the analysis due to contamination, careless collection, or alteration during storage (Takeoka et al. 1986a). Berger et al. (1985) noted that the concentration of 1-(E,Z)-3,5-undecatriene in pentane fractions of apple and pear decreased markedly on prolonged storage. In fruit research the analyst is typically involved in trace analysis, so any surface which our sample contacts becomes a possible source of contamination. Chemical processes occurring in the sample between the time of collection and analysis can also change the analytical results. Examples of such processes are photodecomposition, adsorption, vaporization loss, microbial action, and chemical reaction. Another important factor in fruit analysis is the stage of maturity or ripeness during sampling. With fruit, the formation of flavor constituents is a dynamic process. The concentrations of volatiles are subject to many variations, one of which is the ripeness of the fruit. In their study of nectarine (*Prunus persica* L., Batsch var. Nucipersica Schneid), Engel et al. (1988b) demonstrated that C6 aldehydes and alcohols are the major constituents isolated from "green" fruit. With increasing fruit maturity, the concentrations of these compounds decrease dramatically. In contrast, the level of γ- and δ-lactones increase during nectarine maturation. Another source of error is the incomplete recovery of fruit flavor constituents. Steam distillation is one of the most frequently employed methods of sample preparation. However, quantitative studies have shown that only 33 and 3% of the flavor constituents of strawberries and raspberries, respectively, are recovered by steam distillation (Ohloff et al. 1985). The loss of important flavor constituents can lead to an inaccurate flavor profile. 2,5-Dimethyl-4-hydroxy-3(2*H*)-furanone (*Furaneol*®), a character impact compound of both strawberry and pineapple, does not steam-distill and must be isolated using other sample preparation techniques. Another source of error is the production of secondary volatiles via enzymatic reactions. At the moment the plant material is cut, macerated, or homogenized, certain enzymatic processes begin, some of which are extremely rapid. This leads to the production of many volatiles which are either not present or are present at much lower levels in the intact tissue.

3 Methods of Sample Preparation

The most frequently used methods for isolation and concentration of flavor constituents involve distillation, extraction, and adsorption techniques.

3.1 Extraction

Grosch and coworkers (Guth and Grosch 1989; Jung et al. 1992) have applied direct extraction to isolate flavor constituents from various foods. The exact extraction method has varied with the type of food. Relatively dry products such as cheese (Preininger and Grosch 1994) and popcorn (Schieberle 1995) were frozen with liquid nitrogen and ground in a blender. The powdered material was then extracted with diethyl ether either with stirring or soaking overnight followed by further extraction with a Soxhlet extractor. With boiled trout, the powdered material was mixed with anhydrous Na_2SO_4 (1:1 w/w) prior to extraction with ether. The sodium sulfate binds the water and aids the extraction of highly water-soluble constituents. In their analysis of muskmelons (*Cucumis melo*), Schieberle et al. (1990) incubated the homogenized fruit pulp for 5 min at 22 °C to allow enzymatic development of flavor constituents. Aqueous saturated $CaCl_2$ solution was then added to inhibit further enzyme activity. The suspension was centrifuged and the resulting supernatant was filtered. The filtrate was then extracted with pentane/diethyl ether (1:1 v/v).

Extraction is an effective method of isolating highly water-soluble flavor constituents which are typically poorly recovered by distillation and headspace techniques. Direct liquid-liquid extraction of homogenized fruit pulp has been used to isolate nectarine volatiles (Engel et al. 1988a). Analysis of the highly oxygenated 2,5-dimethyl-4-hydroxy-3(2*H*)-furanone in strawberries, pineapples, and mangoes has been performed by Pickenhagen et al. (1981) using the following extraction technique. The fruit was frozen with liquid nitrogen and blended to a powder. The powder was stirred with water for 1 h at room temperature. The slurry was filtered and the filtrate continuously extracted in a liquid-liquid extractor with ether for 18 h. After drying over sodium sulfate and removal of most of the ether with a Vigreux column, the concentrate was analyzed by gas chromatography.

A problem with direct extraction is that nonvolatiles are also extracted. Introduction of such extracts into gas chromatographic systems would lead to the accumulation of nonvolatile residues in the injector and column inlet. These residues may slowly degrade to produce volatiles or can catalyze the decomposition of subsequent samples injected (Mitzer 1964). Additionally, adsorption effects in a dirty injector can lead to tailing peaks and reduced system efficiency (Grob and Grob 1979). In order not to compromise our analytical results, some method is thus necessary to separate these nonvolatiles from the volatiles. Grosch and coworkers (Sen et al. 1991; Jung et al. 1992) have used high vacuum sublimation to separate volatiles and solvent from the direct extract. The exact method has again varied, depending on the product. The extract is dried over sodium sulfate and concentrated to about 120 ml by distilling of the solvent on a Vigreux column (50 × 1 cm). The sample is placed in a 1-l distillation flask which is connected to a series of traps cooled to ice, dry ice-methanol, and liquid nitrogen temperatures. The sample is frozen for 30 min in liquid nitrogen. A high vacuum source (20 mPa;

consisting of a two-stage rotary vane vacuum pump and an oil diffusion pump) is placed at the end of the series of traps. The volatiles and solvent are sublimed for 3h. The flask is then heated to 50 °C with a water bath and the sublimation is continued for an additional 2h. The condensates of the cold traps are then combined and concentrated using a Vigreux column.

3.2 Headspace Sampling

One of the most popular methods of sample preparation is dynamic headspace sampling. The blended fruit tissue or juice is typically placed in a container and the contents rapidly stirred. The headspace volatiles are purged by air or nitrogen and are trapped by adsorption on porous polymer traps. Various trapping materials have been used such as charcoal, the Porapak series, the Chromosorb series, and Tenax. In a second step the volatiles are recovered by solvent or heat desorption. With thermal desorption the analyst must be constantly alert to the danger of artifact formation. The extent of thermal alteration depends on temperature, duration, and surface catalytic effects (Grob and Habich 1985). The problem may be related to the metal surface which the sample contacts in most commercial thermal desorption devices.

3.2.1 Use of Activated Carbon Traps

Activated carbon is an extremely powerful adsorbent which possesses very high thermal and chemical stability. One problem in using charcoal for analytical work is related to large variations in the chemical and physical properties of different sources of carbon. Until the mid-1970s, activated carbon was infrequently used in flavor research because of fears of incomplete recoveries by solvent extraction or thermal decomposition by applying thermal desorption. Now, to a large extent these difficulties have been overcome by the development of small traps by Grob (1973) and Grob and Zürcher (1976). They employed traps containing only 1–2 mg of charcoal. For recovery of adsorbed volatiles, traps were eluted with 10–50 μl carbon disulfide. This technique has been widely used by Kaiser (1993) to examine flower scents. Kaiser (1993) used traps having 1–5 mg of charcoal and stated that for complete recovery of flower scents containing extremely polar constituents, further extraction with 10–50 μl of ethanol may be required. If one decides to use carbon, it is important to test its activity. This can be accomplished by using model systems. For example, Brunke and coworkers (1993) used a model system of (E) and (Z)-β-ocimenes to check the activity of various carbons. The ocimenes are unstable compounds which are sensitive to oxygen. Some carbons tested were found to produce artifacts, the dimethyloctatrien-2-ols and the octatetraenes. Similar artifact formation was observed on activated carbon by Surburg et al. (1993).

3.2.2 Use of Tenax Traps

Various sizes of Tenax traps have been utilized in flavor research. In contrast to many researchers, Buttery and coworkers (1987a,b) have used relatively large Tenax traps (2.2 cm i.d. × 14 cm length; 10 g Tenax) to collect headspace volatiles. They determined that large traps and high flow rates ($3\,l\,min^{-1}$) were necessary to isolate the low concentrations of volatiles present in most plant material. In their decision to use the above conditions they considered model systems of water solutions containing typical plant volatiles. With such systems, Buttery and coworkers (1987b) calculated the total volume of sweep gas (V_g) needed to transfer a percentage (P) of compound in water solution to the Tenax trap with the following equation:

$$V_g = -V_W K^{-1} \ln\left[(100-P)/100\right],$$

where V_w is the total volume of the water solution and K is the air to water partition coefficient of the compound at the temperature under consideration (usually room temperature, about 25 °C).

For tomato samples, the isolation procedure was typically performed as follows: 100 g of tomato sample was blended for 30 s with a Waring blender. The blended material was allowed to stand for 3 min to allow the enzymatic development of flavor constituents. A saturated solution of $CaCl_2$ (100 ml) was added and the mixture was blended for an additional 10 s. The $CaCl_2$ solution was added to deactivate the enzyme systems, and also aided in the recovery of volatiles through the "salting out" effect. Internal standards were then added as a 5-ml solution containing 20 ppm 3-pentanone, 20 ppm 2-octanone, and 5 ppm anethole. The mixture was added to a 1-l round-bottomed flask, Purified air or nitrogen ($3\,l\,min^{-1}$) was passed over the rapidly stirred sample via a Teflon tube and exited out of the flask through a Tenax trap. The apparatus and all connections were either glass or Teflon. The isolations were usually conducted for 1 h, at which time the trap was removed and eluted with 100 ml of freshly distilled diethyl ether (containing ca. 10 ppm ethyl antioxidant 330). The ether extract was finally carefully concentrated to ca. 50 µl using a Vigreux column. The high-flow Tenax trapping method gave good recoveries of most flavor constituents in the C5 to C13 range (Table 1). A conspicuous exception was 2-phenylethanol, which gave a 2% absolute recovery. This was probably due to its high water solubility and low vapor pressure, producing a low partition coefficient (K). The average variation found for standard solutions of low to medium polarity compounds for this sampling method was less than 8%. The error with poorly recovered compounds such as 2-phenylethanol is probably considerably higher.

Two different headspace methods were utilized by Takeoka et al. (1990a) to sample intact apricot volatiles. The first method used small Tenax traps (0.7 g of Tenax), low purge flow rates ($50\,ml\,min^{-1}$), short sampling periods (15–30 min), and thermal desorption of trapped volatiles (ambient to 220 °C over 15 min). This method was useful for detecting low boiling constituents which would be obscured by solvent peaks. The second method involved large Tenax traps (10 g of Tenax),

Table 1. Absolute Recoveries for Standard Solutions of Various Flavor Volatiles Using the High Flow Tenax Trapping Method of Buttery et al. (1987b)[a]

Compound	Absolute % Recovery
3-methylbutanal	19
1-penten-3-one	31
1-penten-3-ol	52
isobutyl cyanide	96
3-methylbutanol	40
(E)-2-pentenal	43
(Z)-3-hexenal	43
hexanal	54
(E)-2-hexenal	61
(Z)-3-hexenol	50
hexanol	66
6-methyl-5-hepten-2-one	68
6-methyl-5-hepten-2-ol	75
linalool	86
geranial	75
geraniol	57
β-cyclocitral	68
methyl salicylate	35
geranylacetone	36
2-phenylethanol	2
β-ionone	44

[a] Adapted from Buttery et al. 1987b, 1988.

fast purge gas flow rates ($3\,l\,min^{-1}$), long sampling times (22 h), and solvent elution (freshly distilled diethyl ether) of entrained constituents. This method was better suited to the analysis of higher boiling constituents and exposed the sample to less thermal stress. The air flow rate during trapping kept the fruit in good condition, with no condensation observed in the sampling chamber. The importance of high gas flows during fruit headspace sampling has been discussed by Ismail et al. (1980), who felt that these conditions were less conducive to microbial growth (which would contribute its own volatiles). With the exception of the very low boiling compounds, more peaks were observed and could be identified using the second headspace method.

Stromvall and Petersson (1992) have shown that, even on Tenax, monoterpene hydrocarbons such as myrcene, limonene, and β-phellandrene can undergo oxidative decomposition and can be partially lost when the sampling air contains high photooxidant levels. These researchers found that pretreatment of the Tenax cartridges with thiosulfate (antioxidant) prevented oxidative reactions of the terpenes as well as the Tenax polymer, thereby preserving the analytical quality. In a similar fashion, traces of ethyl antioxidant 330 [1,3,5-trimethyl-2,4,6-tris(3,5-di-tert-butyl-4-hydroxybenzyl)benzene; contained in the ether elution solvent] may have had a protective effect on the labile flavor constituents investigated by Buttery et al. (1987b). They found good recovery of aldehyde and terpenoid constituents examined by sampling on Tenax, indicating that no appreciable oxidation was occurring on the polymer.

3.2.3 Use of Solvents as Trapping Material

Solvents have also been used to trap headspace vapors. Jennings (1979) described a cocondensation method to trap headspace volatiles. A round-bottomed flask was attached to a semimicro Vigreux column via a side-arm adapter. Freon 12 (dichlorodifluoromethane, b.p. 30 °C) was added and a cold-finger condenser containing dry-ice/acetone was positioned at the top of the column. The headspace vapors were passed through the Vigreux column of refluxing Freon 12. The Freon containing entrained volatiles was continuously returned to the round-bottomed flask located beneath the Vigreux column. At the end of sampling, a solvent such as ether can be added to the flask and the Freon allowed to completely evaporate. The ether can then be removed from the flask before the ice melts, thereby resulting in a dry sample for GC analysis. Rapp and Knipser (1980) conducted a sweeping gas through wines to a Freon 11 (trichlorofluoromethane) system to selectively concentrate headspace constituents (excluding water and ethanol). Williams et al. (1980) used this method to examine muscat grape juice volatiles. The juice samples (400 ml) were swept with nitrogen at 50 ml min^{-1} for 24 h. Another variation of headspace sampling with solvent trapping was developed by Buttery and Ling (1993). Sweep gas (nitrogen at 3 l min^{-1}) was purged through a flask containing the sample of interest and passed into a three-necked 1-l flask containing 50 ml of diethyl ether (freshly distilled) cooled by dry ice surrounding the flask. The ether was continuously stirred during the isolation. A cold-finger condenser containing dry ice was placed on the opposite neck through which the sweep gas exited (the center neck was stoppered). The sampling was typically performed for 1 h, at which time the trapping flask was removed. The ether solution was freed from water by freezing and then concentrated prior to analysis. These workers found that with food systems such as the tomato this method gave recovery of volatiles very similar to the high-flow Tenax trapping method described earlier. It should be stressed that an inert gas such as nitrogen must be used as the sweep gas in this system. Due to its potential fire hazard, air should not be used because of possible electrostatic effects from the flowing gas and from any escaping ether. Instead of purging headspace volatiles directly into solvent, Umano and Shibamoto (1987) developed an apparatus where the volatiles were purged (nitrogen at 7.2 ml min^{-1}) from the sample flask into a gas-washing bottle containing 250 ml of cooled water (10 °C). The volatiles dissolved by the water were then continuously extracted with dichloromethane (70 ml) for 6 h. These workers also replaced the water in the gas-washing bottle with a aqueous cysteamine solution and recovered and identified trace amounts of various aldehydes as their thiazolidine derivatives. They detected highly volatile aldehydes such as acetaldehyde and propanal, which are very difficult to isolate and identify by conventional headspace techniques.

3.2.4 Effect of Sweep Gas

In their investigation of strawberry foliage headspace volatiles, Hamilton-Kemp et al. (1993) compared the use of air and nitrogen as sweep gases. Nitrogen purging yielded greater amounts of aliphatic alcohols and esters such as 1-hexanol, (Z)-3-hexen-1-ol, (Z)-3-hexenyl acetate, (Z)-3-hexenyl 2-methylbutanoate, 6-methyl-5-hepten-1-ol, and 1-octen-3-ol, and aromatics such as benzyl alcohol, methyl and ethyl salicylate, and 2-phenylethanol. The presence of higher levels of alcohols and esters with nitrogen purging suggests that the anaerobic conditions resulted in increased alcohol dehydrogenase activity. Davies (1980) observed enhanced alcohol dehydrogenase activity in plant tissues under anaerobic conditions. With air purging, higher levels of the terpene hydrocarbons, (E)-β-ocimene, germacrene D, and α-farnesene, were observed compared to nitrogen purging. Hamilton-Kemp et al. (1993) postulated that the anaerobic conditions produced by nitrogen purging probably altered the metabolism of the strawberry foliage during the sampling stage. Based on their observations of different results with nitrogen and air, they concluded that air was the preferred purge gas for strawberry. The results are probably highly dependent on the plant species under investigation. For example, in their study of tomato leaf headspace volatiles, Buttery et al. (1987a) found no significant differences among the constituents isolated when using either air or nitrogen as sweep gases. When investigating fruit possessing active enzyme systems, air may be the preferred sweep gas to minimize problems caused by metabolic changes during sampling.

3.2.5 Closed-Loop Stripping

The use of closed-loop stripping for sampling of headspace volatiles was pioneered by the Grobs (Grob 1973; Grob and Grob 1974; Grob et al. 1975; Grob and Zürcher 1976). Though they used this technique to examine organic substances in water, it is applicable to any type of headspace analysis. Small charcoal traps (1–5 mg) and high gas flow rates (2–3 $l\,min^{-1}$) produced by a pump with stainless-steel bellows were utilized in their procedure. With room temperature sampling stripping was possible for substances up to ca. C20. An additional advantage is that compounds which break through the trap are automatically recycled. For low and medium molecular weight constituents, closed-loop stripping was three to four times more efficient than liquid extraction (Grob et al. 1975). For C20, both methods were equivalent in terms of efficiency. They cautioned against the use of PTFE in the stripping equipment, especially in the fittings and pump, due to contamination with hydrocarbons. Buttery and Ling (1996) adapted this method with the use of Teflon tubing and diaphram pumps with Teflon heads and valves. After careful cleaning, they found the level of hydrocarbon background to be very low. The inertness of Teflon was stressed, since some constituents may interact with metal surfaces. Buttery and Ling (1996) utilized this technique for the analysis of flavor constituents. To sample the highly polar constituents Furaneol and maltol, they used excess Na_2SO_4 to bind water before subjecting the sample to closed-loop stripping. A typical analysis was performed in the following way: a weighted amount (30 g) of sample (strawberries or tomatoes) and internal standard (usually 1 ml of 50.0 ppm maltol) were blended for 30 s. The mixture was then slowly added and stirred into ca. eight times the amount of anhydrous sodium sulfate (240 g) in a 1-l beaker. The resulting powder was packed into a Pyrex tube (30 cm × 3.9 cm o.d.) containing a coarse-fritted glass disk at the bottom. This tube was connected to a Tenax trap via ball joints, and volatiles were collected on the trap using the closed-loop stripping system. Argon or nitrogen were utilized (the system was purged with the appropriate gas for 10 min prior to closing it) as the sweep gas at a flow rate of 6 l min^{-1}. The sampling was performed for 3 h, at which time the trap was removed and eluted with ether. The ether was concentrated on a Vigreux column prior to gas chromatographic analysis. The average absolute recoveries for Furaneol and maltol were 44 and 46%, respectively. For unknown reasons, aldehydes and methyl ketones were not recovered quantitatively with this technique. It is possible that these constituents interacted with the Na_2SO_4.

3.3 Distillation

Distillation is a frequently used technique to isolate volatile from nonvolatile materials. Distillation may be performed at atmospheric or reduced pressure.

High vacuum distillation (40–50 °C/0.1 bar) of homogenized fruit pulp (diluted with water) has been used by Schreier and coworkers (Idstein et al. 1984; Idstein and Schreier 1985a,b,c; Fröhlich and Schreier 1989, 1990) to isolate volatiles. The volatiles were subsequently recovered from the distillates by solvent extraction with pentane-dichloromethane (2:1). The extracts were concentrated and then fractionated by silica gel chromatography. Similarly, vacuum distillation (25–35 °C/1 mmHg) was used by Takeoka et al. (1986b, 1988) to isolate volatiles from kiwifruit and nectarine. The distillates were subjected to continuous liquid-liquid extraction with Freon 11 (trichlorofluoromethane) or Freon 11-dichloromethane (9:1 v/v).

Though atmospheric pressure simultaneous distillation-extraction (SDE) exposes the sample to high temperatures, and is known to produce thermally generated artifacts, it may be sometimes used in flavor research for the following reasons (Takeoka et al. 1992). First, it can simulate the composition of volatiles that would be formed during processing such as canning. Second, it has been demonstrated that there is a low recovery of certain polar constituents such as lactones with vacuum SDE. Third, knowledge of the volatiles formed during thermal treatment gives information on the type and amount of flavor precursors present. For example, the composition of volatiles obtained by thermal treatment at the native fruit pH could be compared to those obtained at a neutral pH of 7.

The maintenance of constant pressure in an SDE system run under vacuum (vacuum SDE) is of great importance if steady distillation at constant temperature is to be achieved. We have used Cartesian manostats to achieve constant pressure in our vacuum SDE systems. We found mechanical stirring to be much more effective than magnetic stirring in agitating the often viscous fruit slurries. Efficient stirring can solve many foaming problems which occur during vacuum SDE.

However, it may be necessary to add antifoam solution to control some problems. We typically prepare antifoam solution in the following way. Ten ml of Hartwick antifoam 50 emulsion is added to 1000 ml of water and boiled until the volume is reduced to ca. 700 ml to remove volatiles. For example, 80 ml of antifoam solution (prepared as above) was added to a blended slurry of nectarine fruit (3.18 kg) and water (3500 ml) to control foaming during vacuum SDE (Takeoka et al. 1992).

3.4 Prefractionation by Column Chromatography and High Performance Liquid Chromatography (HPLC)

The identification of trace constituents in complex fruit flavor mixtures (i.e., extracts) is greatly facilitated by preseparation into different fractions prior to gas chromatographic analysis. Column chromatography on silica gel is a widely used and effective method of fractionation. The hydrocarbons can be easily separated from the whole extract by selective elution with pentane or hexane. The polar, oxygenated constituents can then be eluted using gradient elution with increasing levels of diethyl ether. Schreier and coworkers have fractionated various fruit extracts such as cherimoya (Idstein et al. 1984), guava (Idstein and Schreier 1985c), Alphonso mango (Idstein and Schreier 1985a), papaya (Idstein and Schreier 1985b), carambola (Fröhlich and Schreier 1989), and loquat (Fröhlich and Schreier 1990) using cooled (11–13 °C) glass columns, 2.0 cm i.d. × 30 cm, containing silica gel 60 (Merck), activity grade II, with a pentane-diethyl ether gradient.

Silica gel should be used with caution, since it may cause compositional changes in samples due to isomerization reactions. The isomerization reaction of limonene with silica gel at 100 and 150 °C has been studied by Hunter and Brogen (1963). Limonene initially isomerized to α-terpinene, γ-terpinene, terpinolene, and isoterpinolene, which subsequently underwent disproportionation and polymerization reactions. Wrolstad and Jennings (1965) observed the isomerization of sabinene into α-thujene, α-terpinene, γ-terpinene, limonene, β-phellandrene, and terpinolene during thin layer chromatography. Evidence supported an acid-catalyzed isomerization of sabinene since silica gel chromatostrips prepared by using 0.1N NaOH in place of distilled water resulted in much lower isomerization. It has been suggested that acid-catalyzed reactions observed with silica gel chromatography are probably due to the presence of acidic impurities, since carefully prepared silicas exhibit only a small tendency toward acid-catalyzed sample reaction (Snyder 1975).

The use of water is an effective and simple alternative for deactivation of silica gel. Kubeczka (1973) found no changes in a monoterpene hydrocarbon mixture after column chromatography on silica gel with a water content of 7%. Scheffer et al. (1976) prevented isomerization of various monoterpene hydrocarbons by using an acid- and base-washed silica gel with a 5% water content. These workers reported that the water content of the silica gel should be kept low for the efficient separation of the terpenes. In their fractionation of linden honey volatiles, Black et al. (1989) used silica gel treated with conc. HCl and deactivated with 7% water (by mass).

High performance liquid chromatography (HPLC) is being increasingly used for the prefractionation of flavor mixtures. Due to its greater speed and efficiency, HPLC has advantages over conventional column chromatography. There is less danger of sample rearrangement, since the constituents are in contact with the adsorbent for a much shorter time. The sample throughput is increased and the improved resolution is advantageous in the subsequent GC analyses. Many of the early flavor fractionating studies utilizing HPLC involved the use of synthetic mixtures. Jones et al. (1979) used HPLC to separate a model mixture of terpenes. Reversed-phase HPLC has been utilized to fractionate a mixture of oxygenated terpenoids, monoterpene hydrocarbons, and sesquiterpene hydrocarbons in 12 min (Kubeczka 1981). A general HPLC method for the prefractionation of citrus essential oils was proposed by Chamblee et al. (1985). The essential oils are initially separated by column chromatography into two fractions, hydrocarbons and oxygenated constituents. Each fraction is then further separated by normal phase (silica gel) HPLC using three columns connected in series (Whatman Partisil 5 µm, 10 µm, and a Waters silica radial compression column). The best overall separation was achieved with a ternary solvent system of ethyl acetate:methylene chloride:hexane (8:46:46, v/v). Collected fractions were concentrated and then directly analyzed by GC-MS. The application of the technique to cold pressed lime oil facilitated the identification of 23 new constituents. Schieberle and Grosch (1988) fractionated fresh and aged lemon oil constituents by HPLC (50 × 0.46 cm stainless-steel column filled with 5 µm Shandon Hypersil) with various pentane-diethyl ether mixtures. The use of HPLC with diol-bonded silica to prefractionate flavor extracts has been discussed (Barron 1991). Guth and Grosch (1993) utilized HPLC with diol-bonded silica to fractionate acidic volatiles of stewed beef.

4 Selective Detection for the Analysis of Chiral Fruit Volatiles

4.1 Abbreviations

AED	atomic emission detector
CC	column chromatography
df	film thickness of capillary columns for gas chromatography
FID	flame ionization detector
FPD	flame photometric detector
FTIR	Fourier transform infrared spectroscopy
GC	gas chromatography
HPLC	high performance liquid chromatography
MDGC	multidimensional gas chromatography

MS	mass spectrometry
MS-MS	tandem mass spectrometry
SIM	single ion monitoring
SPE	solid phase extraction
SRM	single reaction monitoring
SWC	selected wavelength chromatogram
TEA	thermal energy analyzer
TIC	total ion chromatogram
LC-GC	on-line coupled liquid chromatography-gas chromatography

4.2 Introduction

Since volatiles extracted from plants, especially fruit, are in most cases very complex mixtures, even separation by capillary gas chromatography is sometimes not powerful enough to achieve resolution of all compounds involved. Therefore, their analysis requires either a more powerful separation or a more selective detection technique.

There are several possibilities to achieve better chromatography:

- off-line fractionation (CC, HPLC, liquid liquid partitioning) of the complex analytes before the final GC analysis step,
- special sample preparation techniques (SPE, affinity chromatography),
- on-line multidimensional chromatographic methods (on-line LC-GC, MDGC).

To achieve selective detection the following detection methods can be useful:

- GC-MS (SIM mode) and GC-MS-MS (SRM mode),
- GC-FTIR (SWC mode),
- GC-FPD (ECD, TEA, AED).

Many of the above methods are frequently used in the analysis of fruit volatiles. The following is focused on the analysis of chiral fruit volatiles using some of the methods mentioned above.

4.3 Approaches in the Analysis of Chiral Fruit Volatiles

Since natural fruit aroma is an ever-growing market, there is also a growing demand for determining authenticity of those products. One possibility to assess this kind of information is to determine the enantiomeric distribution of chiral fruit volatiles present in the aroma. Many fruit volatiles are optically active. Additionally, in many cases, their biosynthesis is stereoselective, rendering optically enriched or pure compounds (Tressl et al. 1988, 1992; Engel et al. 1989). In contrast, many synthesized (nature identical) chiral aroma compounds are racemic. Much work has already been done on the evaluation of the enantiomeric distribution of chiral volatiles in fruits. Examples are γ-lactones (Bernreuther et al. 1989;

Mosandl et al. 1989; Hildenbrand et al. 1990; Kollmansberger et al. 1991), δ-lactones (Bernreuther et al. 1991; Palm et al. 1991), some monoterpenes like α, β-pinene and limonene (Mosandl et al. 1990), linalylacetate (Mosandl and Schubert 1990), linalool (Bernreuther and Schreier 1991), as well as 2-methylbutyric acid and ethyl 2-methylbutanoate (Rettinger et al. 1990, 1991; Takeoka et al. 1990b), and α-ionone (Braunsdorf et al. 1991; Werkhoff et al. 1991b). By comparing the enantiomeric distribution of chiral aroma compounds found in commercial natural aromas with the data found in the literature (for that particular fruit), one is now able to determine the aroma's authenticity in many cases. However, for some chiral flavor compounds there is a wide variation of enantiomeric distribution also in nature. Additionally, in some cases, initially (from biosynthesis) optically enriched compounds may form racemates during sample preparation or due to the existing natural conditions (e.g., keto-enol tautomerism of 3-(2H)-furanones in acidic media, Bruche et al. 1991). In such cases, determination of authenticity has to be done by different techniques, e.g., GC-IRMS (isotope ratio-mass spectrometry) (Mosandl 1992).

In order to evaluate the authenticity of a particular fruit aroma by enantioselective gas chromatography, a chiral compound must be selected which in its natural occurrence has a high optical purity or a well-defined enantiomeric ratio. In addition, the enantiomeric ratio should not change during sample preparation.

The following describes several approaches to analyze chiral fruit volatiles using enantioselective cyclodextrin phases for gas chromatography. Analysis of complex mixtures is achieved either by selective detection or by multidimensional separation techniques or by combining both.

Since the introduction of peralkylated and partially alkylated/acylated derivatives of α–, β- and γ-cyclodextrin phases for gas chromatography (König 1990; Schurig and Nowotny 1990) there have been many publications regarding their applications for the analysis of volatiles. There are some reviews in the field of flavor compounds (Mosandl 1992; Schreier 1993; Werkhoff et al. 1993).

These columns can be used in a gas chromatograph to achieve separation (and thus determination of the enantiomeric distribution) of chiral volatile compounds. However, as mentioned in the introduction, most volatiles extracted from fruits are very complex mixtures, and it is likely that two or more compounds from that mixture coelute with the chiral compound of interest and thus hamper the determination of the correct enantiomeric distribution.

Several solutions for this particular problem are presented in the following sections.

4.3.1 Selective Detection Using GC-FTIR (SWC)

One way to obtain a more accurate determination of the enantiomeric ratio of a target compound is to employ more selective detection than an FID. One example of this is using an FTIR detector. We employed an FTIR detector for the analysis of α-ionone from raspberry fruit. GC-FTIR was introduced to flavor analysis in the early 1980s (Herres 1983, 1987) and is particularly useful for the identification of unknown compounds. For the chirospecific analysis of complex mixtures we utilized the FTIR as a selective detector.

After an analysis is completed, the computer of an FTIR system is able to calculate reconstructed chromatograms on the basis of a specific wavelength window (selected wavelength chromatogram, SWC). By applying this method on a specific absorption wavelength of the target compound, one is able to make interfering peaks invisible and thus calculate the correct enantiomeric ratio of the target compound. We employed this technique for the chirospecific analysis of α-ionone from raspberry fruit. For the analysis of α-ionone the reconstructed chromatogram was based on the C=O stretching band (1680–1730 cm^{-1}) of the molecule.

Figure 1 shows the effect of employing selective detection in the analysis of a raspberry fruit extract. A comparison between FID detection (Fig. 1A) and SWC detection (Fig. 1B) reveals that a correct determination of the enantiomeric distribution of α-ionone is only possible when using selective detection. In Fig. 1C, the result of enantioseparation after spiking the sample with synthetic α-ionone is represented.

The crucial point with this method is that the target molecule needs to have an FTIR-absorption band which does not occur in interfering peaks and vice versa. Furthermore, an FTIR detector is generally (depending on the target compound's chemical structure) not as sensitive as an MS detector, meaning that the absolute detection limit is higher (Schreier and Idstein 1985). We found that the detection limit of about 15 ng/peak is sufficient to perform direct enantiodifferentiation of α-ionone as well as of some other chiral volatiles in fruit extracts and aroma concentrates (Full et al. 1991). However, when the target compounds are present only at trace levels, one has to switch to more sensitive selective detection techniques like GC-MS.

4.3.2 Selective Detection Using GC-MS (SIM)

GC-MS is a more widespread technique than GC-FTIR. Performing an analysis based on the total ion current (TIC) output of a GC-MS results in a selectivity similar to an FID, with the advantage of the additional spectroscopic information. Increasing selectivity and sensitivity can be achieved by using SIM (single ion monitoring) detection. In SIM mode the detector scans the GC effluent only for specific m/z fragments of the compound of interest. By selecting only a few ions (compared to full scan mode) the MS-detector scans the selected ions for a longer period of time, thus resulting in greater sensitivity.

There are many examples in the literature (Krammer et al. 1990b; Haase-Aschoff et al. 1991) and the technique will not be treated in further detail here. As with FTIR, the crucial point with this technique is that the interference could also have the same mass fragments as the target compound, thus again

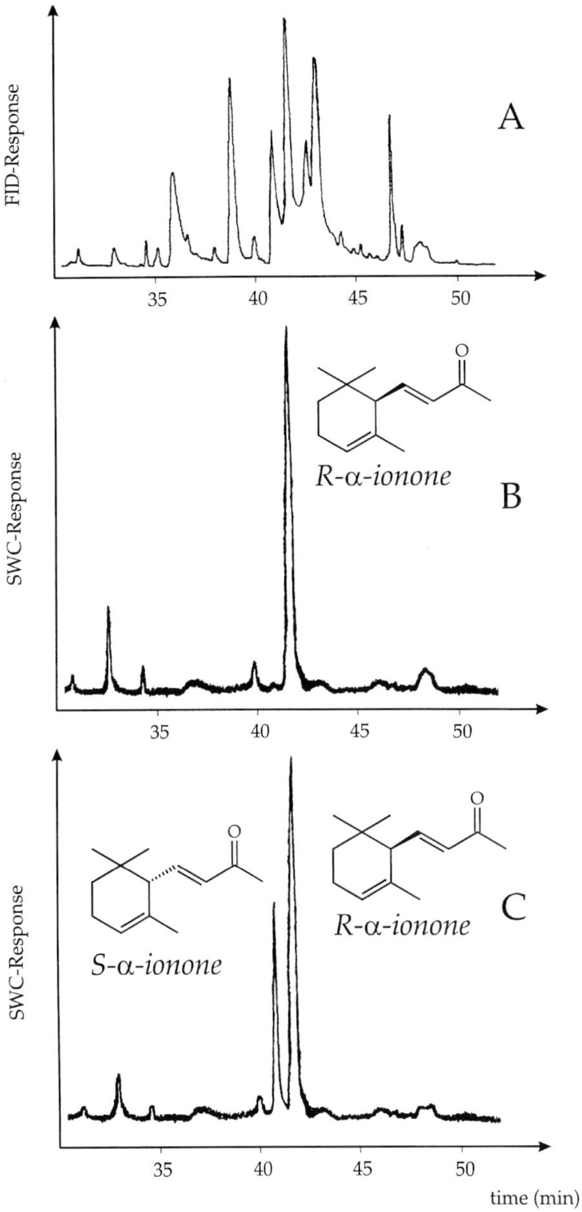

Fig. 1A–C. GC-FTIR analysis of α-ionone from raspberry. **A** FID detection. **B** FTIR-SWC detection (wavelength window 1680–1730 cm^{-1}). **C** As **B**, but spiking the sample with synthetic α-ionone. Separation column: J&W CDex-B (permethylated β-cyclodextrin, 25 m × 0.25 mm. df = 0.25 μm). Temperature programming: 80 °C, 1.5 °C min^{-1} to 280 °C

hampering the determination of the correct enantiomeric ratio (Hener and Mosandl 1993).

4.3.3 Separation Employing Multidimensional Systems (MDGC)

Multidimensional separations are a powerful tool for analyzing complex mixtures. Basically, separations with this kind of setup are done using two different columns. For the analysis of fruit volatiles, multidimensional gas chromatography (MDGC) is an ideal tool. Usually, a preseparation is achieved on the first column and the final separation step is conducted after selective on-line transfer (heart-cut) of the fraction containing the target compounds onto the second column. There are several MDGC systems commercially available (Krammer et al. 1990a). Some of them offer a separate GC oven for each column. The examples of multidimensional GC separations in this chapter were all performed using a pressure-regulated double oven gas chromatograph (Siemens Sichromat 2, Siemens, Germany). For the analysis of chiral fruit volatiles, the first column is a nonenantioselective column and it ideally isolates the compound of interest (Schomburg et al. 1984a). A heart-cut then transfers this isolated compound and possible interfering compounds onto the second, enantioselective column where the final analysis step is

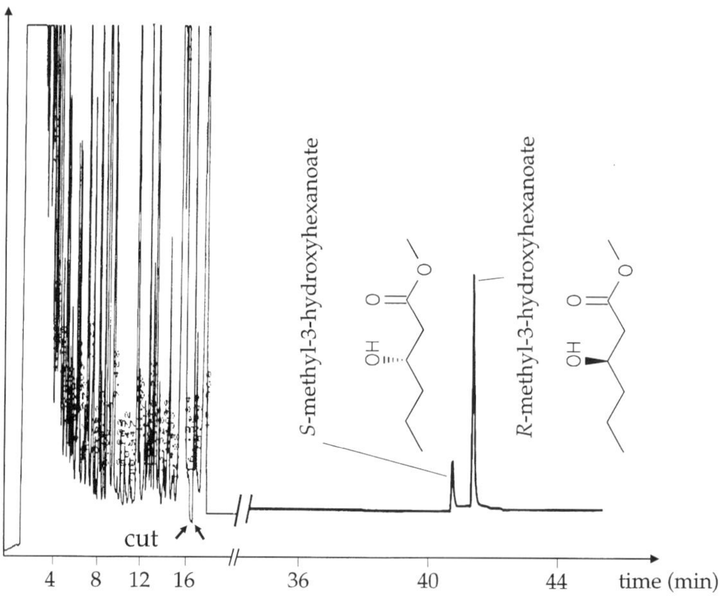

Fig. 2. MDGC analysis of methyl-3-hydroxyhexanoate from peelings of lulo (*Solanum vestissimum* D.). Column 1: J&W DB-Wax (25 m × 0.25 mm, df = 0.25 μm). Temperature programming 60 °C, 4 °C min^{-1} to 240 °C. Column 2: Heptakis-2,6-di-O-methyl-3-O-pentyl-β-cyclodextrin (30 m × 0.25 mm, df = 0.3 μm). Temperature programming 60 °C, 20 min isothermal, 2 °C min^{-1} to 200 °C. Heart-cut from column 1 to column 2: 16.2–16.5 min

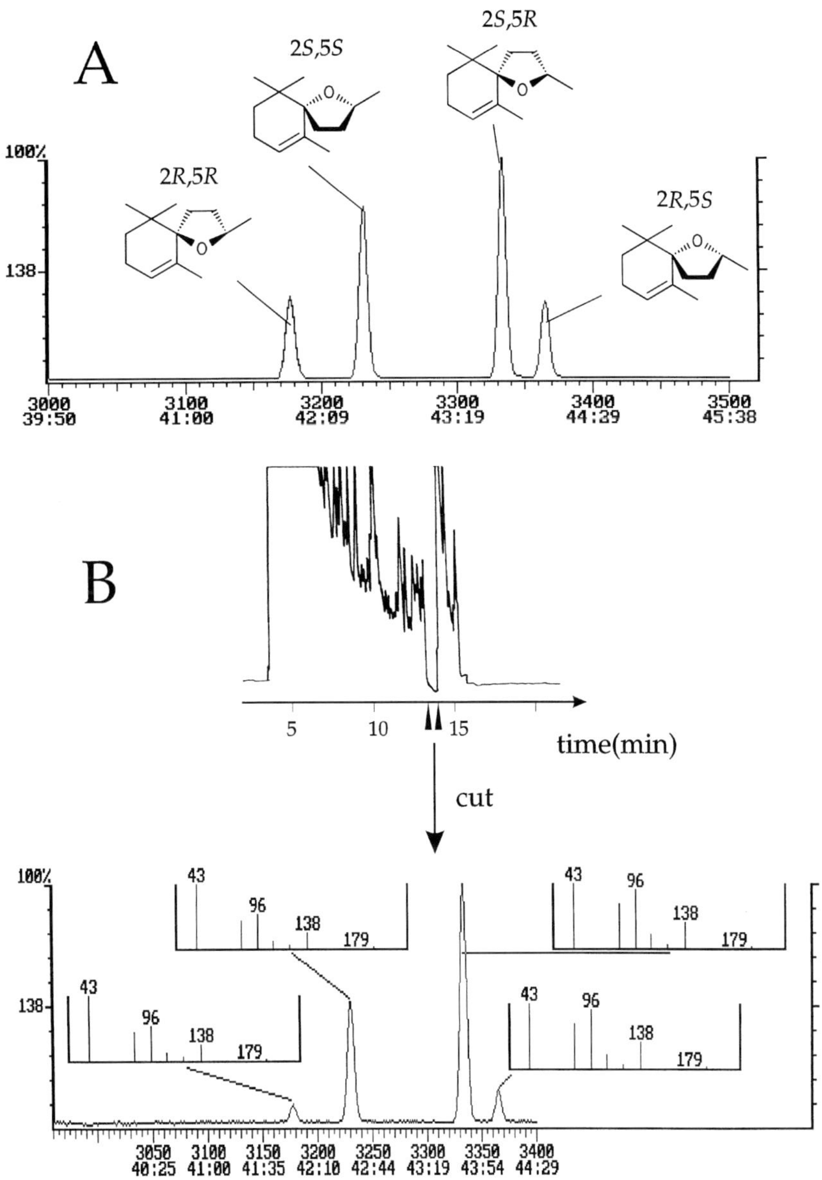

Fig. 3A,B. MDGC-MS analysis of theaspirane enantiomers. **A** Reference compound to determine the order of elution. **B** Theaspiranes in an aglycon fraction from purple passion fruit. Column 1: J&W DB-5(25 m × 0.25 mm,df = 0.25 μm). Temperature programming 60 °C, 5 °C min^{-1} to 300 °C. Column 2: J&W C-Dex B (permethylated β-cyclodextrin) (25 m × 0.25 mm, df = 0.25 μm). Temperature programming 80 °C, 20 min isothermal, 2 °C min^{-1} to 200 °C. Heart-cut from column 1 to column 2: 14.0–14.6 min

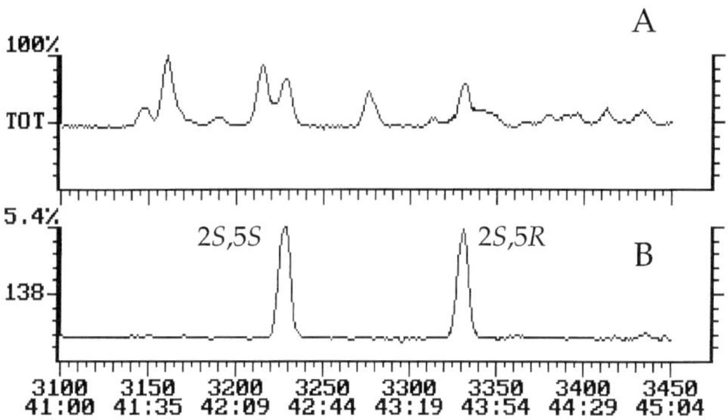

Fig. 4A,B. MDGC-MS analysis of theaspiranes in guava fruit. Comparison of TIC **A** and SIM **B** chromatograms

performed. The analysis of methyl 3-hydroxyhexanoate in lulo (*Solanum vestissimum* D.) peelings is presented (Fig. 2) as an example of using this technique (Suárez et al. 1993).

4.3.4 Combining MDGC and Selective Detection

Despite the high resolving power of MDGC, there still may be samples which have interfering peaks. If this happens, there are two possibilities to solve the problem: a more efficient off-line sample preparation can be performed, or more selectivity can be added to the detection. For the investigation of the enantiomeric distribution of theaspiranes, we employed MDGC-MS coupling, which resulted in excellent selectivity and sensitivity.

Figure 3 shows the enantiodifferentiation of theaspiranes, Fig. 3A order of elution, Fig. 3B analysis of an aglycone fraction of purple passion fruit (Full et al. 1993).

As mentioned above, in some cases, even when using MDGC techniques, some interfering peaks may occur. This was the case when analyzing the volatiles of guava fruits. As seen in Fig. 4, when employing nonselective detection, some interfering peaks occur which are invisible upon selectively detecting the m/z 138 ion.

5 Conclusions

The chapter has discussed two important challenges in the analysis of fruit flavors. The first is the quantitative isolation of flavor compounds which may occur at trace levels. There is no single ideal sample preparation method, and frequently it may be necessary to employ more than one method to obtain accurate qualitiative and quantitative data on a fruit sample. The chemist must consider the formation of artifacts during sample preparation and analysis. Glycosidic flavor precursors occur widely in fruit, and application of heat during sample preparation may release additional flavor constituents, distorting the original profile of volatiles. Labile constituents may also be transformed by even mild sample preparation conditions. The second involves the separation, detection, and quantitation of chiral flavor constituents. There has been great progress in the chirospecific analysis of flavor compounds due to introduction of modified cyclodextrin capillary columns and multidimensional gas chromatography. Knowledge of the enantiomeric composition of flavor constituents gives insight into their biosynthesis. Although selective detection is in many cases a good solution for problems of coeluting interferences, one should still pay attention to sample preparation. More efficient sample preparation will lead to better chromatographic separation. Better chromatography will yield more accurate analytical results. The most powerful combination to achieve accurate results is a combination of a multidimensional separation technique and selective detection. Such a combination is a powerful tool for assessing the authenticity of natural fruit flavors.

References

Barron D (1991) HPLC using diol-bonded silica, an alternative to silica gel in the prefractionation of aroma extracts. Z Lebensm Unters Forsch 193:454–459

Berger RG, Drawert F, Kollmannsberger H, Nitz S (1985) Natural occurrence of undecanes in some fruits and vegetables. J Food Sci 50:1655–1656, 1667

Bernreuther A, Schreier P (1991) Multidimensional gas chromatography/mass spectrometry: a powerful tool for the direct chiral evaluation of aroma compounds in plant tissues. II. Linalool in essential oils and fruits. Phytochem Anal 2:167–170

Bernreuther A, Christoph N, Schreier P (1989) Determination of the enantiomeric composition of γ-lactones in complex natural matrices using multidimensional capillary gas chromatography. J Chromatogr 481:363–367

Bernreuther A, Bank J, Krammer G, Schreier P (1991) Multidimensional gas chromatography/mass spectrometry: a powerful tool for direct chiral evaluation of aroma compounds in plant tissues, I. 5-alkanoides in fruits. Phytochem Anal 2:43–47

Blank I, Fisher KH, Grosch W (1989) Intensive natural odorants of linden honey. Differences from honeys of other botanical origin. Z Lebensm Unters Forsch 189:426–433

Braunsdorf R, Hener U, Lehmann D, Mosandl A (1991) Analytische Differenzierung zwischen natürlich gewachsenen, fermentativ erzeugten und synthetischen (naturidentischen) Aromastoffen Mitt. In: Herkunftsspezifische Analyse des (E)-α(β)-ionons. Dtsch Lebensm Rundsch 87:277–280

Bruche G, Schmarr HG, Bauer A, Mosandl A, Rapp A, Engel L (1991) Stereoisomere Aromastoffe LI: Stereodifferenzierung chiraler Furanone. Möglichkeiten und Grenzen der herkunftsspezifischen Aromastoff-Analyse. Z Lebensm Unters Forsch 193:115–118

Brunke EJ, Hammerschmidt FJ, Schmaus G (1993) Flower scents of some traditional medicinal plants. In: Teranishi R, Buttery RG, Sugisawa H (eds) Bioactive volatile compounds from plants. ACS Symposium Series 525, American Chemical Society, Washington, DC, pp 282–296

Buttery R, Ling L (1993) Enzymatic production of volatiles in tomatoes. In: Schreier P, Winterhalter P (eds) Progress in flavour precursor studies: analysis-generation-biotechnology. Allured Publishing Corporation, Carol Stream, Illinois, pp 137–146

Buttery RG, Ling LC (1996) Methods of isolating food and plant volatiles for biotechnology studies. In: Takeoka G, Teranishi R, Williams P, Kobayashi A (eds) Biotechnology for improved foods and flavors. ACS Symposium Series 637, American Chemical Society, Washington, DC pp 240–248

Buttery RG, Ling LC, Guadagni DG (1969) Food volatiles – volatilities of aldehydes, ketones, and esters in dilute water solution. J Agric Food Chem 17:385–389

Buttery RG, Ling LC, Light DM (1987a) Tomato leaf volatile aroma components. J Agric Food Chem 35:1039–1042

Buttery RG, Teranishi R, Ling LC (1987b) Fresh tomato aroma volatiles: a quantitative study. J Agric Food Chem 35:540–544

Buttery RG, Teranishi R, Ling LC, Flath RA, Stern DJ (1988) Quantitative studies on origins of fresh tomato aroma volatiles. J Agric Food Chem 36:1247–1250

Chamblee TS, Clark BC Jr, Radford T, Iacobucci GA (1985) General method for the high-performance liquid chromatographic prefractionation of essential oils and flavor mixtures for gas chromatographic-mass spectrometric analysis. Identification of new constituents in cold pressed lime oil. J Chromatogr 330:141–151

Davies DD (1980) Anaerobic metabolism and the production of organic acids. In: Stumpf PK, Conn EE (eds) The biochemistry of plants: a comprehensive treatise, vol 2. Academic Press, New York, pp 581–611

Engel KH, Flath RA, Buttery RG, Mon TR, Ramming DW, Teranishi R (1988a) Investigation of volatile constituents in nectarines. 1. Analytical and sensory characterization of aroma components in some nectarine cultivars. J Agric Food Chem 36:549–553

Engel KH, Ramming DW, Flath RA, Teranishi R (1988b) Investigation of volatile constituents in nectarine. 2. Changes in aroma composition during nectarine maturation. J Agric Food Chem 36:1003–1006

Engel KH, Heidlas J, Albrecht W, Tressl R (1989) Biosynthesis of chiral flavor and aroma compounds in plants and microorganisms. In: Teranishi R, Buttery RG, Shahidi F (eds) Flavor chemistry: trends and developments. ACS Symposium Series 388, American Chemical Society, Washington, pp 8–22

Fröhlich O, Schreier P (1989) Additional volatile constituents of carambola (*Averrhoa carambola* L.). Flavour Fragr J 4:177–184

Fröhlich O, Schreier P (1990) Volatile constituents of loquat (*Eriobotrya japonica* Lindl.) fruit. J Food Sci 55:176–180

Full G, Bernreuther A, Krammer G, Schreier P (1991) Lebensmittelanalytik: On-Line-HRGC-FTIR zur direkten Enantiodifferenzierung von Aromastoffen aus komplexen Matrices. LABO 30–34

Full G, Winterhalter P, Schmidt G, Herion P, Schreier P (1993) MDGC-MS: a powerful tool for enantioselective flavor analysis. J High Resolut Chromtogr Chromatogr Commun 16:642–644

Grob K (1973) Organic substances in potable water and its precursor. Part I. Methods of their determination by gas-liquid chromatography. J Chromatogr 84:255–273

Grob K (1985) Headspace gas analysis: the role and the design of concentration traps specifically suitable for capillary gas chromatography. J Chromatogr 321:45–58

Grob K, Grob G (1974) Organic substances in potable water and its precursor. Part II. Applications in the area of Zürich. J Chromatogr 90:303–313

Grob K, Grob G (1979) Practical capillary gas chromatography – a systematic approach. J High Resolut Chromatogr Chromatogr Commun 2:109–117

Grob K, Habich A (1985) Headspace gas analysis: The role and the design of concentration traps specifically suitable for capillary gas chromatography. J Chromatogr 321:45–58

Grob K, Zürcher F (1976) Stripping trace organic substances from water; equipment and procedure. J Chromatogr 117:285–294

Grob K, Grob K Jr, Grob G (1975) Organic substances in potable water and its precursor. III. The closed-loop stripping procedure compared with rapid liquid extraction. J Chromatogr 106:299–315

Guth H, Grosch W (1989) 3-Methylnonane-2,4-dione – an intense odour compound formed during flavour reversion of soya-bean oil. Fat Sci Technol 91:225–230

Guth H, Grosch W (1993) 12-Trimethyldecanal, a species-specific odorant of stewed beef. Lebensm Wiss Technol 26:171–177

Haase-Aschoff K, Haase-Aschoff I, Laub H (1991) Chirale Analyse und Bewertung der 2-Methyl-Buttersäure und ihrer Ester in Früchten und Fruchtprodukten. Lebensmittelchemie 45:107–109

Hamilton-Kemp TR, Loughrin JH, Andersen RA, Rodriguez JG (1993) Volatile compounds from strawberry foliage and flowers. Air versus nitrogen entrainment; quantitative changes and plant development. In: Teranishi R, Buttery RG, Sugisawa H (eds) Bioactive volatile compounds from plants. ACS Symposium Series 525, American Chemical Society, Washington, DC, pp 229–239

Hener U, Mosandl A (1993) Zur Herkunftsbeurteilung von Aromen. Dtsch Lebensm Rundsch 89:307–312

Herres W (1987) HRGC-FTIR: Capillary gas chromatography-Fourier transform infrared spectroscopy - theory and applications. Hüthig, Heidelberg

Herres W, Idstein H, Schreier P (1983) Flavor analysis by HRGC/FT-IR: Cherimoya (*Annona cherimolia*, Mill.) fruit volatiles. J High Resolut Chromatogr Chromatogr Commun 6:590–594

Hildenbrand K, Christoph N, Bernreuther A (1990) Aktuelle Aromastoff-Analytik in der amtlichen Weinüberwachung sowie analytische Charakterisierung von Weinfehlern. Dtsch Lebensm Rundsch 2:39–44

Hunter GLK, Brogen WB Jr (1963) Isomerization and disproportionation of d-limonene on silica gel. J Org Chem 28:1679–1682

Idstein H, Schreier P (1985a) Volatile constituents of Alphonso mango (*Mangifera indica*). Phytochemistry 24:2313–2316

Idstein H, Schreier P (1985b) Aromastoffe der Papaya Frucht (*Carica papaya*, L., var. Solo). Lebensm Wiss Technol 18:164–169

Idstein H, Schreier P (1985c) Volatile constituents from guava (*Psidium guajava*, L.) fruit. J Agric Food Chem 33:138–143

Idstein H, Herres W, Schreier P (1984) High resolution gas chromatography-mass spectrometry and -Fourier transform infrared analysis of cherimoya (*Annona cherimolia*, Mill.) volatiles. J Agric Food Chem 32:383–389

Ismail HH, Tucknott OG, Williams AA (1980) The collection and concentration of aroma components of soft fruit using Porapak Q. J Sci Food Agric 31:262–266

Jennings WG (1979) Vapor-phase sampling. J High Resolut Chromatogr Chromatogr Commun 2:221–224

Jones BB, Clark BC Jr, Iacobucci GA (1979) Semipreparative high performance liquid chromatographic separation of a characterized flavor mixture of monoterpenes. J Chromatogr 178:575–578

Jung HP, Sen A, Grosch W (1992) Evaluation of potent odorants in parsley leaves [*Petroselinum crispum* (Mill.) Nym. ssp. crispum] by aroma extract dilution analysis. Lebensm Wiss Technol 25:55–60

Kaiser RA (1993) On the scent of orchids. In: Teranishi R, Buttery RG, Sugisawa H (eds) Bioactive volatile compounds from plants. ACS Symposium Series 525, American Chemical Society, Washington, DC, pp 240–268

Kollmannsberger H, Nitz S, Drawert F (1991) Über den Nachweis von nicht-natürlichen Aromen in Fruchtschaumweinen. 3. Enantiomere γ-Lactone in Ananasfrüchten und Ananasprodukten. Chem Mikrobiol Technol Lebensm 13:58–63

König WA (1990) Modifizierte Cyclodextrine als chirale Trennphasen in der Gaschromatographie. Kontakte 2:3–14

Krammer G, Bernreuther A, Schreier P (1990a) Multidimensionale Gas-Chromatographie. GIT Fachz Lab 3:306–312

Krammer G, Bernreuther A, Herderich M, Schreier P (1990b) Multidimensionale Gas-Chromatographie-Massenspektrometrie (MDGC-MS): Chirale Analytik mit dem ITD 800. Finnigan MAT Appl Rep 77

Kubeczka KH (1973) Vortrennung ätherischer Öle und ähnlich komplexer Stoffgemische für die GC-Analyse durch modifizierte Trockensäulen-Chromatographie. Chromatographia 6:106–108

Kubeczka KH (1981) Application of HPLC for the separation of flavour compounds. In: Schreier P (ed) Flavour '81, 3rd Weurman Symp. de Gruyter, Berlin, pp 345–359

McFadden WH, Teranishi R, Black DR, Day JC (1963) Use of capillary gas chromatography with a Time-of-Flight Mass Spectrometer. J Food Sci 28:316–319
Mitzer B (1964) Decomposition of oxygenated terpenes in the injection heaters of gas chromatographs. Anal Chem 36:242
Mosandl A (1992) Capillary GC in quality assessment of flavours and fragrances. J Chromatogr 624:267–292
Mosandl A, Schubert V (1990) Stereoisomere Aromastoffe XXXIX. Chirale Inhaltsstoffe ätherischer Öle (I) Stereodifferenzierung des Linalylacetats – ein neuer Weg zur Qualitätsbeurteilung des Lavendelöls. Z Lebensm Unters Forsch 190:506–510
Mosandl A, Palm U, Günther C, Kustermann A (1989) Stereoisomeric flavour compounds XXV. Stereodifferentiation of enantiomeric γ-lactones by HRGC on chiral stationary phases. Z Lebensm Unters Forsch 188:148–150
Mosandl A, Hener U, Kreis P, Schmarr HG (1990) Enantiomeric distribution of α-pinene, β-pinene and limonene in essential oils and extracts. Part 1. Rutaceae and Gramineae. Flavour Fragr J 5:193–199
Ohloff G, Flament I, Pickenhagen W (1985) Flavor chemistry. Food Rev Int 1:99–148
Palm U, Askari C, Hener U, Jakob E, Mandler C, Geßner M, Mosandl A, König W, Evers P, Krebber R (1991) Stereoisomere Aromastoffe XLVII. Direkte stereospezifische HRGC-Analyse natürlicher δ-Lactone. Z Lebensm Unters Forsch 192:209–213
Petro-Turza M (1986–1987) Flavor of tomato and tomato products. Food Rev Int 2:309–351
Pickenhagen W, Velluz A, Passerat JP, Ohloff G (1981) Estimation of 2,5-dimethyl-4-hydroxy-3(2H)-furanone (Furaneol) in cultivated and wild strawberries, pineapples and mangoes. J Sci Food Agric 32:1132–1134
Preininger M, Grosch W (1994) Evaluation of key odorants of the neutral volatiles of Emmentaler cheese by the calculation of odour activity values. Lebensm Wiss Technol 27:237–244
Rapp A, Knipser W (1980) Eine neue Methode zur Anreicherung von Dampfkomponenten. Dargestellt am Beispiel des Weines. Chromatographia 13:698–702
Rettinger K, Weber B, Mosandl A (1990) Stereoisomere Aromastoffe XLII. 2-Methylbuttersäure-Enantiomerenverteilung in Äpfeln und apfelhaltigen Lebensmitteln. Z Lebensm Unters Forsch 191:265–268
Rettinger K, Burschka C, Scheeben P, Fuchs H, Mosandl A (1991) Chiral 2-alkylbranched acids, esters and alcohols, preparation and stereospecific flavour evaluation. Tetrahedron Asymmetry 2:965–968
Scheffer JJC, Koedam A, Baerheim-Svendsen A (1976) Occurrence and prevention of isomerization of some monoterpene hydrocarbons from essential oils during liquid-solid chromatography on silica gel. Chromatographia 9:425–432
Schieberle P (1995) Quantitation of important roast-smelling odorants in popcorn by stable isotope dilution assays and model studies on flavor formation during popping. J Agric Food Chem 43:2442–2448
Schieberle P, Grosch W (1988) Identification of potent flavor compounds formed in an aqueous lemon oil/citric acid emulsion. J Agric Food Chem 36:797–800
Schieberle P, Ofner S, Grosch W (1990) Evaluation of potent odorants in cucumbers (*Cucumis sativus*) and muskmelons (*Cucumis melo*) by aroma extract dilution analysis. J Food Sci 55: 193–195
Schomburg G, Husmann H, Hübinger E, König WA (1984a) Multidimensional capillary gas chromatography-enantiomeric separations of selected cuts using a chiral second column. J High Resolut Chromtogr Chromatogr Commun 7:404–410
Schomburg G, Husmann H, Podmaniczy L, Weeke F, Rapp A (1984b) Coupled GC methods for separation, identification, and quantitative analysis of complex mixtures: MDGC, GC-MS, GC-IR, LC-GC. In: Schreier P (ed) Analysis of volatiles: Methods Appl Proc – Int Worksh Meet Date 1983. de Gruyter, Berlin, pp 121–150
Schreier P (1993) Chirality evaluation in flavour precursor studies. In: Schreier P, Winterhalter P (eds) Progress in flavour precursor studies: analysis-generation-biotechnology. Allured, Wheaton, pp 45–61
Schreier P, Idstein H (1985) HRGC-FTIR in flavour analysis – limits and perspectives. Z Lebensm Unters Forsch 181:183–188

Schurig V, Nowotny HP (1990) Gaschromatographische Enantiomerentrennung an Cyclodextrinderivaten. Angew Chem 102:969–1108

Sen A, Laskawy G, Schieberle P, Grosch W (1991) Quantitative determination of β-damascenone in foods using a stable isotope dilution assay. J Agric Food Chem 39:757–759

Snyder LR (1975) Adsorption. In: Heftmann E (ed) Chromatography. Van Nostrand Reinhold, New York, 54 pp

Stromvall A-M, Petersson G (1992) Protection of terpenes against oxidation and acid decomposition on adsorbent cartridges. J Chromatogr 589:385–389

Suárez M, Duque C, Bicchi C, Wintoch H, Full G, Schreier P (1993) Volatile constituents from the peelings of lulo (*Solanum vestissimum* D.) fruit. Flavour Fragr J 8:215–220

Surburg H, Güntert M, Harder H (1993) Volatile compounds from flowers – analytical and olfactory aspects. In: Teranishi R, Buttery RG, Sugisawa H (eds) Bioactive volatile compounds from plants. ACS Symposium Series 525, American Chemical Society, Washington, DC, pp 168–186

Takeoka GR, Güntert M, Smith SL, Jennings W (1986a) Changes in aroma concentrates during storage. In: Parliment TH, Croteau R (eds) Biogeneration of aromas. ACS Symposium Series 317, American Chemical Society, Washington, DC, pp 65–74

Takeoka GR, Güntert M, Flath RA, Wurz RE, Jennings W (1986b) Volatile constituents of kiwi fruit (*Actinidia chinensis* Planch.). J Agric Food Chem 34:576–578

Takeoka GR, Flath RA, Güntert M, Jennings W (1988) Nectarine volatiles: vacuum steam distillation versus headspace sampling. J Agric Food Chem 36:553–560

Takeoka GR, Flath RA, Mon TR, Teranishi R, Güntert M (1990a) Volatile constituents of apricot (*Prunus armeniaca*). J Agric Food Chem 38:471–477

Takeoka G, Flath RA, Mon TR, Buttery RG, Teranishi R, Güntert M, Lautamo R, Szejtli J (1990b) Further applications of permethylated β-cyclodextrin capillary gas chromatographic columns. J High Resolut Chromtogr Chromatogr Commun 13:202–206

Takeoka GR, Flath RA, Buttery RG, Winterhalter P, Güntert M, Ramming DW, Teranishi R (1992) Free and bound flavor constituents of white-fleshed nectarines. In: Teranishi R, Takeoka GR, Güntert M (eds) Flavor precursors – thermal and enzymatic conversions. ACS Symposium Series 490, American Chemical Society, Washington, DC, pp 116–138

Teranishi R, Corse JW, McFadden WH, Black DR, Morgan AI Jr (1963) Volatiles from strawberries. I. Mass spectral identification of the more volatile components. J Food Sci 28:478–483

Tressl R, Heidlas J, Albrecht W, Engel KH (1988) Biogenesis of chiral hydroxyacid esters. In: Schreier P (ed) Bioflavour '87. de Gruyter, Berlin, pp 221–236

Tressl R, Albrecht W, Heidlas J (1992) Chiral flavour compounds – Properties and analysis. In: Patterson RLS, Charlwood BV, MacLeod G, Williams AA (eds) Bioformation of flavours. Royal Society of Chemistry, Cambridge, pp 109–122

Umano K, Shibamoto T (1987) Analysis of headspace volatiles from overheated beef fat. J Agric Food Chem 35:14–18

Werkhoff P, Brennecke S, Bretschneider W (1991a) Fortschritte bei der chirospezifischen Analyse natürlicher Riech- und Aromastoffe. Chem Mikrobiol Technol Lebensm 13:129–152

Werkhoff P, Bretschneider W, Güntert M, Hopp R, Surburg H (1991b) Chirospecific analysis in flavor and essential oil chemistry. Part B. Direct enantiomer resolution of trans-α-ionone and trans-α-damascone by inclusion gas chromatography. Z Lebensm Unters Forsch 192:111–115

Werkhoff P, Brennecke S, Bretschneider W, Güntert M, Hopp R, Surburg H (1993) Chirospecific analysis in essential oil, fragrance and flavor research. Z Lebensm Unters Forsch 196:307–328

Williams PJ, Strauss CR, Wilson B (1980) Hydroxylated linaool derivatives as precursors of volatile monoterpenes of muscat grapes. J Agric Food Chem 28:766–771

Wrolstad RE, Jennings W (1965) Chromatostrip isomerization of terpenes. J Chromatogr 18:318–324

Gas Chromatographic Techniques for the Measurement of Isoprene in Air

R.R. Arnts, E.L. Singsaas, and T.D. Sharkey

1 Introduction

Isoprene emission from plants was discovered by Sanadze in 1966 (reviewed in Sanadze 1991) while studying plant-insect interactions. Isoprene, 2-methyl-1,3-butadiene, appears to have no role in plant-insect interactions but the amount of isoprene emitted is so great that it affects atmospheric chemistry and may be important to the carbon balance of some plants. Rasmussen and Went (1964) independently discovered isoprene emission from plants while studying plant hydrocarbon emissions (Went 1960). Isoprene is the root member of a large class of compounds, many of which are important in plant fragrance. Isoprene itself is not universally appealing when in high enough concentration to be detected by smell.

2 Types of Measurement

Measurements of isoprene are commonly made in studies of (1) the biochemical reactions involved in its synthesis, (2) emission from intact leaves, (3) fluxes from forest canopies, and (4) concentration in the atmosphere. These measurements increase in difficulty as concentrations descend from the ppm (parts per million by volume, a measure of mole fraction of the gas in air) range of static enclosed atmospheres to the ppb (parts per billion) to ppt (parts per trillion) levels of the free atmosphere above many ecosystems. Table 1 provides a rough guide to the concentrations typically expected in these measurement situations.

The volume of sample available for analysis and its concentration provide primary information for selecting the proper analytical scheme. However, frequency of measurement and speed of analysis will determine whether sample integrity between sampling and analysis need to be considered. All the methods described here are discrete measurements of a sampled volume. Analysis times can be as short as a few minutes or as long as an hour to make a single measurement. If there is a need for continuous, near-instantaneous measurement, the reader should consult the chemiluminesence method of Hills et al. (1991) and Hills and Zimmerman (1990).

Modern Methods of Plant Analysis, Vol. 19
Plant Volatile Analysis
Edited by H.-F. Linskens and J.F. Jackson

Table 1. Typical concentrations of isoprene (in ppb) in different measurement situations

Biochemical reactions	50–500
Emission from intact leaves	50–500
Fluxes from forest canopies	2–20
Concentration in the atmosphere	0.5–20

2.1 Subleaf Level Analyses

Isoprene emission from leaf pieces, leaf extracts, and enzyme preparations are useful in correlating isoprene production with enzyme amount or activity. Such measurements are usually carried out in a sealed vial followed by analysis of the headspace gas or by solvent layering. Isoprene will build up to high levels with no apparent feedback on its own synthesis (Sharkey 1991), making subsequent measurement easier. Vial size can vary over a wide range; we have used 1- to 10-ml vials with good results. We use vials suitable for immersing in a water bath so that temperature can be controlled, a critical parameter in many isoprene studies. Given the small volume of these vials and the high concentrations of isoprene, small samples are withdrawn for assay.

2.1.1 Headspace Analysis

Headspace analysis is often the method of choice when subleaf level measurements are made. For headspace analysis only a small amount of gas is withdrawn with a gas-tight syringe from the vial used for the reaction. When a significant portion of the gas of the headspace is withdrawn, the pressure inside the vial and syringe can be significantly below atmospheric. When the syringe is withdrawn, outside air can infiltrate the syringe, diluting the sample. Using a syringe with a valve at the opening, and closing the valve before pulling the needle out of the vial can overcome this problem. Once outside the vial, the pressure in the syringe can be increased by pressing the plunger, then the pressure is relieved by opening the valve. While some gas is lost, the concentration of isoprene in the syringe should be the same as in the headspace. Another method to overcome this problem is replacement of the volume withdrawn from the vial. If air is introduced, dilution of the isoprene in the head space must be calculated. We have used two syringes, one filled with water which is injected at the same rate that the head space gas is withdrawn with the other syringe. This analysis relies on the assumption that all the isoprene will be in the gas phase and none left in the aqueous phase.

2.1.2 Solvent Layering

Another method for collecting isoprene from reactions carried out in liquid phase in vials is to layer an organic solvent such as heptane on the aqueous phase. This technique requires that isoprene partition strongly into the organic solvent, otherwise the technique will be more qualitative than quantitative. If the partitioning between air and heptane is high enough, practically all the isoprene will be in the heptane, which is then removed by syringe for injection into a gas chromatograph. Since a liquid is being withdrawn, the loss of pressure in the vial is not a problem. Since this is liquid sample, standard liquid injection techniques may be employed to analyze for isoprene.

2.2 Leaf Level Sampling

One of the most commonly performed measurements is to estimate the flux of isoprene from an intact leaf. One method is simply taking a small sample of the air around a leaf. Near the leaf, the isoprene concentration will be higher than in the bulk air if the leaf is emitting substantial amounts of isoprene. This method is useful for qualitative surveys of plants to determine which might emit isoprene (M. Lerdau, pers. comm.). However, in most cases, the leaf is enclosed in a cuvette and the air supply is controlled. This reduces contamination from the surrounding

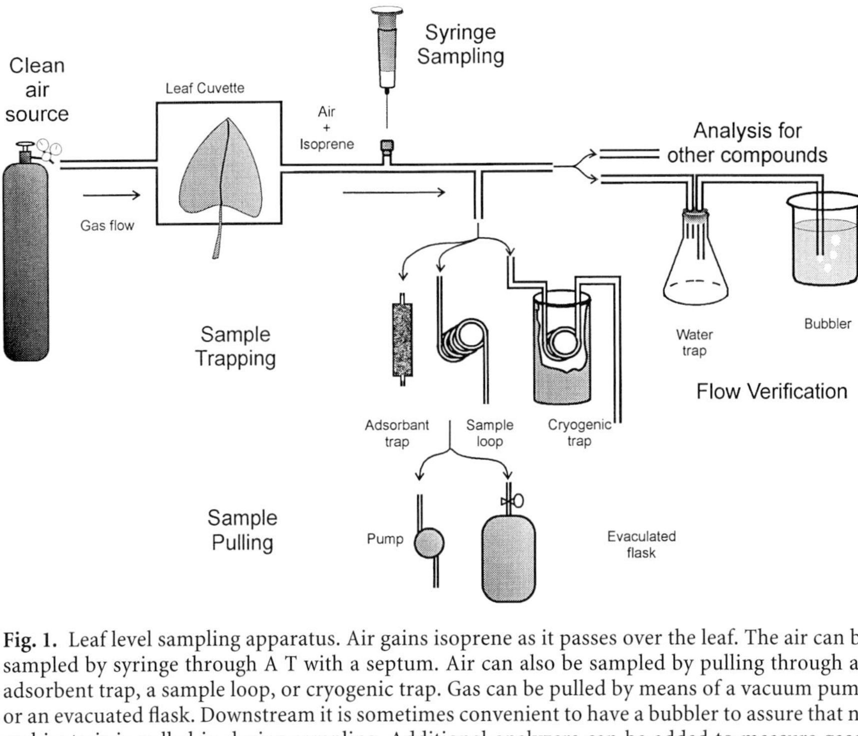

Fig. 1. Leaf level sampling apparatus. Air gains isoprene as it passes over the leaf. The air can be sampled by syringe through A T with a septum. Air can also be sampled by pulling through an adsorbent trap, a sample loop, or cryogenic trap. Gas can be pulled by means of a vacuum pump or an evacuated flask. Downstream it is sometimes convenient to have a bubbler to assure that no ambient air is pulled in during sampling. Additional analyzers can be added to measure gases such as carbon dioxide

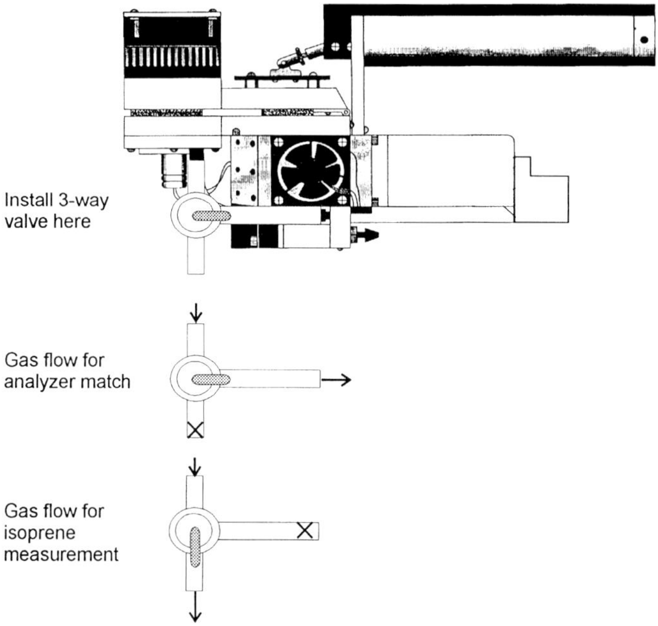

Fig. 2. Installation of a three-way valve on the LiCor 6400 leaf cuvette to allow sampling gas after it has passed over the leaf. The valve is switched to allow gas analyzer matching or isoprene measurement

air and allows the isoprene to reach higher concentrations, making measurement easier.

2.2.1 Dynamic Gas Exchange Cuvettes

Most measurements of isoprene from intact leaves make use of sophisticated gas exchange system. Many gas-exchange cuvettes do not have temperature regulation capabilities, making isoprene measurements difficult. The best leaf level measurements are made with gas-exchange systems with temperature-controlled cuvettes. In many of these systems, leaf temperature can be controlled to within 0.1 °C. We have used a number of methods for transferring the gas leaving the gas-exchange cuvette to a gas chromatograph. We have removed gas directly from a gas-exchange cuvette on occasion but it is much more common, and in most cases recommended, to sample the gas that has left the cuvette (Fig. 1). In many systems it is easy to obtain access to the gas stream leaving the cuvette, but in the new LiCor 6400 gas-exchange system, this requires installing a three-way valve in the short piece of tubing under the chamber. The gas flow must be directed into the lower part of the chamber during the gas analyzer match procedure, then can be directed past collection ports for the rest of the time (see Fig. 2).

2.3 Branch and Plant Enclosures

2.3.1 Bag Enclosures

Another method that has been used in the past is enclosure of a branch or small seedling in bags, followed by evacuation, refilling the bag with clean air, then analyzing that air after some period of time (usually about 10–15 min). While this technology played an important role in establishing early inventories of isoprene emission (Zimmerman 1979), there are now better methods widely available and the bag enclosure method is no longer recommended except for qualitative surveys of isoprene-emitting species. The bag enclosure method suffers from several drawbacks: (1) lack of temperature control results in rapid rise in air temperature, particularly when the bag is exposed to direct sunlight, (2) despite refilling the bag with dry air, humidity often rises to 100% within a few minutes, resulting in condensation on the bag walls, and (3) difficulties in making a meaningful radiation measurement(s) of a branch with widely varying degrees of exposure to direct sunlight. Since isoprene emission is particularly sensitive to temperature and radiation (Sanadze and Kursanov 1966; Rasmussen and Jones 1973; Tingey et al. 1979; Monson and Fall 1989; Loreto and Sharkey 1990; Guenther et al. 1991; Sharkey and Loreto 1993), variations in leaf temperature and radiation will result in a large uncertainty in isoprene emission. Additionally, the collapsing bag may damage plant tissue, resulting in altered emission rates (Juuti et al. 1990). Measurements of isoprene from bags is relatively easy with concentrations in the range of 0.01 to 1 ppm. Volumes up to several liters can be withdrawn easily if needed.

2.3.2 Growth Chambers and Other Rigid Structures

In an effort to alleviate some of the problems encountered with bag enclosure methods, Teflon film rigid chambers were built (Tingey et al. 1979; Juuti et al. 1990; Janson 1993). Generally on the order of less than 1 m^3 in volume, these chambers minimize contact of foliage with the walls and are continuously purged with clean air (often prehumidified). By purging continuously, buildup of heat and water vapor is minimized, and a steady-state concentration of isoprene is achieved. Also, enclosure volume is more stable and more easily measured than with collapsible bags. These chambers have proven useful in field and laboratory experiments. While the concentrations of isoprene are lower in these purged systems compared to the bag systems, the steady-state concentration is more accurately measured than the constantly increasing concentration found in the bags. Furthermore, large samples may be collected over long periods without upsetting the equilibrium of the plant atmosphere.

2.4 Ambient Air Measurements

Isoprene emissions from plants are rapidly diluted in the atmosphere. Concentrations as low as 0.1 ppb and up to 100 ppb are important for ambient air monitoring, while leaf cuvette isoprene measurements are typically in the range of 10 to 500 ppb (Warneck 1988). To make an accurate isoprene measurement nearly always requires some method of concentrating the sample. In contrast to laboratory measurements, where charcoal-filtered air or bottled air is used, ambient measurements are more difficult because of potential sampling losses due to the presence of reactive species such as ozone. Sample volumes are not usually limited. Recently, isoprene flux measurements in the mixing layer above forest canopies have been made by integrating 10–15-l air samples into solid adsorbent collection tubes (see Sect. 3.3; Nie et al. 1995).

3 Sample Collection Methods and Storage

3.1 General Considerations

3.1.1 Perturbation of Conditions

When planning sample collection and storage, consideration must be given to the collection of a sample that is sufficient to quantify accurately and be representative of the atmosphere under evaluation. If the rate at which a sample is withdrawn is a significant fraction of the natural air exchange near the plant, then the equilibrium established by the natural emission and ventilation of the plant may be disturbed, resulting in possible perturbation of the process of interest. These effects are generally of most concern in tissue level and leaf level studies. Lastly, sufficient sample must also be allowed to permit adequate flushing of syringes, transfer lines, and sample loops. If replicate analysis of the sample is required, the volume collected must be adjusted accordingly.

3.1.2 Contamination Control

Although isoprene is not a particularly adsorptive compound, most analysts use Teflon (FEP or PFA) or stainless steel materials for transfer lines and valves. If stainless steel tubing is used, we recommend that the lines be warmed to 40–60 °C to minimize adsorption, particularly when using long runs of tubing. Non-Teflon polymers are usually avoided because of the widespread use of plasticizers and residual monomers which may interfere with analysis. Similarly, isoprene may be passed through most oil-less pumps without problem. However, wetted stainless steel or Teflon surfaces are usually preferred in order to avoid introducing contaminants which may complicate analysis.

3.1.3 Reactive Losses

Isoprene in humid air has been demonstrated to store for periods as long as 31 days in passivated (electropolished) stainless steel cans (Pate et al. 1992). However, isoprene is easily oxidized by atmospheric levels of ozone and hydroxyl radicals. For example, in the presence of a typical summer value of 100 ppb of ozone, isoprene has a lifetime of about 8 h (Atkinson and Carter 1995). Similarly, in the presence of an average hydroxyl radical concentration of $1 \times 10^{-6}\,cm^{-3}$, isoprene has a lifetime of 3 h (Atkinson 1986). While hydroxyl radical will not store in any of the sampling containers addressed here, ozone may be captured by the whole-air sampling containers. Ozone will decay over time depending on the surface of the container, temperature, and the presence of other reactive species such as nitric oxide. Laboratory experiments conducted with bottled air or carbon-filtered air will not contain ozone or hydroxyl radicals.

3.2 Whole-Air Collection

3.2.1 Syringe Sampling

Generally, syringe sampling is limited to grab sampling, where a point measurement in time is suitable. Volumes ranging from 10^{-6} to 2 l can be sampled with a gas-tight syringe. We have used 10 and 100 ml gas-tight glass syringes and 1 to 30 ml plastic syringes. We have found that inexpensive disposable plastic syringes are satisfactory for sampling isoprene. Syringes with valves that can be closed after collection are available. We have used syringes without valves when the samples is immediately transferred to the analytical instrument. In those instances where the sample could not be analyzed immediately, we use valved syringes or we cap the needle with a gas chromographic grade septum. We have stored isoprene at concentrations between 50 and 500 ppb without losses. The isoprene is stable for at least an hour in the syringe, whether plastic or glass gas-tight. Syringes find their greatest utility in sampling leaf cuvettes, solvent layering in vials (liquid-type syringes), and head space measurements where volumes are limited and concentrations of isoprene are generally high.

3.2.2 Polymeric Bags

For larger sample volumes (1–10^{2} l), plastic bags (usually Tedlar or Teflon) are relatively noncontaminating containers. Prefabricated bags are available from various vendors (SKC; Alltech Associates). A noncontaminating pump (i.e., metal bellows or Teflon-lined diaphragm) is usually used to pump the sample into the bag (KNF Neuberger, Inc; Thomas Industries; Metal Bellows Co.). Samples can be integrated over periods of minutes or hours if a constant flow device is used (i.e., critical orifice, needle valve, mass flow controller). Bags are rather bulky, especially when dealing with numerous samples. They are prone to develop pin-hole leaks that are difficult to detect, and analytes and contaminants can permeate into and out of the walls of the bag (Kelly 1983). Thus, bags are not appropriate for long-term storage of samples.

3.2.3 Electropolished Stainless Steel Flasks

These are popular vessels for ambient air collection and are available in sizes from about 1 to 25 l (Graseby Anderson; B R C/Rasmussen; Scientific Instrumentation Specialists; Valin Corp). They can be pressurized to 40–50 psi using a non-contaminating pump. Alternatively, they can be evacuated and opened to collect sample without a pump. Constant flow devices are available for filling cans. High pressures are not recommended since water condensation may occur and safety limits may be exceeded. They have been shown to be good storage containers for isoprene in concentrations as low as 6 ppb (Pate et al. 1992). A considerable investment is required for a baking and vacuum system for reproducibly cleaning cans after each use. They usually find their greatest utility in ambient air sampling where isoprene concentrations are low to sub-ppb (Rudolph et al. 1990).

3.3 Solid Adsorbents

For air samples ranging in volume from 100 ml to 15 l solid adsorbents offer a compact, reusable means of collecting isoprene. Sampling tubes made of stainless steel or glass are packed with coarse-grade (usually 20/40 mesh) adsorbent through which sample air is drawn. Target analytes are adsorbed by the solid substrate while the bulk components of air (oxygen, nitrogen, carbon dioxide, water) pass through substantially unretained. After sampling, the analytes are released from the adsorbent by passing an inert carrier gas through the tube (usually helium) and heating the tube until the analytes are released. Details of the process will be described in Section 4.3. Adsorbent tubes are ideal for collecting time-averaged integrated samples since they operate best in a constant flow mode. Also, because flow can be effected by applying a vacuum downstream of the tube, contamination due to pumps and control devices can be avoided.

Numerous solid adsorbents are available that can be used for collecting isoprene. However, the most useful of these is probably the graphitized carbon black (i.e., Carbotrap, Supelco). It is capable of withstanding the high bake-out temperatures necessary for cleaning the tube, while retaining its mechanical and chemical integrity (Betz and Lambiase 1989, 1991). It does not retain significant amounts of water, while possessing a significant affinity for isoprene (Cao and Hewitt 1993a; Helmig and Vierling 1995). Other popular adsorbents which do not perform as well for isoprene include Tenax-GC, Tenax-TA, Tenax-GR, and the carbon molecular sieves (i.e., Carbosieve S-III, Carboxen 569, etc.; Riba et al. 1985). The Tenax materials do not adequately retain isoprene (retention volumes 1.5 to 2.9 l g^{-1}) at room temperature (Riba et al. 1985; Maier and Fieber 1988). The carbon molecular sieves do retain isoprene; however, they can retain sufficient water to cause chromatographic column obstruction and interfere with detectors.

Adsorbent sampling tubes are particularly suited to integrating large samples; however, they require attention to a number of variables. Breakthrough volume is the volume of sample that can be passed through the tube before the analyte (i.e., isoprene) emerges from the downstream side of adsorbent bed. This term is imprecisely applied in many publications. When it is used interchangeably with gas chromatographic retention time, it represents 50% breakthrough of the concentration applied. However, it is more usefully defined by the volume that can be passed without 1 or 5% breakthrough (see Brown and Purnell 1979; Namiesnik 1988 for detailed discussions of breakthrough volume and its proper application). In practice, no more than about one half of the 1% breakthrough volume should be sampled. Breakthrough volume is also strongly affected by temperature. A rough guide is that an increase in temperature of 10 °C causes a 50% decrease in breakthrough volume. Breakthrough volumes can also be decreased by high concentrations (generally reported in the range of 100 ppm or higher) of the target analyte.

At high levels, the isoprene can saturate the available adsorptive surfaces, causing it to pass unadsorbed through the bed. Similarly, other compounds in the sample may compete for the adsorptive sites along with isoprene, and decrease the breakthrough volume (Hyver and Parcher 1984; Raymer et al. 1989; Schoene et al. 1990). Breakthrough volume is not very sensitive to flow rate except that very low flows should be avoided. At low flow, diffusional sampling (passive) becomes important relative to turbulent sampling. The most efficient sampling occurs at the chromatographic minimum in the van Deemter curve (theoretical plates vs. gas velocity) for the adsorbent. This minimum is about 11 cm s^{-1} for Carbotrap (Betz and Lambiase 1991). The adsorbent gradually becomes less efficient at higher flows, but generally allows a wide range of flows to be used.

The potential for on-sorbent reactions or sampling artifacts should also be considered when using solid adsorbents to sample ambient air (Pellizari et al. 1984). Adsorbed analytes can react (i.e., isomerize, oxidize) on the surface of the adsorbent. These effects can be reduced by selecting the appropriate adsorbent. Adsorbed analytes may also react with common ambient air species (particularly oxidizers such as ozone, nitrogen dioxide, etc.). Ozone has been observed to react with monoterpenes which have been adsorbed on Tenax (Roberts et al. 1984; Stromvall and Petersson 1992; Peters et al. 1994). However, Carbotrap tubes loaded with isoprene were challenged with up to 240 ppb of ozone and not found to lose isoprene (Van Eijk and Kotzias 1994). Their tubes were stored for up to 30 days without measurable losses. Lastly, another potential problem can occur when a cosampled analyte (such as ozone) attacks the adsorbent or a coadsorbed species, yielding a product which subsequently elutes with the target analyte. This has been observed to occur with Tenax and terpenes where the ozone attack on the Tenax yields benzaldehyde, acetophenone, and phenol (Roberts et al. 1984). The benzaldehyde often coelutes with α-pinene on a nonpolar column.

Adsorbent tubes need to be evaluated for potential increases in background over anticipated storage periods. Contaminants adsorbed within deep pores, residual monomers in the adsorbent matrix, or degradation of the adsorbent all can contribute to the background of the tube. Also the integrity of the sealing system should be evaluated to ascertain that room air is not infiltrating the tube during storage. Refrigeration (freezing can cause problems with ice formation causing glass tubes to crack) may be useful in reducing increases in background (Hazard and Brown 1993). Subambient temperature storage may also prevent adsorbed analyte from migrating into deep pores where thermal desorption may not readily release them.

Adsorbents may change their adsorptive characteristics over time. Tenax has been reported to turn from white to amber with repeated use. This is accompanied by partial fusing of the particles and contraction of the absorbent bed, causing channeling of the sample air around the bed. We have not observed similar effects with Carbotrap or Carbosieve SIII, probably because these materials have already been subjected to high temperatures during their manufacture. We have, however, noted that Carbotrap will pulverize if packed too tightly. For this reason, we recommend sampling with tubes in a vertical position so that channeling will be avoided if this occurs.

Goldan et al. (1993) have reported that 2-methyl-3-buten-2-ol (MB) may be a major emission from Ponderosa pine. In the course of evaluating recovery efficiencies of MB collected on Carbotrap C/Carbotrap adsorbent tubes, we have found that MB apparently dehydrates to form isoprene. This may be caused by acidic surfaces on some batches of Carbotrap C (W.R. Betz, Supelco Co., Bellefonte, Pennsylvania, pers. comm.). R.A. Rasmussen (pers. comm.) has observed that air containing MB passed through a Nafion drier (an acidic ion exchange media) also dehydrates to form isoprene. If plant emissions are suspected of including MB, analysts should evaluate their procedures for possible transformation of MB to isoprene.

4 Sample Introduction Methods

4.1 Direct Injection

Before the sample can be delivered to the detector for quantitation, the sample must be injected into and separated by the chromatographic column. Injection can be as simple as a direct introduction into the carrier stream of a measured volume of the bulk gas sample (by syringe or through a fixed volume sample loop with a gas sampling valve; see Fig. 3).

Alternatively, the injection can include steps to effect elimination of bulk components (i.e., oxygen, water, carbon dioxide) from the sample matrix.

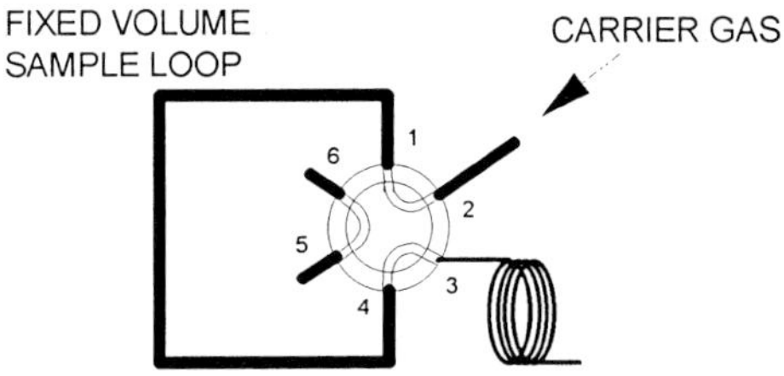

Fig. 3. Gas valve injector for gas chromatography. To load sample loop (valve rotated to place ports *1–6*, *2–3*, and *4–5* in common) syringe sample is introduced through port *6*. Sufficient volume is introduced to flush sample loop and excess gas exits through port *5*. The valve (as shown) is switched (ports *1–2*, *3–4*, and *5–6* in common) and the carrier gas sweeps the contents of the loop into the column. For system automation, samples can be drawn directly from test atmosphere through loop by applying a vacuum to port *5*. Vacuum is then disconnected before injection to allow loop to fill to atmospheric pressure

Injections must be rapid and the total volume must be compatible with the column. Direct injections of whole air samples (direct sampling, bags, cans, etc.) via syringe or fixed volume loops with gas sampling valves can be used with volumes up to about 5 cc with 2 mm i.d. packed columns and down to 0.1 to 2 cc with 0.53 mm i.d. capillary columns. Presumably smaller i.d. columns could be used with smaller injected volumes. However, syringe-injected samples would not be advisable, since the injectors of smaller columns operate at higher carrier pressures, thus making accurate injections from small-volume syringes difficult (due to pressurization of the syringe and potential leakage around the syringe barrel). Syringe injection requires development of a consistent smooth injection technique and therefore is operator-dependent. Gas sampling valves give more reproducible results than syringes and are more easily automated. Gas sampling valves are required when withdrawing sample from a subatmospheric pressure container. A wide range of gas sampling valves and associated fittings are available (Valco Instruments).

4.2 Preconcentration

If whole-air injection will not introduce sufficient analyte for detection, a preinjection concentration step is necessary. At a minimum, this step allows for the target analyte to be concentrated while allowing the oxygen and nitrogen to pass unretained. This is usually accomplished by drawing a metered volume of sample air through a cryogenically cooled trap (usually −196 °C liquid nitrogen, −185 °C liquid argon, or −182 °C liquid oxygen) packed with glass beads, or a trap packed with an adsorbent (cooled or ambient temperature; see Fig. 4).

With cryogenically cooled traps, sample volume is usually limited to 300–500 cc because of flow obstruction due to ice formation at ambient humidities. Larger samples with adsorbent(s) can be drawn at room temperature if the appropriate hydrophobic adsorbent can be identified (Riba et al. 1988; Cardin and

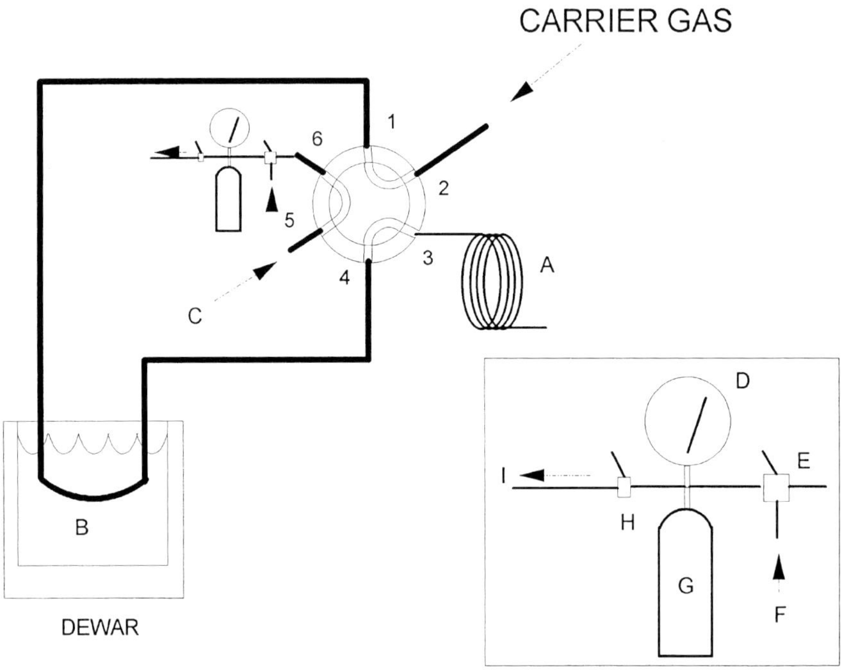

Fig. 4. Cryogenic preconcentration system (shown inject mode). Before loading a sample vacuum ballast *G* (approx 6 l) is evacuated by opening valve *H*. Three-way toggle valve *E* is closed to the vacuum side during this operation (note: this allows clean helium to enter through *F* and purge the cryogenic sampling loop *B* of semivolatiles). The cryotrap (usually about 10″ of 1/8″ o.d. stainless steel tubing packed with glass beads) is submerged in a Dewar flask of liquid argon. To load a sample through port 5, valve *H* is closed and valve *E* is opened to allow the vacuum to draw on port 5 (with 6-port valve as shown). After allowing the lines to flush, the 6-port valve is switched allowing the vacuum to draw sample through the cryotrap (the vacuum gauge reading is noted at this point). After sufficient sample has been drawn through the cryotrap (as determined by the pressure change and the combined volume of the vacuum system), the 6-port valve is switched to place the cryotrap in line with the GC column. The cryogenic Dewar is removed and is replaced by a Dewar of boiling water. Simultaneously, the data acquisition system is initiated. After about 2 min, the 6-port valve is switched again allowing *F* to backflush the cryotrap of any residual semivolatiles

Deschenes 1993; Yokouchi et al. 1993; Helmig and Greenberg 1994; Riemer et al. 1994). Several approaches to water reduction prior to preconcentration have been reported. These include passage of the sample stream through a tube in a dry ice/ethanol bath, through a desiccant, or through a dryer consisting of a section of Nafion tubing (a perfluorosulfonic acid membrane) surrounded by a desiccant (Burns et al. 1983). Water reduction by compression sufficient to cause condensation or expansion to cause adiabatic cooling have also been reported to be effective water-management techniques. If isoprene is the only analyte of interest, these methods can be successfully applied. It should also be noted that the water-management goal of reduction to avoid trap or column flow obstruction is the minimum. Liquid water deposited on the column in quantities not sufficient to cause flow obstruction can still cause retention time shifts due to the formation of films on the column surfaces. These quantities can also cause detector problems such as flame ionization detector flameout or excessive pressures in mass spectrometers.

In order to transfer the preconcentrated sample from the cryotrap or adsorbent trap to the gas chromatographic column, the trap needs to be heated to release the isoprene. Good chromatographic peak shapes require that the sample be introduced into the carrier stream as a distinct plug. Because the traps have larger inside diameters and a large degree of interstitial volume between the beads or adsorbent material, several minutes are required at the low capillary flows, after heating the trap, to effectively sweep the analytes into a capillary column. In order to refocus the analytes so that a plug injection is achieved, the head of the column or the entire chromatographic oven is cooled (i.e., −50 °C) to deposit them in a narrow band at the head of the column. Upon temperature programming, the analytes revolatilize and can be separated. Because the optimum flows through a glass bead cryotrap and a packed column are more compatible, a refocusing step may not be needed. The rate of heating and the end temperature of the trap have been reported to vary widely, depending on the method. By controlling the rate of heating, temperature, and the volume of carrier used to perform the transfer, isoprene can be released and leave much of the water in the trap. We have found that only 25 ml of helium carrier flowing through a glass bead cryotrap warmed to 5 °C is sufficient to transfer isoprene to the column. Adsorbent type traps require the application of higher temperatures to effect release of isoprene (i.e., 180 °C for Carbotrap).

4.3 Adsorbent Tube Desorption and Refocusing

Selection of a method for the analysis of a sample collected on an adsorbent tube is dependent on the adsorbent, the target analyte, and co-sampled components. When returned from the field, the tube contains air in the void between the adsorbent particles and dead volume on both ends of the adsorbent bed. With an easily oxidizable compound such as isoprene, it is essential to purge this residual air to avoid oxidizing the isoprene when the tube is thermally desorbed. In addition, this step removes water which could cause the problems noted above. To accomplish this, carrier gas (usually helium) is passed through the tube to sweep the air from it. It is preferable to purge in the same direction of the sample flow. This will move the analyte deeper into the adsorbent bed, whereas purging in the reverse flow mode will cause some loss of the target analyte into the purge stream. It is necessary to account for this additional volume when calculating necessary quantities of adsorbent so as not to exceed breakthrough volumes.

Adsorbent tube samples usually cannot be directly injected into a gas chromatographic column in a well-defined plug. Because of poor heat transfer into an adsorbent tube, resulting temperature gradients across the adsorbent bed produce nonuniform release of adsorbed analytes. Also the large volume required to sweep the analytes from the tube to a smaller diameter gas chromatographic column produces a nonideal injection (rapid increase in concentration followed by an exponential decrease). These difficulties can be overcome by cryofocusing on the head of the column or cooling the oven (Helmig and Arey 1992) or cooling the oven, as discussed in Section. 4.2. Lewis et al. (1995) used very small adsorbent traps which they were able to flash-heat and achieve a plug injection without refocusing. These types of simple system have the advantage of minimizing contact with transfer lines and valves. The disadvantages are that there is not much flexibility for performing predesorption steps, as mentioned above, because carrier flow rates and direction of flow are limited by the GC carrier gas system. These systems are generally manually operated because the adsorbent tube is placed in line with the column.

A more flexible procedure for analyzing tubes consists of desorbing the sample to a cryotrap or small adsorbent trap, as described above for processing whole air samples. By using a helium source independent of the GC carrier gas and requisite valving, the adsorbent tubes can be purged of air and desorbed in the optimum directions. Higher purge rates can be used to transfer to a trap than are possible when directly transferring to a capillary column. This "off-line" processing of tubes permits the introduction of multiport valves which can be used to automate the method. A number of automated multitube desorbers are available from various vendors (Perkin-Elmer, Tekmar, Entech, OI Analytical, Dynatherm, Envirochem).

5 Separation by Gas Chromatography

Gas chromatographic columns are available in a range of characteristics that allow separation of complex mixtures of widely varying polarities and volatilities. The reader can find in-depth guidance elsewhere (Grob 1977; Supina 1977; Jennings 1980) and from column vendors in selecting the best column for a particular application (Supelco 1995). A number of decisions need to be made when selecting a column. These include: polarity of the liquid phase, film thickness, and length. In considering the initial criteria for column selection, the minimum and maximum quantities of analyte that need to be delivered to the detector and the complexity of the sample matrix help determine the most suitable column size (inside diameter). Table 2 shows the most common sizes of gas chromatographic columns available along with the maximum quantity of analyte that can be loaded onto them without overload and the optimum carrier flow rate.

In general, the larger diameter columns tolerate a larger sample amount, albeit at the expense of efficiency (ability to resolve similar compounds). The capacity to handle larger quantities of sample can be useful when the sample contains water that is sufficient to plug a smaller column or when wide variation in isoprene

Table 2. Effect of column diameter on column loading capacity, efficiency and optimum flow. (Supelco 1995)

Internal diameter (mm)	Sample capacity (ng) (each component)	Efficiency (theorectical plates m^{-1})	Optimum flow rate (cc min^{-1})
0.20	5–30	5000	0.4
0.25	50–100	4170	0.7
0.32	400–500	3330	1.4
0.53	1000–2000	1670	2.5
0.75	10000–15000	1170	5.0
2 (packed column)	20000	2000	20

concentration is expected. In addition to the primary air components of nitrogen, oxygen, carbon dioxide, and water vapor, the sample may contain constituents that interfere with the accurate measurement of isoprene. With the possible exception of the chemiluminescence detector, the available detectors require chromatographic separation of these other substances from the isoprene in the sample in order to perform properly. Depending on the detection system, it may be necessary to reduce some of the major gases prior to separation in order to avoid overloading the chromatographic column or the detector. Water and carbon dioxide can obstruct flow and degrade column and detector performance. Elimination of most of the carbon dioxide and water may be incorporated in the sampling procedure, or it may be implemented as part of the pre-separation sample enrichment and injection procedures.

If the sample contains significant concentrations of higher molecular weight components, temperature-programmed chromatographic runs may need to be extended past the elution of isoprene. This is necessary to ensure that slow elution of these components does not cause noise and baseline drift. The reduction gas detector, for example, will drift and become noisy if monoterpenes are allowed to deposit on the mercuric oxide bed. Therefore, chromatographic run times can be as short as a few minutes or up to 40–50 min for temperature-programmed, high resolution, capillary column analysis. High resolution runs may be necessary to resolve isoprene from compounds that are of similar volatility to isoprene. On our nonpolar columns, the following common air pollutants elute within 1 min of isoprene: acetone, 2-methyl-1-butene, furan, isopentane, n-pentane, t-2-pentene, c-2-pentene, 1-pentene, propanal, 2 propanol, and trichlorofluoromethane. In general, reduction of possible interferences requires longer analysis time and therefore often limits the frequency of measurements. The use of a selective detector, however, may reduce the need for high resolution separations.

Gas chromatographic columns generally separate compounds based on their volatility and polarity. Because more than one compound may elute from the column at the same time, interferences are always a concern. Performing the analysis of a sample on two different columns of differing polarity is a useful means of looking for coeluting peaks (if the compounds are of significantly different polarity). In practice, the analyst rarely has the luxury of running two analytical systems. Changing columns is usually avoided due to the inconvenience of making and remaking capillary connections. Sample splitting or column diversion through use of valves has also been reported. If the analyst expects that most samples will contain significant concentrations of other species (i.e., other C_5 species) that will coelute with or elute close to isoprene (especially at higher concentrations), then the selectivity of the detector should be considered.

6 Detection Methods

A number of detectors are available that are capable of quantifying isoprene. Based on different operating principles, each has different sensitivities, linear dynamic ranges, and selectivities. Low sensitivity detectors may be suitable for high concentration work or in cases where large volumes of low concentration air can be enriched. However, in the latter case, enrichment of isoprene can also cause accompanying enrichment of water, carbon dioxide, and other undesirable species. A high-sensitivity detector may be needed for low-concentration measurements or for high-concentration measurements where the sample volume available is restricted. Linear dynamic range may need to be considered since isoprene levels may span orders of magnitude. Some detectors are only linear over three decades. Calibrations above the linear range become more difficult, incurring larger errors in the nonlinear regions of operation. Of course, to bring the concentrations within range, dilutions can also be performed at the expense of some loss in accuracy. Lastly, selectivity can be very useful, reducing the need for separation.

The minimum quantity of isoprene that needs to be injected is dictated by the sensitivity of the detector. Table 3 gives a brief summary of the approximate sensitivities of a several detectors which have been applied to the measurement of isoprene.

Detector sensitivity for all those listed, except for the flame ionization and mass spectrometer, are properly defined in terms of their concentration sensitivity (Sullivan and O'Brien 1977). The FID and MS are mass-flow-sensitive detectors. Mass-flow-sensitive detectors increase their signal-to-noise ratio when the flow is increased. Concentration-sensitive detectors, conversely, increase their signal-to-noise ratio when flow is decreased. Expression of the sensitivities as shown in Table 3 removes the influence of the chromatographic conditions on sensitivity.

6.1 Flame Ionization Detector (FID)

The FID responds to any substance that will ionize in a hydrogen/air flame. It is often the detector of choice for organic analysis because it responds to the carbon content of the molecule. Thus, one mole of ethane produces roughly twice the response of 1 mol of methane. The presence of other elements in the molecule,

Table 3. Gas Chromatographic detectors useful for isoprene measurements

Detector	Minimum detectable level (S/N = 2)	Linear dynamic range
Flame ionization	$10^{-12}\,g\,s^{-3}$[a]	10^{7}[b]
Photoionization	$10^{-14}\,g\,cm^{-3}$[c]	10^{7}[d]
Argon ionization	$10^{-9}\,g\,cm^{-3}$[e]	10^{3}[a]
Reduction gas	$10^{-13}\,g\,cm^{-3}$[f]	10^{3}[f]
Mass spectrometer	$10^{-10}\,g\,s^{-1}$(scan)[g]	10^{5}[g]
(Electron impact)	$10^{-13}\,g\,s^{-1}$(SIM)	
Chemiluminescence	Not available	10^{3}[h]

[a] Sullivan and O'Brien (1977).
[b] Willard et al. (1965).
[c] R.A. Rasmussen (pers. comm.) Minimum detectable level calculation based on isoprene retention time of 0.8 min with a 3.5-s peak half-width produced with a 1-cc injection and a column flow of 40 ml min^{-1} using a 11-eV lamp.
[d] Driscoll (1977).
[e] T.D. Sharkey, Sentex Scentoscreen (unpubl.).
[f] Greenberg et al. (1993).
[g] McFadden (1973); Buffington (1988).
[h] Hills and Zimmerman (1990).

such as oxygen, reduces the response. It does not respond to inorganic substances, including water. However, large quantities of water injected as a result of preconcentrating large volumes of air can extinguish the flame if steps are not implemented to limit water injection as described above. Portable FID-equipped gas chromatographs are available. However, their use is limited to direct injection via syringe or gas-sampling valve. Because FIDs can require hydrogen, air, helium (carrier), and nitrogen (makeup gas for capillary column operation), portability is limited. In addition, portable GCs usually have fairly rudimentary column oven-temperature capabilities which can limit temperature-programming options and reduce reproducibility. Laboratory GCs with FID have good sensitivity, very wide linear range, and a very stable response.

6.2 Argon Ionization Detector (AID)

The AID responds to any substance that will ionize as a result of collision with metastable argon atoms (produced by exposure to a source of β particles). The ionized compounds are measured as an increase in current between an applied electric field (500–1100 V). As can be seen in Table 3, the AID has fair linear dynamic range (10^3) and sensitivity (Sevcik 1976; Sullivan and O'Brien 1977). It does not respond to water, methane, oxygen, carbon dioxide, nitrogen, carbon monoxide, ethane, acetonitrile, or flurocarbons (Willard et al. 1965). However, the presence of impurities in the carrier gas such as water or oxygen can impair the performance of the detector by increasing the background signal. Thus, care must be exercised in selecting the proper chromatographic conditions so that water and air peaks do not tail into the isoprene peak. We have used a field-portable GC equipped with AID with good success. Its field portability is good since it requires only one support gas.

6.3 Photoionization Detector (PID)

The PID uses ultraviolet radiation to ionize chromatographic effluents and then measures the current that results from the products by applying a voltage downstream of the irradiation zone (Sevcik 1976; Sullivan and O'Brien 1977). By selecting the energy of ionization (through use of UV source lamps), the PID can be selective. With isoprene, selection of a 10-eV source will make it possible to detect isoprene but not detect any C_5 paraffins that may be eluting nearby. Use of a 9.5-eV source will still allow isoprene detection but also exclude acetone, which usually elutes close to isoprene. Additional information regarding selectivity of the PID can be found in articles by Cox and Earp (1982), Rasmussen and Khalil (1988), and Pierotti (1990). The PID is a concentration-sensitive detector, as is noted in Table 3 (Driscoll 1977; Freedman 1980). The PID is insensitive to water, oxygen, nitrogen, carbon dioxide, and carbon monoxide, as well as a number of other molecules which require higher energies to ionize them (Driscoll et al. 1978). However, in our experience, we found that both oxygen and water will quench the signal from a PID. The PID has a linear dynamic range of 10^7. Its use for measuring isoprene and terpenes at ambient concentrations without preconcentration has been reported (Das 1992). Its field portabilty is excellent since it does not require supporting gases (Sharkey et al. 1991). Field-portable GCs with PID are available, Rasmussen and Khalil (1988) measured isoprene with a field-portable instrument. C.D. Geron (US EPA, Research Triangle Park, North Carolina, pers. comm.) found a problem with condensation in a portable GC-PID with an unheated detector when using highly humidified air coming from a leaf cuvette.

6.4 Reduction Gas Detector (RGD)

The RGD operates by passing the column effluent through a heated bed of HgO and detecting elemental mercury that is released when analytes react with the HgO (Trace Analytical Corp.). The elemental mercury is detected by passing through a UV absorption cell. Because of the very strong UV absorption by Hg, the sensitivity for isoprene is excellent. The RGD is concentration-sensitive and has a fairly low linear dynamic range of about 10^2, which is extended to 10^3 via electronic compensation. It is not sensitive to water, oxygen, carbon dioxide, or nitrogen. It is sensitive to hydrogen, carbon monoxide, acetaldehyde, methanol, and acetone (Schmidt and Seiler 1970; O'hara and Singh 1988; Foulger and Simmonds 1993; Greenberg et al. 1993; Singh et al. 1994). The detector has a significant dead volume, which causes considerable chomatographic peak tailing. If other species elute close to isoprene, baseline resolution of isoprene can become difficult. The

HgO bed is subject to contamination, and attention must be paid to the prevention of the accumulation of high-molecular-weight species and column bleed (non-bonded liquid phase columns). The accumulation of such species can cause long-term baseline drift and noise. A precolumn separation may need to be performed to prevent high-molecular-weight material from accumulating on the HgO bed (Singh et al. 1994). The HgO bed and UV lamp will degrade over time, causing gradual baseline drift and loss of sensitivity. The RGD has been reported to measure ambient isoprene at levels as low as 0.3–0.4 ppb using a 1-ml direct injection, (Greenberg et al. 1993). Cao and Hewitt (1993a) used adsorbent tubes in conjunction with the RGD to measure isoprene.

6.5 Mass Spectrometric Detection (MS)

Mass spectrometry requires sample ionization, separation of resulting ions, and detection of the resulting mass/charge ratio. A wide variety of ionization and separation methods are available. Because of their complexity and high cost, mass spectrometers are not generally employed for routine quantitation of relatively simple structures such as isoprene. However, given the widespread use of lower-cost benchtop mass spectrometers for identification and quantitation of organics, we have included it for consideration (Bishop and Valis 1990; Almasi et al. 1993; Ciccioli et al. 1993; Yokouchi et al. 1993; Helmig and Greenberg 1994). Benchtop mass spectrometers use an electron beam to ionize analytes and a quadrupole mass filter to separate the ions. (Note: chemical ionization sources have recently become available on benchtop systems). Operated in a full scan mode, these units can produce mass spectra that can be used to help identify eluting peaks. Sensitivity in the full scan mode is poor compared to the above detectors. Sensitivity of mass spectrometers operated in a full scan mode is specific to the analyte and the operating conditions. Excellent discussions of the meaning of sensitivity for mass spectrometers can be found elsewhere (McFadden 1973). The mass flow rate *roughly* required for full scan identification of relatively low molecular weight species (<500 amu) is given in Table 3. However, these systems are also capable of operating in a selected ion mode (SIM). By detecting only ions specific to the fragmentation of the target analyte (e.g., 67 for isoprene), as much as three orders of magnitude better sensitivity can be gained. This also is a very powerful tool for rejecting adjacent or coeluting peaks. Benchtop mass spectrometers have insufficient pumping to accept more than about 1 to 2 ml min^{-1} of carrier gas into the ion source. Thus 0.25 mm i.d. capillary columns are usually used to separate samples because their optimum operating flow is close to this flow. Larger columns can be used if part of the flow is split away before introduction into the mass spectrometer. This causes loss of sample unless a postcolumn analyte/carrier separator is used (see McFadden 1973). Because sample utilization is usually a high priority and because the post-column separators discriminate against low molecular weight species, most researchers use the small capillary column introduced directly into the mass spectrometer. Use of the 0.25-mm columns produces maximum chromatographic resolution, albeit at the expense of lower tolerance for large samples and more susceptibility to icing, as mentioned earlier. In addition to the physical blockage caused by icing, the introduction of microgram or higher quantities of water, carbon dioxide, nitrogen, and oxygen must be avoided, or excessive ion source pressures will result. Large amounts of these can cause instrument shutdown or electrical arcing within the source. Often the high ratio of these gases relative to light analytes such as isoprene causes long tailing of these gases into the analyte peak. This often causes the analyst to scan only for mass peaks above the tailing peaks, i.e., above 44 amu. This can cause loss of significant mass fragmentation information when analyzing relatively simple molecules such as isoprene.

6.6 Chemiluminescence Detector

Various investigators have reported the construction and testing of gas chromatographic chemiluminescence detectors for measurement of olefins (Seila 1976). However, we are not aware of any commercially available detectors. Hills and Zimmerman (1990) have reported the development of a *continuous* chemiluminescence detector that operates without any separation preprocessing step. The advantage of this method is that it is capable of fast response (0.1 s) with 0.4-ppb detection limit and it is linear over three orders of magnitude. It will respond to other olefins (C_2–C_4), with only propene displaying a comparable response to isoprene (Hills and Zimmerman 1990). Changing concentrations of water vapor will cause a baseline shift. Therefore, sample stream drying is recommended, i.e., passage through a Nafion dryer. Although this instrument is not available commercially, its construction is straightforward (see also Hills et al. 1991).

7 Calibration

Both liquid and gas standards have been used to calibrate for gas phase determinations. Liquid standards are the easiest to prepare. Concentrations of isoprene in a solvent such as methanol are mixed to bracket the expected isoprene amounts to be injected. Since isoprene is not strongly retained on many columns, the choice of solvent is important to ensure that the solvent peak does not tail into the isoprene peak. Solvent-based calibrations are acceptable for direct injections of whole-air samples. They may not be acceptable when a solvent-splitting injector is used since isoprene may not be quantitatively transferred to the column under these conditions.

If a preconcentrating or refocusing system is employed, a gas standard is required. Isoprene standards prepared in high pressure nitrogen have been reported (Monson and Fall 1989), although long-term stability has not been established. Short-term standards (hours) of isoprene in air or nitrogen in Teflon bags or glass mixing flasks have also been reported (Loreto and Sharkey 1990). We have made isoprene-diffusion tubes (Altshuller and Cohen 1960) as well. Other methods of preparing gas standards may be applicable (Nelson 1971; Namiesnik 1984).

8 Conclusion

Measurement of isoprene in air can be roughly considered in three parts: (1) collection of sufficient sample volume for representative and accurate quantitation, (2) separation (if necessary) of isoprene from interfering substances in the sample matrix, and (3) detection of isoprene. Selection of the gas sampling, separation, and detection combination appropriate for a given experiment should be based on consideration of the following factors: (1) expected concentration range, (2) limitations of sample volume and sampling rate, (3) frequency of measurement, (4) complexity of sample matrix, (5) field or laboratory measurement (portability), and (6) cost.

Acknowledgments. The authors gratefully acknowledge useful discussions with William A. Lonneman and Robert L. Seila of the National Exposure Research Laboratory US Environmental Protection Agency.
Disclaimer. The information in this document has been funded wholly or in part by the United States Environmental Protection Agency under Cooperative Agreement CR 823791-01 to the University of Wisconsin at Madison. It has been subjected to Agency review and approved for publication. Mention of trade names or commercial products does not constitute endorsement or recommendation for use.

References

Almasi E, Kirshen N, Kern H (1993) The determination of sub part-per-billion levels of volatile organic compounds in air by preconcentration from small sample volumes. Int J Environ Anal Chem 52:39–48
Altshuller AP, Cohen IR (1960) Application of diffusion cells to the production of known concentrations of gaseous hydrocarbons. Anal Chem 32:802–810
Anonymous (1995) Analytical chemistry lab guide edition. Anal Chem 74-75-75
Atkinson R (1986) Kinetics and mechanisms of the gas-phase reactions of hydroxyl radical with organic compounds under atmospheric conditions. Chem Rev 86:69–201
Atkinson R, Carter WPL (1995) Kinetics and mechanisms of the gas-phase reactions of ozone with organic compounds under atmospheric conditions. Chem Rev 84:437–470
Berkley RE, Miller M, Chang JC, Oliver K, Fortune C (1981) Evaluation of commercially available portable gas chromatographs. Proc EPA/AWMA Symp on Toxic and Related Air Pollutants, 413 pp. Air and waste management association; Pittsburgh, 1992
Betz WR, Lambaise SJ (1989) Characterization of carbon molecular sieves and activated charcoal for use in airborne contaminant sampling. Am Ind Hyg Assoc J 50:181–187
Betz WR, Lambiase SJ (1991) Dynamic gas solid chromatographic techniques for characterizing carbon molecular-sieves. J Chromatogr 556:433–440

Bishop RW, Valis RJ (1990) A laboratory evaluation of sorbent tubes for use with a thermal-desorption gas chromatography-mass selective detection technique. J Chromatogr Sci 28:589–593

Brown RH, Purnell CJ (1979) Collection and analysis of trace organic vapour pollutants in ambient atmospheres: the performance of a Tenax-GC adsorbant tube. J Chromatogr 28:589–593

Buffington R (1988) GC-Atomic emission spectroscopy using microwave plasmas, Hewlett-Packard Co.

Burns WF, Tingey DT, Evans RC, Bates EH (1983) Problems with a Nafion membrane for drying chromatographic samples. J Chromatogr 269:1–9

Cao XL, Hewitt CN (1993a) Thermal-desorption efficiencies for different adsorbate adsorbent systems typically used in air monitoring programs. Chemosphere 27:695–705

Cao XL, Hewitt CN (1993b) Passive sampling and gas-chromatographic determination of low concentrations of reactive hydrocarbons in ambient air with reduction gas detector. J Chromatogr 648:191–197

Cardin DB, Deschenes JT (1993) A cryogenless autogc system for enhanced ozone monitoring using a simplified, single detector approach. Air and Waste Management Association, Pittsburgh, pp 282–288

Ciccioli P, Brancaleoni E, Cecinato A, Sparapani R, Frattoni M (1993) Identification and determination of biogenic and anthropogenic volatile organic compounds in forest areas of northern and southern Europe and a remote site of the Himalaya region by high-resolution gas-chromatography mass-spectrometry. J Chromatogr 643:55–69

Cox RD, Earp RF (1982) Determination of trace level organics in ambient air by high-resolution gas chromatography with simultaneous photoionization and flame ionization detection. Anal Chem 54:2265–2270

Das TN (1992) Terpenes in a conifer forest – detection at sub-parts per billion level without sample preconcentration. Atmos Environ Pt A Gen Top 26:2853–2857

Driscoll JN (1977) Evaluation of a new photoionization detector for organic compounds. J Chromatogr 134:49–55

Driscoll JN, Ford J, Jaramillo LF, Gruber ET (1978) Gas chromatographic detection and identification of aromatic and aliphatic hydrocarbons in complex mixtures by coupling photoionization and flame-ionization detectors. J Chromatogr 158:171–180

Foulger BE, Simmonds PG (1993) Ambient-temperature gas purifier suitable for the trace analysis of carbon-monoxide and hydrogen and the preparation of low-level carbon-monoxide calibration standards in the field. J Chromatogr 630:257–263

Freedman AN (1980) The photoionization detector: theory, performance, and application as a low-level monitor of oil vapor. J Chromatogr 190:263–273

Goldan PD, Kuster WC, Fehsenfeld FC, Montzka SA (1993) The observation of a C5 alcohol emission in a North American pine forest. Geophys Res Lett 20:1039–1042

Greenberg JP, Zimmerman PR, Taylor BE, Silver GM, Fall R (1993) Sub-parts per billion detection of isoprene using a reduction gas detector with a portable gas-chromatograph. Atmos Environ Pt A Gen Top 27:2689–2692

Grob RL (1977) Theory of gas chromatography. John Wiley, New York, pp 40–112

Guenther AB, Monson RK, Fall R (1991) Isoprene and monoterpene emission rate variability: observations with *Eucalyptus* and emission rate algorithm development. J Geophys Res 96:10, 799–10, 808

Hazard SA, Brown JL (1993) Recovery after storage and desorption efficiencies for volatile organic compounds spiked on thermal desorption tubes. Air and Waste Management Association, Pittsburgh, pp 571–578

Helmig D, Arey J (1992) Organic chemicals in the air at Whitaker's Forest/Sierra Nevada Mountains, California. Sci Total Environ 112:233–250

Helmig D, Greenberg JP (1994) Automated in-situ gas-chromatographic mass-spectrometric analysis of ppt level volatile organic trace gases using multistage solid-adsorbent trapping. J Chromatogr 677:123–132

Helmig D, Vierling L (1995) Water adsorption capacity of the solid adsorbants Tenax TA, Tenax GR, Carbotrap, Carbotrap C, Carbosieve SIII, and Carboxen 569 and water management techniques for the atmospheric sampling of organic gases. Anal Chem 67:4380–4386

Hills AJ, Zimmerman PR (1990) Isoprene measurement by ozone-induced chemiluminescence. Anal Chem 62:1055–1060

Hills AJ, Fall R, Monson RK (1991) Methods for the analysis of isoprene from leaves. Springer, Berlin Heidelberg New York, pp 288–315

Hyver KJ, Parcher JF (1984) Nonreactive coadsorption of solutes on a sampling adsorbent. Anal Chem 56:274–278

Janson RW (1993) Monoterpene emissions from Scots pine and Norwegian spruce. J Geophys Res 98:2839–2850

Jennings W (1980) Gas chromatography with glass capillary columns. Academic Press, New York, pp 1–18

Juuti S, Arey J, Atkinson R (1990) Monoterpene emission rate measurement from Monterey pine. J Geophys Res 95:7515–7520

Kelly NA (1983) The contamination of fluorocarbon-film bags by hydrocarbons and nitrogen-oxides. J Air Pollut Control Assoc 33:120–125

Lewis AC, Seakins PW, Denha AM, Bartle KD, Pilling MJ (1995) Programmed temperatire vapoization injection (PTV) for in situ field measurements of isoprene, and selected oxidation products in a eucalyptus forest. Atmos Environ 29:1871–1875

Loreto F, Sharkey TD (1990) A gas-exchange study of photosynthesis and isoprene emission in *Quercus rubra* L. Planta 182:523–531

Maier I, Fieber M (1988) Retention characteristics of volatile compounds on Tenax-TA. J High Resolut Chromatogr Commun 11:566–576

McFadden WH (1973) Techniques of combined gas chromatography/mass spectrometry: applications in organic analysis. John Wiley, New York, p 1

Monson RK, Fall R (1989) Isoprene emission from aspen leaves. The influence of environment and relation to photosynthesis and photorespiration. Plant Physiol 90:267–274

Namiesnik J (1984) Generation of standard gaseous mixtures. J Chromatogr 300:79–108

Namiesnik J (1988) Preconcentration of gaseous organic pollutants in the atmosphere. Talanta 35:567–587

Nelson GO (1971) Controlled test atmospheres. Ann Arbor Science Publishers, Ann Arbor

Nie D, Kleindienst TE, Arnts RR, Sickles JE (1995) The design and test of a relaxed eddy accumulation system. J Geophys Res 100:11415–11423

O'hara D, Singh HB (1988) Sensitive gas chromatographic detection of acetaldehyde and acetone using a reduction gas detector. Atmos Environ 22:2613–2615

Pate B, Jayanty RKM, Peterson MR, Evans GF (1992) Temporal stability of polar organic-compounds in stainless-steel canisters. J Air Waste Manage Assoc 42:460–462

Pellizari E, Demian B, Krost K (1984) Sampling of organic compounds in the presence of reactive inorganic gases with Tenax GC. Anal Chem 56:793–798

Peters RJB, Duivenbode ADVRV, Duyzer JH, Verhagen HLM (1994) The determination of terpenes in forest air. Atmos Environ 28:2413–2419

Pierotti D (1990) Analysis of trace oxygenated hydrocarbons in the environment. J Atmos Chem 10:373–382

Rasmussen RA, Jones CA (1973) Emission isoprene from leaf discs of *Hamamelis*. Phytochemistry 12:15–19

Rasmussen RA, Khalil MAK (1988) Isoprene over the Amazon Basin. J Geophys Res 93:1417–1421

Rasmussen RA, Went FW (1964) Volatile organic material of plant origin in the atmosphere. Proc Natl Acad Sci USA 53:220

Raymer JH, Cooper SD, Pellizzari ED (1989) Analyte competition on polyimide adsorbents studied by deuterated tracer pulse chromatography. ACS Monograph, American Chemical Society, Washington, DC

Riba ML, Randrianalimanana E, Mathieu J, Torres L (1985) Preconcentration of atmospheric terpenes on solid adsorbants. Int J Environ Anal Chem 19:133–143

Riba ML, Tsiropoulos N, Clement B, Golfier A, Torres L (1988) Preconcentration and analysis of atmospheric isoprene and monoterpenes: system automation. J Chromatogr 456:165–173

Riemer DB, Milne PJ, Farmer CT, Zika RG (1994) Determination of terpene and related compounds in semi-urban air by GC-MSD. Chemosphere 28:837–850

Roberts JM, Fehsenfield FC, Albritton DL, Sievers RE (1984) Sampling and analysis of monoterpene hydrocarbons in the atmosphere with Tenax gas chromatographic porous polymer. Butterworth, Boston, pp 371–387
Rudolph J, Muller KP, Koppmann R (1990) Sampling of organic volatiles in the atmosphere at moderate and low pollution levels. Anal Chim Acta 236:197–211
Sanadze GA (1991) Isoprene effect – light dependent emission of isoprene by green parts of plants. In: Sharkey TD, Holland EA, Mooney HA (eds) Trace gas emissions by plants. Academic Press, San Diego, pp 135–152
Sanadze GA, Kursanov AL (1966) On certain conditions of the evolution of the diene C5H8 from poplar leaves. Sov Plant Physiol 13:184–189
Schmidt U, Seiler W (1970) A new method for recording molecular hydrogen in atmospheric air. J Geophys Res 75:1713–1716
Schoene K, Steinhanses J, Konig A (1990) The sorptive efficiency of Tenax for binary vapor mixtures. Fresenius J Anal Chem 336:114–119
Seila RL (1976) GC-chemiluminescence method for the analysis of ambient terpenes. In: Proc Int Conf on Photochemical oxidant pollution and its control, vol 1, EPA-600/3-77-00a, US Environmental Protection Agency, Research Triangle Park, North Carolina, pp 41–50
Sevcik J (1976) Detectors in gas chromatography. Elsevier, New York, pp 87–144
Sharkey TD (1991) Stomatal control of trace gas emissions. In: Sharkey TD, Holland EA, Mooney HA (eds) Trace gas emissions by plants. Academic Press, San Diego, pp 335–339
Sharkey TD, Loreto F (1993) Water stress, temperature, and light effects on the capacity for isoprene emission and photosynthesis of kudzu leaves. Oecologia 95:328–333
Sharkey TD, Loreto F, Delwiche CF (1991) High carbon dioxide and sun/shade effects on isoprene emission from oak and aspen tree leaves. Plant Cell Environ 14:333–338
Singh HB, O'hara D, Herlth D, Sachse W, Blake DR, Bradshaw JD, Kankidou M, Crutzen PJ (1994) Acetone in the atmosphere: distribution, sources, and sinks. J Geophys Res 99:1805–1819
Stromvall AM, Petersson G (1992) Protection of terpenes against oxidative and acid decomposition on adsorbent cartridges. J Chromatogr 589:385–389
Sullivan JJ, O'Brien MJ (1977) Detectors. John Wiley, New York, pp 213–288
Supelco (1995) Supelco Co. Chromatographic products catalogue, Belletonte, Pennsylvania
Supina WR (1977) Columns and column selection in gas chromatography. John Wiley, New York, pp 114–148
Tingey DT, Manning M, Grothaus LC, Burns WF (1979) The influence of light and temperature on isoprene emission rates from live oak. Physiol Plant 47:112–118
Van Eijk J, Kotzias D (1994) Sampling and analysis of isoprene in ambient air. Fresenius Environ Bull 3:220–225
Warneck P (1988) Chemistry of the natural atmosphere. Academic Press, San Diego, 241 pp
Went FW (1960) Blue hazes in the atmosphere. Nature 187:641–643
Willard HH, Merritt LL Jr, Dean JA (1965) Instrumental methods of analysis. Van Nostrand, Princeton, pp 503–504
Yokouchi Y, Bandow H, Akimoto H (1993) Development of automated gas-chromatographic mass-spectrometric analysis for natural volatile organic-compounds in the atmosphere. J Chromatogr 642:401–407
Zimmerman PR (1979) Tampa Bay area photochemical oxidant study – determination of emission rates of hydrocarbons from indigenous species of vegatation in the Tampa/St. Petersburg Florida area. EPA 904/9-77-028, US Environmental Protection Agency, Atlanta

Appendix. Company Addresses and Phone Numbers

Alltech Associates, Inc., 2051 Waukegan Rd., Deerfield, IL 60015, (800)-255-8324, fax (708)-948-1078
BRC/Rasmussen, 1121 N.W. Hillsboro, Oregon 97124, (503)-648-2264
Chrompack Inc., 1130 Route 202 South, Raritan, NJ 08869, (800)-526-3687, fax (908)-722-8365

Dynatherm Analytical Instruments, Inc., Kelton, PA 19346, Tel (215)-869-8702
Entech Laboratory Automation, Inc., 950 Enchanted Way #101, Simi Valley, CA 91365, (805)-527-5939, fax (805)-527-5687
Envirochem, Inc, Route 896, Box 180, Kemblesville, PA 19347
Grasbey Andersen, 4801 Fulton Industrial Blvd., Atlanta, GA 30336-2003, (800)-241-6898, fax (404)-691-6315
J & W Scientific, 91 Blue Ravine Rd, Folsom, CA 95638-4714, (800)-223-3424, fax (916)-985-1101
KNF Neuberger, Inc., P. O. Box 4060, Princeton, NJ 08543, (609)-799–4350
LI-COR, Inc., 4421 Superior Street, Lincoln, NE 68504, (402)-467-3576
Metal Bellows Co., 1075 Providence Highway, Sharon, MA 02067, (617)-668-3050
OI Analytical, Graham Rd @ Wellborn Rd, P O Box 2980, College Station, TX 77841–2980, (409)-690-1711, fax (409)-690-0440
Perkin-Elmer Corp., 761 Main Ave., Norwalk, CT 06859, (800)-762-4000
Photovac Inc., 330 Cochrane Dr., Markham, Ontario L3R8E5, Canada, (905)-477-8088, fax (905)-477-8220
Restek Corp., 110 Benner Circle, Bellefonte, PA 16823-8812, (800)-356-1688, fax (814)-353-1309
Scientific Instrument Services, Inc., 1027 Old York Rd., Ringoes, NJ 08551, (908)-788-5550, fax (908)-806-6631
Scientific Instrumentation Specialists, P. O. Box 8941, 815 Courtney St., Moscow, Idaho 83843, (202)-882–3860
Sentex Systems Inc., 553 Broad Avenue, Ridgefield, NJ 07657, (201)-945-3694, fax (201)-941-6064
Shimadzu Scientific Instruments, Inc., Chicago Office, 175 Hansen Court, Suite 108W, Wood Dale, IL 60191, (708)-860-7094, fax, (708)-860-7031 Corporate offices (301)-381-1227, fax (301) 381-1222
SKC Inc., RR 1 Box 334, Eighty Four, PA 15330-9614, (412)-941-9701, fax (412)-941-1369
Supelco Inc., Supelco Park, Bellefonte, PA 16823-0048, (800)-247-6628, fax (800)-349-3044
Tekmar, Corp., P.O. Box 429576, Cincinnati, OH 45242-9576, (513)-247-7000
Thomas Industries Inc., P. O. Box 29, 1419 Illinois Ave., Sheboygan, WI 53082-0029, (414)-457-4891, fax (414)–457-2389
Trace Analytical Inc., 3517-A Edison Way, Menlo Park, CA 94025, (415)-364-6895, fax (415)-364-6897
Valin Corp., 209 N Fair Oaks Avenue, Sunnyvale, CA 94086, (408)-730-9850, fax (408)-730-1363
Valco Instruments Co., P. O. Box 55603, Houston, TX 77255, (800)-367-8424, fax (713)-956-1649

Measurement of Ammonia Gas Emission from Plants

J.K. SCHJOERRING and S. HUSTED

1 Introduction

It is now well established that plants have a compensation point for NH_3, implying that they emit NH_3 at atmospheric NH_3 concentrations below the compensation point and absorb NH_3 at concentrations in excess (Farquhar et al. 1980; Langford and Fehsenfeld 1992; Husted et al. 1996; Schjoerring et al. 1996). The compensation point normally ranges between 0.5 and 20 nmol NH_3 mol^{-1} air and is thus of a magnitude similar to the NH_3 concentrations occurring in the atmosphere over terrestrial ecosystems (Sutton et al. 1995). Ammonia exchange over natural land is probably the most uncertain factor in the compilation of a global NH_3 budget (Dentener and Crutzen 1994), and little is as yet known about the interrelationships between NH_3 compensation points, nitrogen metabolism and environmental factors. One major reason for the limited knowledge of plant-atmosphere NH_3 fluxes is analytical difficulties.

The existence of a NH_3 compensation point in plant foliage close to naturally occurring atmospheric NH_3 concentrations makes it essential to carry out experiments at realistic atmospheric NH_3 concentrations. Several requirements are thereby imposed on the experimental equipment in order to minimize NH_3 adsorption and to control and monitor NH_3 concentrations precisely. This is a big challenge because NH_3 readily sticks to solid surfaces in, e.g., plant cuvettes, tubes and monitors. In addition to being affected by the external NH_3 concentration, plant-atmosphere NH_3 exchange is very sensitive to climatic parameters such as air temperature, air humidity and light intensity (van Hove 1989; Husted and Schjoerring 1996), making it necessary to work under well-defined environmental conditions.

In this chapter, we describe recent advances in techniques and equipment for quantifying ammonia emission from the shoots of living plants. The description will predominantly focus on aspects which are relevant for experiments under controlled environmental conditions including atmospheric NH_3 concentrations corresponding to those found naturally. In order to design laboratory systems suitable for the investigation of plant-atmosphere NH_3 exchange, and to understand the behavior of NH_3 in such systems, some basic aspects of NH_3 chemistry with relevance for surface adsorption will also be presented.

Modern Methods of Plant Analysis, Vol. 19
Plant Volatile Analysis
Edited by H.-F. Linskens and J.F. Jackson

2 Basic Physicochemical Properties of Ammonia

Ammonia adsorption to solid surfaces can be divided into three broad categories, physical adsorption, chemical adsorption and reaction with surface bound water.

Physical adsorption (physisorption) is associated with the relatively weak intermolecular attractions that exist between molecules (van der Waals forces). Physisorption has relatively low adsorption enthalpies of $-38\,kJ\,mol^{-1}$, which is comparable to the enthalpy of vaporisation for NH_3 ($\Delta H = +34\,kJ\,mol^{-1}$) and extensive physisorption of NH_3 is observed only at low temperatures.

Chemisorption is associated with the exchange of electrons and the formation of a chemical bond between NH_3 and the surface. Ammonia consists of three bonding electron pairs and one non-bonding electron pair (lone pair) which gives NH_3 some unique chemical properties:

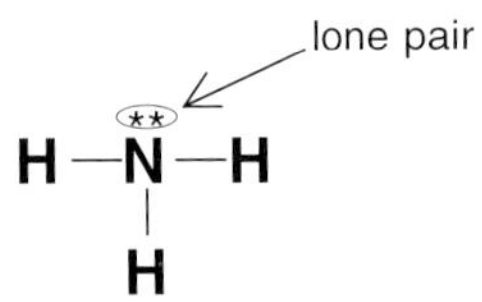

The lone pair in NH_3 is used in the formation of covalent bonds of highly variable strength, depending on the chemical structure of the surface to which it is bound. The enthalpy of NH_3 chemisorption to e.g. steel is around $-188\,kJ\,mol^{-1}$.

The lone pair of NH_3 is also involved in the formation of hydrogen bonds with solid surfaces and compounds already adsorbed to the surface such as water and organic compounds with ability of forming hydrogen bonds. Thus, NH_3 is extremely soluble in water (Table 1) and formation of a water film on the surface of plant cuvettes at high air humidities may lead to serious memory effects. Moreover, the air used to ventilate plant cuvette systems must be filtered and tubings and cuvette walls carefully cleaned for organic residues on a regular basis in order to reduce NH_3 adsorption.

Adsorption of NH_3 is an exothermic process (negative ΔH), suggesting that a temperature increase would decrease NH_3 adsorption and the NH_3 solubility in the water film that might be present on walls of the cuvette system. The equilibrium between adsorption and desorption may be predicted by the van't Hoff equation:

Table 1. The solubility of different gases in water (mol gas l^{-1} H_2O) at a total pressure of 1 atm above the water surface

	10 °C	15 °C	20 °C	25 °C
CH_4	0.002	0.001	0.001	0.001
CO_2	0.048	0.041	0.035	0.030
SO_2	2.5	2.1	1.8	1.5
NH_3	40.5	35.6	31.5	28.0

$$\frac{d(\ln K)}{dT} = \frac{\Delta H}{RT^2}. \tag{1}$$

Figure 1 shows the effect of temperature on NH_3 adsorption/desorption by cuvette walls covered by a water film, as calculated by use of Eq. (1).

The difficulties in working with NH_3 due to surface adsorbance effects are illustrated by the data in Fig. 2, comparing the response time to NH_3 and NO in a plant cuvette system. Even though the two gases were present at similar concentrations, the response to changes in NO concentration was about nine times faster than that of NH_3, permitting a much higher time resolution of NO fluxes compared to NH_3 fluxes.

3 NH_3 Detectors

Continuous measurements of low NH_3 concentrations require highly precise, sensitive, and stable analytical equipment. These requirements are currently best fulfilled by the flow denuder system developed by Wyers et al. (1993). Also chemiluminescent NO_x monitors equipped with an NH_3 converter can, after proper modifications, be utilized, provided atmospheric NO_x concentrations are relatively low and constant (van Putten et al. 1994; Husted and Schjoerring 1995a).

Automatic monitors have not been available in most studies of plant-atmosphere NH_3 exchange. Instead, alternatives such as acid scrubbers have been used (Farquhar et al. 1980; Hooker et al. 1980; O'Deen and Porter 1986; Parton et

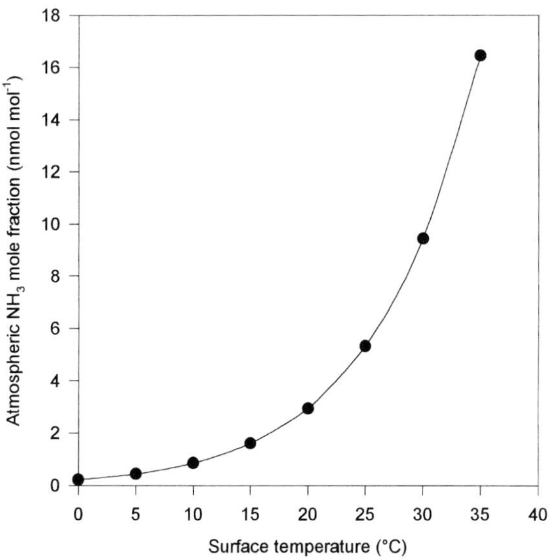

Fig. 1. Effect of temperature on the atmospheric NH_3 mole fraction above a water film equilibrated with 10 nmol NH_3 mol^{-1} air at 30 °C. The enthalpy of NH_3 vaporization in the NH_3/NH_4^+ system is +86.4 kJ mol^{-1} and calculations were made using Eq. (1)

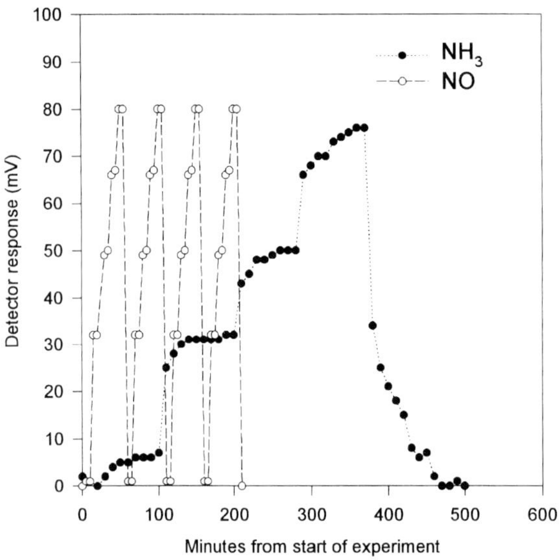

Fig. 2. Response times for NH_3 and NO in plant cuvettes made of coated polycarbonate and connected to a chemiluminescence NH_3/NO_x monitor by use of high-density polyethylene tubing. Cuvettes were ventilated continuously at a rate of 0.45 vol min^{-1} with air containing 0, 5, 15, 20, or 25 nmol NH_3 or NO mol^{-1} air. Air temperature and humidity were 20 °C and 60% RH, respectively

al. 1988). Determination of NH_3 fluxes with acid scrubbers requires a sampling period sufficiently long to ensure that enough NH_3 is collected to exceed the detection limit of the analytical method used for subsequent analysis of the trapped NH_4^+. Consequently, studies of diurnal variations in plant-atmosphere NH_3 fluxes are not possible. Furthermore, determination of NH_3 fluxes by the use of acid scrubbers is very labor-intensive.

3.1 Continuous Flow Denuder

A continuous flow denuder is a new type of automatic NH_3 monitor developed by Wyers et al. (1993). The monitor is available commercially and provides the currently best option for measurements of atmospheric NH_3 concentrations under controlled conditions (Olsen et al. 1995; Husted and Schjoerring 1996; Mattsson and Schjoerring 1996) as well as over intact ecosystems (Sutton et al. 1995).

The working principle of the continuous flow denuder is illustrated in Fig. 3. Ambient air is pumped at a rate of 30 l min^{-1} through an annular denuder, continuously rotating around its axis. The NH_3 in the air is trapped in a $NaHSO_4$ absorption solution, covering the walls of the annular space in the rotating denuder. The absorption solution is pumped out of the denuder and mixed downstream of the denuder with NaOH solution in a mixing chamber, resulting in formation of aqueous NH_3. The solution then passes a gas-permeable Teflon (PTFE) membrane

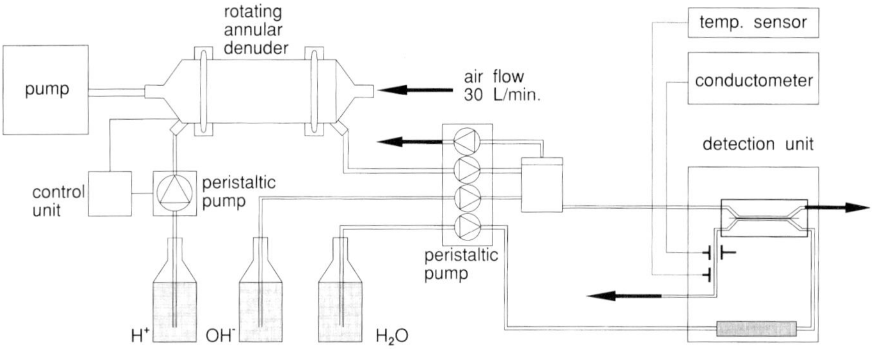

Fig. 3. Schematic diagram of a continuous flow denuder for analysis of atmospheric NH_3 concentrations. (After Wyers et al. 1993)

allowing approx. 30% of the NH_3 to diffuse across the membrane and to become dissolved in an acceptor stream of double-demineralized water from which all traces of bicarbonate and carbonate have been previously removed on an ion exchange column. The NH_4^+ concentration in the water stream is determined by measuring the change in conductivity with corrections for temperature effects. Data acquisition is achieved by a programmable data logger.

The detection limit of the continuous flow denuder is lower than 0.02 nmol NH_3 mol^{-1} air (Wyers et al. 1993; van Putten et al. 1994). The instrument is free from interferences by aerosols as well as by other atmospheric gases with the exception of volatile amines. The precision, i.e. the relative standard deviation of the average of successively measured values at a constant NH_3 concentration, is less than 2% for concentrations below 20 nmol NH_3 mol^{-1} air, and the instrument is linear over a wide span of NH_3 concentrations. The response time is 31 min when the flow denuder is connected to 10 m of clean high density polyethylene (HD-PE) or Teflon FEP tubing in which 55 nmol NH_3 mol^{-1} air is transported at 50% RH (Mennen et al. 1996).

The liquid flow rates must be frequently checked since the calculated NH_3 concentration is proportional to them. A drawback of the continuous flow denuder when used in connection with single plant or leaf cuvettes is the large air flows required. The air flow can probably be reduced, although it is still to be tested how this will affect the performance of the instrument.

3.2 Chemiluminescent NH_3 Monitor

The analytical principle of chemiluminescent NH_3 monitors is based on concomitant conversion of NH_3 and NO_2 to NO in a stainless steel (Fe) converter at 800 °C:

$$2NH_3 + Fe_2O_3 + O_2 \rightarrow 2NO + 3H_2O + 2Fe$$

$$3NO_2 + 2Fe \rightarrow 3NO + Fe_2O_3,$$

whereas only NO_2 is converted to NO in a molybdenum (Mo) converter at 350 °C:

$$3NO_2 + Mo \rightarrow 3NO + MoO_3.$$

The nitric oxide (NO) produced in the two converters then reacts, together with the NO that is already present in the air prior to entering the monitor, with ozone (O_3) producing NO_2 in an exited state (NO_2°):

$$NO + O_3 \rightarrow NO_2^{\circ} + O_2.$$

Upon return to the ground state, NO_2° emits a photon which is detected by a photomultiplier tube:

$$NO_2^{\circ} \rightarrow NO + h\nu.$$

By subtracting the Mo-converter signal (NO_2 + NO) from the Fe-converter signal ($NH_3 + NO_2$ + NO), the concentration of NH_3 can be obtained.

Determination of NH_3 as a difference between two electrical signals from two different converters can be done with sufficient accuracy only under conditions where NO_2 and NO concentrations are relatively low and constant, thus limiting the use of chemiluminescent monitors to studies involving purified air. The latter also reduces problems with interference from NH_4^+ containing aerosols. Errors may also occur as a result of differences in the converter efficiencies of the Mo- and the Fe-converter.

Working with a commercially available type of chemiluminescent NH_3 monitor (Thermo Environmental Instruments, Franklin, Massachusetts, USA), it is the experience of the present authors that a thorough modification is required in order to lower the detection limit and response time to an acceptable level. Firstly, all metal parts in contact with the gas sample must be spray-coated with Teflon (PTFE) and gas lines must be shortened. Secondly, an oxalic acid-coated filter, mounted in, e.g., a vacuum filter holder made of polyoxymethylene (POM), must be inserted in front of the molybdenum converter in order to prevent NH_3 from entering the converter as it reduces some of the NH_3 to NO. Thirdly, the chemiluminescence monitor is also very sensitive to electronic noise generated by, for example, magnetic valves and relays, and therefore a galvanic separation must be inserted in order to increase the signal-to-noise ratio for the monitor. For a further improvement of the signal-to-noise ratio, it may also be useful to mount an oxalic acid-coated filter after the Fe-converter in order to prevent nonconverted NH_3 from reaching the detector.

Provided the above improvements are carried out, the detection limit for a chemiluminescent NH_3 monitor, defined as two times the standard deviation of the intercept of the calibration curve for NH_3 concentrations in the range of 0–35 nmol NH_3 mol^{-1} air, can be brought down to around 1–2 nmol NH_3 mol^{-1} air. However, the response time remains relatively high, around 55 min, as shown in Fig. 4. The precision of the monitor measured continuously over 24 h (determined as the standard deviation of the mean) for an NH_3 concentration of 6 nmol mol^{-1} was 6.0%. The monitor also shows some drift, implying that daily calibrations are necessary.

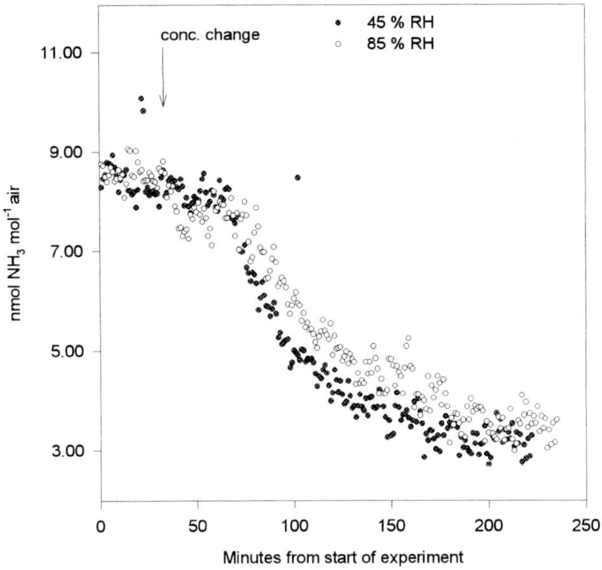

Fig. 4. Influence of air humidity (45 and 85% RH at 20 °C) in response to a concentration change from 7.8 to 3.5 mol NH_3 mol^{-1} air in plant cuvettes made of coated polycarbonate and connected to a chemiluminescence NH_3/NO_x monitor by use of high-density polyethylene tubing. Cuvettes were ventilated continuously at a rate of 0.45 vol min^{-1}.

4 Cuvette Systems

Even the most precise NH_3 monitor would never be able to compensate for errors induced by poor cuvette conditions. Ammonia adsorption must, of course, be kept at a minimum. Furthermore, since plant-atmosphere NH_3 exchange is very sensitive to variations in climatic parameters such as temperature, humidity and light intensity (Husted and Schjoerring 1996), these parameters should also be carefully measured and controlled. However, many studies have been performed under inadequate conditions. O'Deen and Porter (1986) measured NH_3 emission from senescing winter wheat in Plexiglas cuvettes located in growth chambers at 23 °C, but trapping of heat by the cuvette walls in combination with a low air exchange rate (0.02 min^{-1}) caused the temperature to peak daily at 30.5 °C. The studies of Farquhar et al. (1980) were conducted at a relatively low photosynthetic photon flux density (250 $\mu mol\, m^{-2} s^{-1}$) in order to avoid problems with condensation of transpiration water, but under these experimental conditions the NH_3 exchange is likely to be affected by low photosynthetic activity. Stutte and Weiland (1978), Weiland and Stutte (1979), and Stutte and da Silva (1981) studied the NH_3 emission from single leaves by absorbing NH_3 and other heteroatomic N gases in condensation traps in a closed system. This experimental design may have increased the NH_3 emission because NH_3-depleted air was drawn through the leaf cuvette. In addition, CO_2 was removed from the recirculating air, affecting the ratio between photosynthesis and photorespiration.

4.1 Materials

4.1.1 Cuvettes

A very important criterion for materials used for construction of plant and leaf cuvettes to be used for studies of plant-atmosphere NH_3 exchange is that adsorption of NH_3 must be minimal. As described in Section 2, NH_3 is extremely soluble in water and its solubility increases with decreasing temperatures. Consequently, materials that adsorb water will also adsorb NH_3, thereby causing an artificial humidity dependency of NH_3 fluxes.

A material with a relatively slight water adsorption is coated polycarbonate (Margard PC; Trademark of General Electric Company, Pittsfield, Massachusetts, USA), which adsorbs only half the amount of water that acrylic plastic (Plexiglas) does (Knight 1992). In addition, polycarbonate has a relatively high thermal conductance, reducing problems with excessive trapping of IR radiation and temperature increase of the cuvette air. Spectral properties of polycarbonate allow uniform transmittance of more than 90% of incident radiation in the photosynthetically active range as well as of near-IR radiation. Polycarbonate does not allow transmittance of UV-A and UV-B (Knight 1992).

4.1.2 Tubing

High-density polyethylene (HD-PE) tubing has a low gas permeability (Table 2) and the NH_3 adsorption to HD-PE is lower than that to Teflon FEP tubing (Mennen

Table 2. CO_2 permeability of selected polymers at 25 °C. (Calculated on basis of information from Cole-Parmer Instrument Company, Illinois, USA, 1996)

Formulation codes	Gas permeability (10^{-9} cm^2 s^{-1} atm^{-1})
Teflon FEP (fluorinated ethylene propylene)	4.6
Teflon PTFE (polytetrafluoroethylene)	5.3
HD-PE (high density polyethylene)	8.4[a]
PVC (polyvinyl chloride)	14
TYG (Tygon)	29
VIT (Viton)	59
PP (polypropylene)	68
Rubber	996
SIL(Silicone)	15300

[a] Measured at 30 °C.

et al. 1996). In addition, the cost of HD-PE is much lower compared to Teflon polymers like, e.g., PTFE and FEP. No detectable losses of ammonia occurred when 55 nmol NH_3 mol^{-1} air was transported through 2 and 10 m of HD-PE or FEP tubings (Mennen et al. 1996). Other common polymers like PP, PVC, and Tygon are relatively permeable to NH_3. Stainless steel tubes have a very high water and NH_3 adsorption and are, accordingly, inacceptable for use in cuvette systems.

Plasticizers are often added to a number of polymers in order to convert them to elastomers, and these compounds are often loosely linked and are slowly vaporized from the materials. Several plasticizers (e.g., phthalates) are phytotoxic to a large number of plant species in extremely low concentrations and may result in chlorosis and severe growth retardation (Hardwick and Cole 1986). Polymers like PVC, PP, and low-density PE often contain a substantial amount of plasticizers and should therefore be used with extreme caution in plant cuvette systems.

4.1.3 Air Purification

Air can be cleaned for NH_3 by use of activated carbon coated with 5% sulfuric acid. This material effectively absorbs NH_3 in the incoming air, while NO and NO_2 is not trapped effectively. The latter two gases can be removed in a filter of Purafil (Purafil Inc., Atlanta, Georgia, USA), which is expanded aluminum hydroxide coated with potassium permanganate. In the Purafil filter, NO becomes oxidized to NO_2, which absorbs strongly to the filter material. Moreover, it is advisable to incorporate a 1-μm air filter in order to reduce accumulation of dust particles in the cuvette system.

4.1.4 Control of Atmospheric NH_3 Concentrations

Ammonia concentrations in the incoming air can be regulated by use of certified NH_3 standards taken from a pressurized cylinder and fed into the main air stream by programmable digital mass flow controllers. Alternatively, much cheaper needle valves of good quality (e.g., KDG Mobrey LTD, Crawley, UK) can be used, but careful calibration needs to be done on a regular basis in order to obtain accurate NH_3 concentrations.

4.1.5 Air Distribution and Air Flow Within Cuvettes

A slight overpressure of a few hundred Pa in the plant cuvettes is advantageous in order to avoid penetration of external air into the cuvette. The air in the plant cuvette must be exchanged sufficiently rapid to keep the air humidity below a level where liquid water will condensate on the cuvette walls. A high air exchange rate in combination with a uniform mixing of the incoming air with the cuvette air is also necessary in order to minimize leaf boundary layer resistances to NH_3 diffusion.

4.1.6 Sampling of Air

Automatic sampling of gas from the inlet and outlet of several plant cuvette air streams can be done by time-programmed, Teflon-coated valves. In order to reduce response times, sampling lines can be gathered in a heated Teflon gas manifold connected to the NH_3 monitor (see Husted and Schjoerring 1995a).

4.2 Determination of Steady-State NH_3 Concentrations

During a short time interval (dt), the amount of NH_3 emitted or absorbed by the plant foliage causes a change in the cuvette NH_3 concentration which can be calculated using the equation:

$$\frac{dX_{cuv}}{dt} = \frac{F \times A + X_{in} \times Q - X_{out} \times Q}{V}, \tag{2}$$

Where X_{cuv}, X_{in} and X_{out} denote the NH_3 mole fraction (nmol NH_3 mol^{-1} air) in the cuvette, inlet and outlet air, respectively, F is the flux in nmol NH_3 m^{-2} leaf s^{-1}, A is the leaf area (m^2), Q is the air flow rate (mol s^{-1}), and V is the cuvette volume (mol). Equation (2) does not correct for the change in NH_3 concentration due to simultaneous exchange of water vapor and CO_2. However, omitting these corrections leads to errors of less than 1% when normal ratios of transpiration, photosynthesis, and NH_3 exchange are used (calculations not shown; see Leuning 1983).

Equation (2) may be integrated and rewritten as:

$$X_{cuv} = X_o + X_t \times \left(1 - \exp\left(\frac{-t \times Q}{V}\right)\right), \tag{3}$$

where X_o and X_t is the NH_3 mole fraction in the cuvette at time zero and at time t, respectively. As $t \to \propto$ the steady-state concentration equals $X_o + X_t$, and the time needed for reaching steady-state conditions can be calculated. For cuvettes operated with an air renewal rate of 0.25 to 0.5 vol min^{-1}, between 10 to 15 min would be required for establishment of steady-state NH_3 concentrations. However, due to memory effects in cuvette and NH_3 monitor and due to plant adaptation, it usually takes longer to reach steady state in the whole experimental system than predicted on the basis of Eq. (3), which only gives the absolute minimum time lag before measurements can be initiated (see also Sect. 4.3).

The preferable unit in studies of plant gas exchange is mole fraction because it is insensitive to temperature and pressure changes. In order to convert the output from monitors measuring the NH_3 concentration in µg NH_3 m^{-3} air, as is, e.g., the case for the continuous flow denuder described in Section 3.1, to nmol mol^{-1} air, it is first necessary to correct for temperature and pressure differences between monitor and cuvette:

$$[NH_3]_{cuv} = [NH_3]_{monitor} \times \frac{T_{monitor}}{T_{cuv}} \times \frac{p_{cuv}}{p_{monitor}}, \tag{4}$$

where $[NH_3]_{cuv}$ and $[NH_3]_{monitor}$ is the NH_3 concentration in μg NH_3 m^{-3} air in the cuvette and in the monitor, respectively, T is the absolute temperature, and p is the pressure (atm). The true mole fraction, X_{NH3}, in the cuvette can subsequently be calculated as:

$$X_{NH3} = \frac{[NH_3]_{cuv} \times R \times T}{p \times 17}, \quad (5)$$

where R is the gas constant.

4.3 Performance of Plant Cuvette Systems

Some adsorption of NH_3 in cuvette systems is inevitable even though the cuvettes are constructed of materials with a low adsorption of water. This is, e.g., shown by a slower response to a concentration change from 7.8 nmol NH_3 mol^{-1} air to 3.5 nmol NH_3 mol^{-1} at 85% RH than at 45% RH, when measuring on an empty plant cuvette made of coated polycarbonate (Fig. 4). Consequently, at 45% RH a stable reading of 3.5 nmol mol^{-1} was obtained after 70 min, while at 85% RH it took 105 min before a stable signal was achieved, indicating that NH_3 solubilized in moisture on cuvette and tube walls, thus damping short-term oscillations in NH_3 exchange. However, minimizing water absorption by constructing plant cuvettes of coated polycarbonate (Margard PC), having a low water adsorbtion capacity, and using high air flow rates (air renewal every 2 to 4 min), lead to negligible irreversible adsorption of NH_3, and thereby a true steady-state NH_3 flux, as evidenced by no significant differences between inlet and outlet NH_3 concentrations (slope of regression line 1.07 ± 0.02) in a cuvette without plants (Fig. 5).

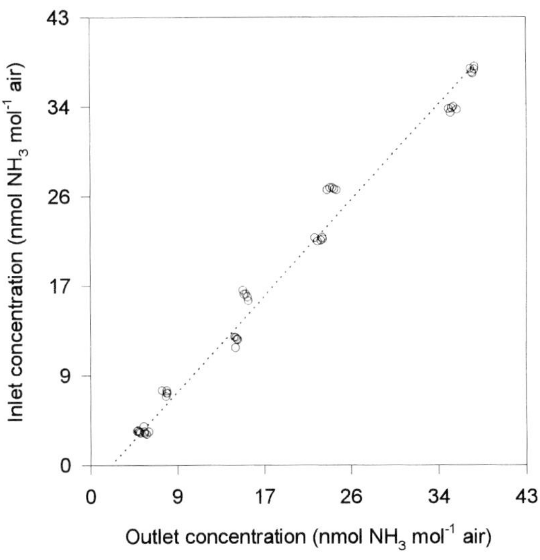

Fig. 5. Relationship between inlet and outlet NH_3 concentrations in plant cuvettes made of coated polycarbonate and connected to a chemiluminescence NH_3/NO_x monitor by use of high-density polyethylene tubing. Cuvettes were ventilated continuously at a rate of 0.45 vol min^{-1}. Regression line: $y = -2.23 + 1.07x$; $r = 0.992$

5 Determination of Ammonia Compensation Points in Plants

The NH_3 compensation point (χ_{NH3}), defined as the mole fraction of gaseous NH_3 in the intercellular cavities of leaves at which the net flux of NH_3 equals zero, is a key parameter in the regulation of plant-atmosphere NH_3 exchange. It has been attempted to measure it by use of plant cuvettes under controlled conditions (Farquhar et al. 1980; Morgan and Parton 1989; Husted and Schjoerring 1995a,b, 1996) and under field conditions (Kesselmeier et al. 1993), as well as by use of micrometeorological techniques under field conditions (Dabney and Bouldin 1990; Langford and Fehsenfeld 1992; Sutton et al. 1992, 1993a,b; Schjoerring et al. 1993; Yamulki et al. 1996). Cuvette measurements are advantageous when the purpose is to investigate the physiological regulation of plant NH_3 compensation points, while micrometeorological measurements are valuable in quantifying NH_3 fluxes on ecosystem level and in deriving NH_3 compensation points on the basis of these fluxes. A major drawback of micrometeological determinations of NH_3 compensation points is the highly skewed distribution of fluxes, with most recorded values scattering around χ_{NH3} and a few observations at atmospheric NH_3 concentrations deviating considerably from the compensation point. This very often leads to errors exceeding 100% (see Sutton et al. 1995; Yamulki et al. 1996). Contributing to the high errors are also diurnal oscillations in air temperature, leaf humidity, and atmospheric stability, which may lead to difficulties in interpretating such data physiologically. NH_3 compensation points in laboratory systems can usually be determined with a much higher accuracy than in field experiments due to the possibilty of controlling the environmental parameters.

5.1 Determination of NH_3 Compensation Points by Exposing Plants to Different Atmospheric NH_3 Concentrations

The NH_3 exchange flux F (nmol NH_3 m^{-2} leaf s^{-1}) can under steady-state conditions be calculated as:

$$F = \frac{Q \times (X_{in} - X_{out})}{A}, \tag{6}$$

where X_{in} and X_{out} denote the NH_3 mole fraction in the cuvette inlet and outlet air, respectively, A is the leaf area (m^2), and Q is the air flow rate (mol s^{-1}).

Ammonia exchange between plants and the atmosphere takes place as a result of a concentration gradient between the atmosphere and the intercellular spaces in the leaves. An integrated form of Fick's first law may therefore be applied in order to deduce χ_{NH3} from flux measurements at a known atmospheric NH_3 concentration:

$$F = g_{NH3} \times (\chi_{NH3} - X_{cuv}), \tag{7}$$

where F is the NH_3 net flux (nmol m^{-2} s^{-1}), g_{NH3} the NH_3 conductance (nmol m^{-2} s^{-1}), and X_{cuv} the external NH_3 mole fraction. According to Eq. (7), a straight line should be obtained when the NH_3 flux is measured in response to changes in the atmospheric NH_3 concentration, assuming g_{NH3} is independent of the external NH_3 mole fraction. Such a straight line is also observed experimentally over a wide range of NH_3 concentrations, as shown by the data in Fig. 6. Linear regression analysis of F on X_{cuv} provides estimates for the slope and intercept, being equal to g_{NH3} and $g_{NH3} \times \chi_{NH3}$, respectively, in Eq. (7). A predicted value for the NH_3 compensation point can subsequently be obtained following division of $g_{NH3} \times \chi_{NH3}$ by g_{NH3}.

In order to incorporate errors associated with both the slope and intercept, standard errors for χ_{NH3} can be calculated on the basis of the following equation (Miller and Miller 1988):

$$s_{x_{NH3}} = \frac{s_y}{g_{NH3}} \left(\frac{1}{m} + \frac{1}{n} + \frac{y^2}{g^2{}_{NH3} \sum_i (x_i - \bar{x})^2} \right)^{0.5}, \tag{8}$$

where s_x is the standard error of the NH_3 compensation point; $s_y = \Sigma(y_i - \bar{y})^2$ the flux residuals; m the total number of NH_3 fumigations; n the number of NH_3 fumigation levels; y the mean NH_3 flux; and $\Sigma(x_i - \bar{x})^2$ the NH_3 mole fraction residuals. All these parameters may be calculated using simple programming in statistical software programs like SAS (SAS Institute, Raleigh, North Carolina, USA).

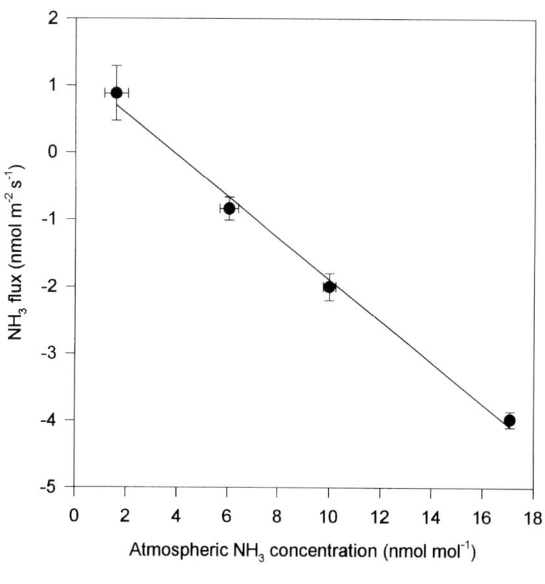

Fig. 6. NH_3 response curves for *Hordeum vulgare* cv. Golf plants exposed to four different NH_3 concentrations during a period of 7 days in the grain-filling period. The photon flux density was 500 μmol m^{-2} s^{-1} in a photoperiod of 16 h, air temperature 20 °C, and relative air humidity 60%

The procedure described above was used for determining ammonia compensation points in two different cultivars of barley at five different growth stages (Fig. 7). The estimated values of χ_{NH3} ranged from 3.0 to 6.4 nmol NH_3 mol^{-1} air in the modern cultivar Golf and from below 0.9 to 5.3 nmol NH_3 mol^{-1} air in the old and primitive cultivar Laevigatum. Averaged over all growth stages, the standard error of χ_{NH3} was 13.1% if values below the detection limit (0.9 nmol NH_3 mol^{-1} air) were omitted. In both cultivars a marked decrease in χ_{NH3} was observed in the period from early vegetative growth to anthesis, and during generative growth χ_{NH3} peaked at approx. 5.3 nmol NH_3 mol^{-1} air. Golf had significantly higher NH_3 compensation points than Laevigatum ($P < 0.01$), except during the middle of the grain-filling period. Golf also had higher shoot NH_4^+ and total nitrogen concentrations than Laevigatum (Husted et al. 1996).

5.2 Determination of NH_3 Compensation Points by Controlled Increments of Leaf Temperature

Ammonia exchange is very sensitive to changes in leaf temperature. An example of a such temperature dependency is given in Fig. 8, showing that NH_3 absorption in *Brassica napus* plants exposed to 15 nmol NH_3 mol^{-1} air decreased with increasing temperatures until, at a leaf temperature of 31 °C, the plants started to emit NH_3, thus indicating an NH_3 compensation point of 15 nmol NH_3 mol^{-1} air at this temperature.

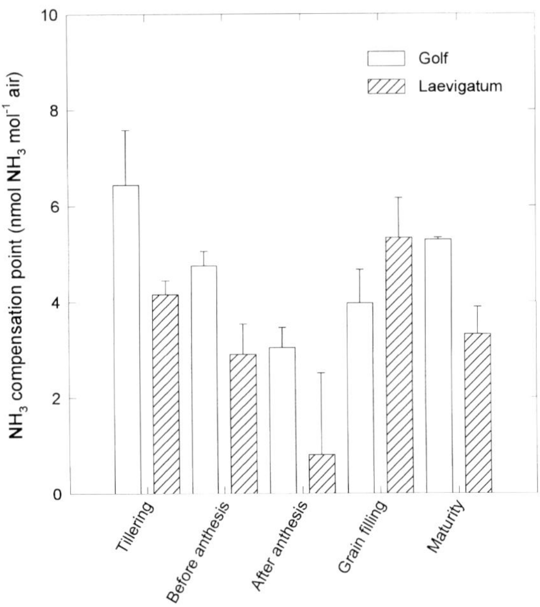

Fig. 7. NH_3 compensation points (20 °C, 60% RH) at five different growth stages of *Hordeum vulgare* cvs. Golf and Laevigatum. Data are means ± SE (n = 7)

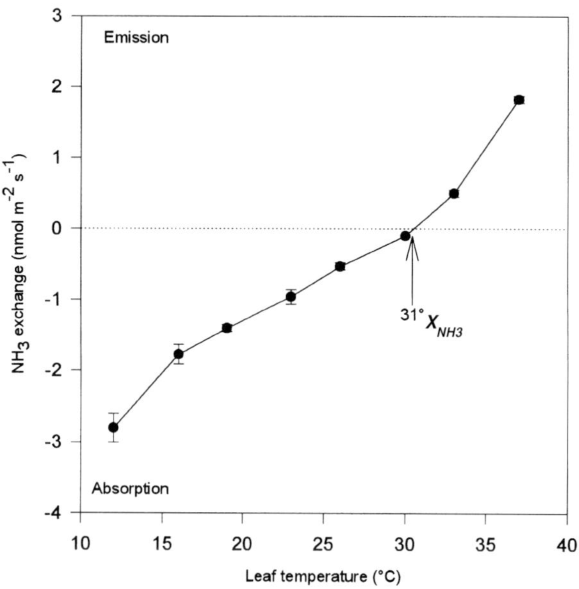

Fig. 8. Effect of increasing leaf temperatures on NH_3 exchange in vegetative, high N status *Brassica napus* plants exposed to 15 nmol NH_3 mol^{-1} air. Photosynthetic photon flux density was 550 µmol m^{-2} s^{-1}, and the relative air humidity 20% at 25 °C. The experiments were done in duplicate over 2 consecutive days

Assuming that χ_{NH3} is a result of a simple chemical equilibrium between gaseous NH_3 in the intercellular air spaces and aqueous NH_3/NH_4^+ in the apoplast (see Sect. 5.3), χ_{NH3} measured at one leaf temperature can be extrapolated to any other given leaf temperature using thermodynamic calculations based on the following equilibria:

$$NH_4^+ \leftrightarrow NH_{3(aq)} + H^+ \qquad \Delta H_{dis}{}^{\circ} = 52.21\,\mathrm{kJ\,mol^{-1}}; \qquad K_d^{25^{\circ}C} = 10^{-9.25} \tag{9}$$

$$NH_{3(aq)} \leftrightarrow NH_{3(g)} \qquad \Delta H_{vap}{}^{\circ} = 34.18\,\mathrm{kJ\,mol^{-1}}; \qquad K_H^{25^{\circ}C} = 10^{-1.76}. \tag{10}$$

The enthalpy (ΔH) of these reactions [Eqs. (9) and (10)] is essentially independent of temperatures below 40 °C, and the variation in ΔH with ionic strength is insignificant below 100 mM (Stumm and Morgan 1981). The mole fraction of gaseous NH_3 may therefore be corrected to any given temperature using a slightly modified Clausius-Clapeyron equation (Atkins 1990):

$$\ln\frac{\chi_2}{\chi_1} = \frac{\Delta H^0_{dis} + \Delta H^0_{vap}}{R} \times \left(\frac{1}{T_1} - \frac{1}{T_2}\right), \tag{11}$$

where χ_1 is the actual NH_3 compensation point at temperature T_1 and χ_2 is the requested NH_3 compensation point at a new temperature T_2. R denotes the gas constant (8.31 J K^{-1} mol^{-1}).

Using Eq. (11), an NH_3 compensation point of 15 nmol NH_3 mol^{-1} air at 31 °C (Fig. 8) corresponds to 7.7 nmol NH_3 mol^{-1} air at 25 °C and 4.2 nmol NH_3 mol^{-1} at 20 °C. This example illustrates the importance of a precise determination of the leaf temperature when NH_3 flux data are reported, because even small errors may result in substantial differences in χ_{NH3}. It must be noted that the extrapolation to other temperatures is only valid if the concentrations of NH_3 and NH_4^+ in the apoplast remain constant in the applied temperature range. In *Brassica napus* this seems to be the case over periods up to about 4 h at 35 °C, as evidenced by stable ammonia emissions (data not shown).

5.3 Determination of NH_3 Compensation Points on the Basis of Apoplastic H^+ and NH_4^+ Concentrations in Leaves

From equilibria (9) and (10), it is evident that χ_{NH3} depends on the H^+ and NH_4^+ concentrations in the apoplast. Provided apoplastic fluid can be isolated without cytoplasmic contamination, it would therefore be possible to determine the NH_3 compensation point on the basis of chemical analyses as a supplement to the more technically and labor-demanding fumigation and temperature-increment methods described in Sections 5.1 and 5.2. Such a technique for isolation of apoplastic solution has recently been developed in the authors' laboratory and is described in the following sections.

5.3.1 Extraction of Apoplastic Solution

The technique is based on vacuum infiltration of small leaf segments (25 × 50 mm) with a 350 mOsm isotonic sorbitol solution (280 mM) at 4°C in a 50-ml plastic syringe mounted on a hydraulic arm that automatically moves the plunger up and down in order to infiltrate the leaf under pressure and vaccum, respectively. The infiltrator exposes the leaf discs to 4 atm pressure and vacuum for 10 s, respectively, and repeats the procedure eight times, ensuring full infiltration. Immediately after infiltration, the leaf apoplast is extracted by centrifugation at 2000 *g* for 10 min at 8 °C and the apoplastic solution (approx. 75 µl) is collected in small test tubes. The concentration of NH_4^+ and NH_3 in the isolated solution is thereafter measured by flow injection analysis (Tecator FIA Star 5020, Höganäs, Sweden) and pH measured with a microelectrode (type 60236, Methrom, Herisau, Schwitzerland). The apoplastic extracts can be screened for cytoplasmic contamination by measuring their activity of malate dehydrogenase (MDH) relative to bulk leaf extracts (Husted and Schjoerring 1995b). Alternatively to MDH, we have tested hexose phosphate isomerase (HPI) and PEP-carboxylase as cytosolic marker enzymes, but assessments of cytoplasmic contamination based on these enzymes seem to be less sensitive compared to MDH.

5.3.2 Apoplastic Air and Water Volumes

The apoplastic solution is diluted when leaf segments are infiltrated with the sorbitol solution and, in order to correct apoplastic H^+ and NH_4^+ concentrations for this dilution, both the apoplastic air and water volume have to be quantified. This is done by weighing and infiltrating leaf segments with high-viscosity silicon fluid (viscosity 5 cgs; density 0.904 g cm^{-1}; Dow Corning, Poole, UK) which is impermeable to the plasma membrane. The segments are thereafter blotted dry with thin paper tissues and the weight increase, corrected for the density of silicone fluid, used for calculation of the apoplastic air volume (V_{air}, cm^3 air cm^{-3} tissue). The amount of silicone fluid adsorbed to the cuticle has been shown to make an insignificant weight increase.

The volume of apoplastic water (V_{apo}) is assessed on the basis of the dilution of the blue dye indigo carmine, which does not adsorb to the cell walls: leaf segments are weighed and infiltrated with 50 µM indigo carmine dissolved in 50 mM phosphate buffer at pH 6.2. Immediately after infiltration, leaves are reweighed and the weight increase is equal to the infiltration volume (V_i). The leaf segments are then put into a small, zippered polyethylene bag to eliminate water vaporization, and stored for a maximum of 15 min at 4 °C before extraction by centrifugation. After extraction, the decrease in absorbance at 610 nm ($\Delta A610$) is measured spectrophotometrically.

The decrease in absorbance, $\Delta A610$, of the indigo carmine solution after vacuum infiltration is given by:

$$\Delta A_{610} = \frac{V_{apo}}{V_{air} + V_{apo}}, \quad (12)$$

which can be rearranged to:

$$V_{apo} = \frac{\Delta A_{610} \times V_{air}}{1 - \Delta A_{610}}. \quad (13)$$

If the osmolarity of the indigo carmine solution does not match the cytoplasmic osmolarity, then water will be transported across the plasmalemma and affect the concentration of the dye. In order to correct for this effect, Eq. (13) is adjusted to:

$$V_{apo} = \frac{\Delta A_{610} \times V_{air}}{1 - \Delta A_{610}} \times \frac{V_i}{V_{air}}, \quad (14)$$

which can be reduced to:

$$V_{apo} = \frac{\Delta A_{610} \times V_i}{1 - \Delta A_{610}}. \quad (15)$$

Finally, the apoplastic dilution factor (f_{dil}) can be calculated as:

$$f_{dill} = \frac{V_{apo} + V_{air}}{V_{apo}}. \tag{16}$$

5.5.3 Calculation of χ_{NH3} Based on Apoplastic pH and NH_4^+ Concentration

Knowing the apoplastic NH_4^+ and H^+ concentrations, the following equations, based on equilibria (9) and (10), can be used to calculate the resulting NH_3 compensation point:

$$[NH_3]_{aq} = C_{tot} \times \frac{K_d}{K_d + (H^+)} \tag{17}$$

$$\chi_{NH3} = [NH_3]_{aq} \times K_H. \tag{18}$$

Combining Eqs. (17) and (18) gives:

$$\chi_{NH3} = K_H \times \frac{K_d \times C_{tot}}{K_d + (H^+)}, \tag{19}$$

where $C_{tot} = [NH_4^+] + [NH_3]_{(aq)}$. Since the physiological ionic strength in the apoplast normally ranges between 14 and 28 mM (Cosgrove and Cleland 1983; Speer and Kaiser 1991), K_d must be adjusted by the extended Debye-Hückel equation (Atkins 1990), giving a $K_d = 10^{-9.32}$ at an ionic stregth of 20 mM.

5.4 Comparison of Methods

The apoplastic pH and NH_4^+ concentration in leaves of *B. napus* plants previously exposed to controlled tempeature increments in cuvette studies (Fig. 8; Sect. 5.2) was found to be 5.90 ± 0.1 and 0.88 ± 0.05 mM, respectively. The NH_3 compensation point calculated on the basis of these values was 5.80 ± 1.24 nmol NH_3 mol^{-1} air at 25 °C, which was not significantly different from the value 7.7 ± 0.8 nmol NH_3 mol^{-1} air obtained by temperature ramping (Fig. 8).

A comparison of NH_3 compensation points in *B. napus* plants, estimated respectively on the basis of apoplastic pH and NH_4^+ concentrations and measured by exposing plants to different atmospheric NH_3 concentrations, is shown in Table 3. In the vegetative growth stage there was close agreement between predicted and experimentally determined χ_{NH3} irrespective of plant nitrogen status. In later growth stages, predicted and measured χ_{NH3} deviated in low-nitrogen plants, while good agreement was still obtained in medium N plants.

Apoplastic pH and NH_4^+ concentrations are promising tools for derivation of NH_3 compensation points. The method is much more rapid than cuvette measurements and permits comparison of a large number of experimental treatments under both controlled-environment and field conditions. The method is quite robust and has been applied successfully in a primitive field laboratory.

Table 3. Apoplastic H^+, NH_4^+ concentrations and NH_3 compensation points in leaves of *Brassica napus* plants growing at low (2N), medium (4N), or high (7N) level of N supply

	Apoplastic NH_4^+ (mM)	Apoplastic pH	Calculated χ_{NH3} at 25 °C (nmol mol^{-1})	Measured χ_{NH3} at 25 °C (nmol mol^{-1})
Vegetative stage				
2N	0.29 ± 0.03	5.28 ± 0.20	0.46 ± 0.17	0.44 ± 0.30
4N	0.64 ± 0.05	5.55 ± 0.12	1.88 ± 0.47	1.72 ± 0.87
7N	2.03 ± 0.30	5.52 ± 0.10	5.58 ± 1.45	4.10 ± 0.20
Anthesis				
2N	0.47 ± 0.04	5.84 ± 0.06	2.70 ± 0.43	5.30 ± 0.80
4N	0.50 ± 0.05	5.87 ± 0.01	3.07 ± 0.31	3.32 ± 0.90
7N	1.05 ± 0.16	5.68 ± 0.12	4.17 ± 1.17	ND
Senescense				
2N	0.71 ± 0.11	5.73 ± 0.09	3.16 ± 0.76	<0.90
4N	0.76 ± 0.04	5.68 ± 0.07	3.02 ± 0.48	2.60 ± 0.50
7N	1.36 ± 0.10	6.10 ± 0.14	14.20 ± 4.12	ND

ND: not determined.

6 Prediction of NH_3 Fluxes on the Basis of Leaf Conductance and Apoplastic H^+ and NH_4^+ Concentrations

Mathematical models of NH_3 surface exchange are essential to analysis of air pollution control options because of the impossibility of measuring NH_3 fluxes over entire regions and in order to integrate standard meteorological information with information on vegetation type and land use.

Several experiments have shown an insignificant NH_3 exchange across the cuticle and epidermal cell layers (van Hove 1989; Husted and Schjoerring 1995a, 1996). Thus, NH_3 seems only to be exchanged through stomata. According to Eq. (7), this implies that the main factors governing NH_3 exchange are the leaf conductance for NH_3 and the NH_3 mole fraction gradient between the apoplast and the atmosphere surrounding the leaf. Theoretically, it should therefore be possible to predict plant-atmosphere NH_3 fluxes on the basis of leaf conductances and the ammonia compensation point.

6.1 Prediction of Leaf Conductance

The conductance for NH_3 (g_{NH3}) can be calculated on the basis of the conductance for water vapor (g_{H2O}) by use of the following equation:

$$g_{NH_3} = g_{H_2O} \times \frac{D_{NH_3}}{D_{H_2O}}, \qquad (20)$$

where D_{NH3} ($2.18 \times 10^{-5}\,m^2\,s^{-1}$) and D_{H2O} ($2.42 \times 10^{-5}\,m^2\,s^{-1}$) are the diffusion coefficients in air for NH_3 and water vapor at 293 K, respectively. The diffusivity of NH_3 in air is thus 0.92 times that of water vapor. The experimentally determined relationship between leaf conductances for NH_3 and water vapor in *Brassica napus* leaves at different levels of light intensity and relative humidity, shown in Fig. 9, is in accordance with the theoretical ratio of 0.92 (slope of line in Fig. 9 is 0.95 ± 0.08). However, in other cases, the ratio may deviate considerably from 0.92 as described below.

6.2 Prediction of NH_3 Fluxes

Equations (7) and (20) may be combined and rearranged to give the following equation, which can be used to predict NH_3 fluxes on basis of g_{H2O} obtained in measurements of the transpiration rate at a given air humidity and the NH_3 compensation point obtained from analyses of apoplastic H^+ and NH_4^+ concentrations:

$$F_{NH3} = g_{H2O} \times 0.92 \times \left(\chi_{NH3} - X_{cuv}\right). \tag{21}$$

It must be noted that this equation is only valid under the assumption that the NH_3 flux is independent of interactions with other atmospheric gases (mainly CO_2, H_2O, O_2, and N_2) diffusing into or out of the intercellular cavity. If extremely high

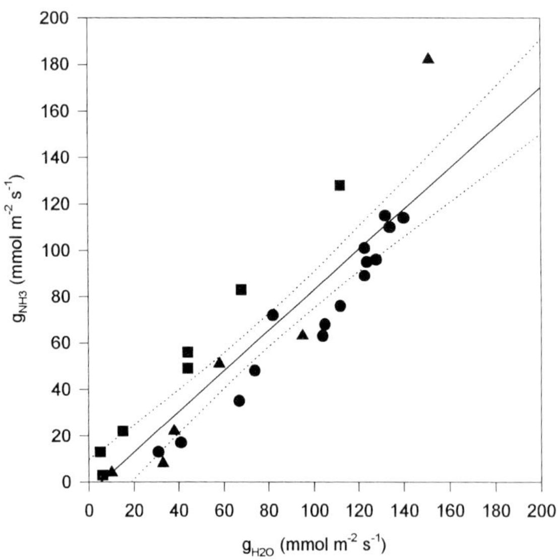

Fig. 9. Leaf conductances for water vapor and NH_3 in vegetative *Brassica napus* plants exposed to 15 nmol NH_3 mol^{-1} air at varying photosynthetic photon flux densities and air humidities at an air temperature of $25 \pm 1\,°C$. The experiments were done in duplicate over a period of 6 days with relative standard errors <10%. ● 20% RH; ■ 60% RH; ▲ 80% RH

accuracy is required in determination of NH_3 fluxes, these molecular interactions have to be taken into account by correcting for multicomponent diffusion (see Jarman 1974; von Caemmerer and Farquhar 1981; Leuning 1983; Parkhurst 1994). Moreover, using Eq. (21), it is assumed that the path length for diffusion from the intercellular cavity to the leaf surface is similar for NH_3 and H_2O, which, as will be shown below, does not always seem to hold true.

NH_3 emissions predicted on the basis of NH_3 compensation points and the leaf conductance for water vapor, adjusted for the difference in diffusivity between NH_3 and H_2O [Eq. (21)], were strongly correlated with experimentally determined NH_3 fluxes in *Brassica napus* grown at different nitrogen levels (Fig. 10). However, it was only in medium N (4N) plants that the slope between predicted and measured NH_3 fluxes was close to the theoretical value of 1; in low and high N plants, the slope was 2.3 and 0.23, respectively, showing that NH_3 emission was overestimated in low N plants and underestimated in high N plants. Thus, the ratio between leaf conductances for NH_3 and H_2O depends on the N status of the plants with a relatively higher H_2O conductance in low N plants, a similar conductance in medium N plants, and a much higher relative NH_3 conductance in high N plants (Fig. 10).

The reasons for deviations in leaf conductances to NH_3 and H_2O transport still remain to be fully resolved (van Hove 1989; Husted and Schjoerring 1995a; Husted et al. 1996). The varying ratio between H_2O and NH_3 conductances may be due to differences in the diffusional path length of the two gases, with NH_3 absorption or emission occurring in different depths of the stomatal cavity depending on the N

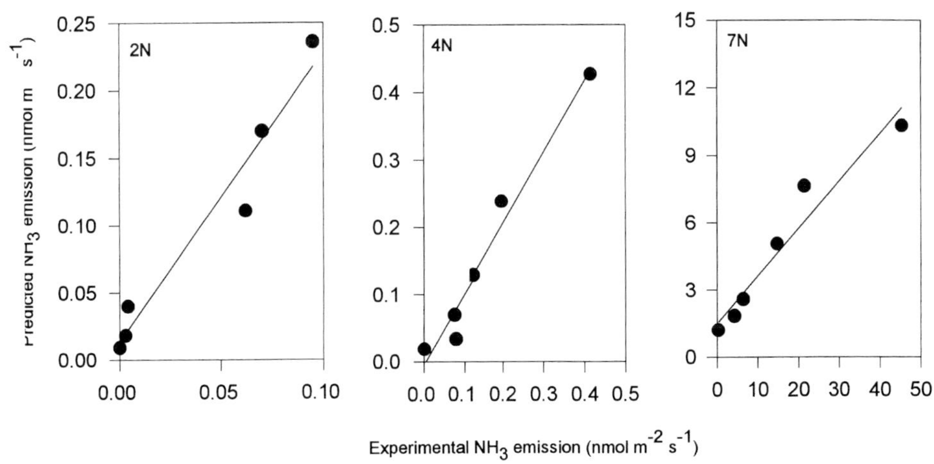

Fig. 10. Predicted versus experimentally determined NH_3 emissions from low, medium, and high N *Brassica napus* plants in vegetative growth stage. Ammonia emissions were predicted on the basis of apoplastic NH_4^+ and H^+ concentrations, and leaf conductances calculated from transpiration data. Photosynthetic photon flux density was 250 $\mu mol\, m^{-2} s^{-1}$ and relative air humidity was 20% at 25 °C. The experiments were done in duplicate over 2 consecutive days with relative standard errors <15%

status of the plants. A similar difference in path length was proposed for SO_2, which, due to its high solubility and reactivity with water, may be predominantly absorbed in the water film in the stomatal cavity, while 70–80% of the water vapor molecules are evaporated from mesophyll cells in the substomatal cavity (Taylor and Tingey 1983). It is very likely that the same mechanism is valid for exchange of NH_3, because the solubility of NH_3 is much higher than that of SO_2.

Acknowledgments. Part of this work was supported by grants from the Danish Agricultural and Veterinary Research Council and The European Commission DG XII/D (Environment Program).

References

Atkins PW (1990) Physical chemistry, 4th edn. Oxford University Press, Oxford, pp 249–252

Cosgrove DJ, Cleland RE (1983) Solutes in the free space of growing tissues. Plant Physiol 72:326–331

Dabney SM, Bouldin DR (1990) Apparent deposition velocity and compensation point of ammonia inferred from gradient measurements above and through alfalfa. Atmos Environ 24A:2655–2666

Dentener FJ, Crutzen PJ (1994) A three-dimensional model of the global ammonia cycle. J Atmos Chem 19:331–369

Farquhar GD, Firth PM, Wetselaar R, Weir B (1980) On the gaseous exchange of ammonia between leaves and the environment: determination of the ammonia compensation point. Plant Physiol 66:710–714

Hardwick R, Cole R (1986) Plastics that kill plants. New Sci Jan:38–40

Hooker ML, Sander DH, Peterson GA, Daigger LA (1980) Gaseous N losses from winter wheat. Agron J 72:789–792

Husted S, Schjoerring JK (1995a) A computer-controlled system for studying ammonia exchange, photosynthesis and transpiration of plant canopies growing under controlled environmental conditions. Plant Cell Environ 18:1070–1077

Husted S, Schjoerring JK (1995b) Apoplastic pH and ammonium concentration in leaves of *Brassica napus* L. Plant Physiol 109:1453–1460

Husted S, Schjoerring JK (1996) Ammonia flux between *Brassica napus* plants and the atmosphere in response to changes in leaf temperature, light intensity and air humidity: interactions with leaf conductance and apoplastic NH_4^+ and H^+ concentrations. Plant Physiol 112:67–74

Husted S, Mattsson M, Schjoerring JK (1996) Ammonia compensation points in two cultivars of *Hordeum vulgare* L. during vegetative and generative growth. Plant Cell Environ 19:1299–1306

Jarman PD (1974) The diffusion of carbon dioxide and water vapour through stomata. J Exp Bot 25:927–936

Kesselmeier J, Merk L, Bliefernicht M, Helas G (1993) Trace gas exchange between terrestrial plants and the atmosphere: carbon dioxide, carbonyl sulfide and ammonia under the rule of compensation points. In: Slanina J, Angeletti G, Beilke S (eds) General assessment of biogenic emissions and deposition of nitrogen compounds, sulfur compounds and oxidants in Europe. CEC Air Pollution Research Report 47. E Guyot SA, Brussels, pp 71–80

Knight SL (1992) Constructing specialized plant growth chambers for gas-exchange research: considerations and concerns. HortScience 27:767–769

Langford AO, Fehsenfeld FC (1992) Natural vegetation as a source or sink for atmospheric ammonia: a case study. Science 255:581–583

Leuning R (1983) Transport of gases into leaves. Plant Cell Environ 6:181–194

Mattsson M, Schjoerring JK (1996) Ammonia emission from young barley plants: influence of N source, light/dark cycles and inhibition of glutamine synthetase. J Exp Bot 47:474–484

Mennen MG, van Putten EM, van Elzakker BG (1996) Effects of sampling tube on measurements of ammonia concentrations in ambient air. Poster manuscript, National Institute of Public Health and Environmental Protection (RIVM), Bilthoven

Miller JC, Miller JN (1988) Statistics for analytical chemistry, 2nd edn. Ellis Horwood, Chichester, pp 101–136

Morgan JA, Parton WJ (1989) Characteristics of ammonia volatilization from spring wheat. Crop Sci 29:726–731

O'Deen WA, Porter LK (1986) Continuous flow system for collecting volatile ammonia and amines from senescing wheat. Agron J 78:746–749

Olsen C, Mattsson M, Schjoerring JK (1995) Ammonia volatilization in relation to nitrogen nutrition of young *Brassica napus* plants growing with controlled nitrogen supply. J Plant Physiol 147:306–312

Parkhurst D (1994) Tansley Review No 65. Diffusion of CO_2 and other gases inside leaves. New Phytol 126:449–479

Parton WJ, Morgan JA, Altenhofen JM, Harper LA (1988) Ammonia volatilization from spring wheat plants. Agron J 80:419–425

Schjoerring JK, Kyllingsbaek A, Mortensen JV, Byskov-Nielsen S (1993) Field investigations of ammonia exchange between barley plants and the atmosphere. I. Concentration profiles and flux densities of ammonia. Plant Cell Environ 16:161–167

Schjoerring JK, Husted S, Mattsson M (1996) Physiological parameters controlling plant-atmosphere ammonia exchange. Atmos Environ (in press)

Speer M, Kaiser WM (1991) Ion relations of symplastic and apoplastic space in leaves from *Spinacia oleracea* L. and *Pisum sativum* L. under salinity. Plant Physiol 97:990–997

Stumm W, Morgan JJ (1981) Aquatic chemistry, 2nd edn. Wiley, New York, pp 68–78

Stutte CA, da Silva PRF (1981) Nitrogen volatilization from rice leaves. I. Effects of genotype and air temperature. Crop Sci 21:596–600

Stutte CA, Weiland RT (1978) Gaseous nitrogen loss and transpiration of several crop and weed species. Crop Sci 18:887–889

Sutton MA, Moncrieff JB, Fowler D (1992) Deposition of atmospheric ammonia to moorlands. Environ Pollut 75:15–24

Sutton MA, Fowler D, Moncrieff JB (1993a) The exchange of atmospheric ammonia with vegetated surfaces. I. Unfertilized vegetation. Q J R Meteorol Soc 119:1023–1045

Sutton MA, Fowler D, Moncrieff JB, Storeton-West RL (1993b) The exchange of atmospheric ammonia with vegetated surfaces. II. Fertilized vegetation. Q J R Meteorol Soc 119:1047–1070

Sutton MA, Schjoerring JK, Wyers GP (1995) Plant-atmosphere exchange of ammonia. Philos Trans R Soc Lond A 351:261–278

Taylor GE, Tingey DT (1983) Sulfur dioxide flux into leaves of *Geranium carolinianum* L. Plant Physiol 72:237–244

van Hove LWA (1989) The mechanism of NH_3 and SO_2 uptake by leaves and its physiological effects. PhD Thesis, Wageningen Agricultural University, Wageningen, pp 97–131

van Putten EM, Mennen MG, Regts T, Uiterwijk JW (1994) Performance study of four automatic ammonia monitors under controlled conditions. Report 723101004, National Institute of Public Health and Environmental Protection (RIVM), Bilthoven, pp 1–79

von Caemmerer S, Farquhar GD (1981) Some relationships between the biochemistry of photosynthesis and the gas exchange of leaves. Planta 153:376–387

Weiland RT, Stutte CA (1979) Pyro-chemiluminescent differentation of oxidized and reduced N forms evolved from plant foliage. Crop Sci 19:545–547

Wyers GP, Otjes RP, Slanina J (1993) A continuous-flow denuder for the measurement of ambient concentrations and surface-exchange fluxes of ammonia. Atmos Environ 27A:2085–2090

Yamulki S, Harrison RM, Goulding KWT (1996) Ammonia surface-exchange above an agricultural field in southeast England. Atmos Environ 30:109–118

Use of GC-FID, GC-FTIR-MS, and Olfactory Characterisation in the Analysis of Essential Oils and Plant Extracts

G. Buchbauer, L. Jirovetz, and A. Nikiforov

1 Introduction

The combined use of GC-FID and the coupling of GC with FTIR spectroscopy and mass spectrometry (MS) as well as with olfactory evaluation (GC sniffing technique) renders the analysis of complex natural product systems more efficient. Additional information from GC (e.g. Kovats indices and use of chiral phases), FTIR spectra (functional groups), MS spectra (structure and isotope information as well as molecular weight) and from olfactory detection (qualitative and quantitative odour value) allows the determination of the identity of single compounds even in complex mixtures more effectively and rapidly, and the analytical data can be further evaluated in additional steps (e.g. library search and multivariate data analysis). The basic principles of these methods, together with the examples of application of the chromatographic-spectroscopic and of the chromatographic-olfactory combinations for investigation of aroma compound in headspace samples, essential oils and plant extracts will be presented in this chapter.

2 Methods

2.1 General Remarks, GC Detection

2.1.1 GC-FID

The most important recent developments in the field of gas chromatography (GC) with flame ionisation detection (FID) led to spectacular results especially in two directions: in access to higher molecular compounds (molecular weight beyond of 1500 amu), and in the availability of chiral phases.

The use of capillary columns (up to 500 000 theoretical plates) with phenyl silicon phases allows the separation of derivatised cyclodextrins at a temperature of about 450 °C (Blum 1991). However, this method is not further discussed in this chapter, because for all applications presented the molecular weight of the odour compounds are in the range from 100 to 300 amu and, generally, the components with a molecular weight higher than 500 amu play no role in aroma analysis.

Modern Methods of Plant Analysis, Vol. 19
Plant Volatile Analysis
Edited by H.-F. Linskens and J.F. Jackson

For investigation of complex odour compound systems, the developments in the field of the chiral-phase GC are more interesting. Enantiomers often show different odour impressions [e.g. (+)-carvone with caraway aroma and (–)-carvone with spearmint aroma] and constitute in many cases so-called character impact compounds. The may therefore be responsible for the total odour of the analysed samples (König 1992; Mosandl 1992; Jirovetz et al. 1994a; Maas et al. 1994). The use of enantioselective GC as well as retention time correlations (Kovats indices and published data) will be discussed later (Sects. 4.1 and 4.3).

2.1.2 Specific GC Detection (N-, S- and P-Specific Detection)

The FID is generally used as an all-round detection device with extensive dynamic and linearity range, as the response coefficients for the most compounds involved in aroma analysis, with few exceptions, are of comparable order. Many analytical problems in aroma analysis, however, require more specific component information for their solution. As more extended data structures stem from coupled methods, which, in turn, require more expensive system components than the GC equipment alone, many simple, selective GC detectors have been developed in the past and are commonly used today. Their additional selective information, in combination with the sole quantitative information of FID-GC, may be efficiently incorporated in the strategy of aroma analysis. Apart from electron capture detector (ECD), which accentuates aromatic components and is seldom used in aroma analysis, most GC-specific detectors are available and suitable rather for specific element detection and are not built for target analysis of any particular class of compounds. The most-used types of nitrogen-, sulphur- or phosphorus-specific detectors (or combinations of them) preferably indicate components with these heteroatoms and may therefore deliver the decisive information to solve any problems.

An example for key information supplied by a nitrogen-specific detector is shown in Fig. 1, giving a comparison of FID- and nitrogen-specific detection of opium headspace. The positions of the chromatogram that exhibit strong olfactory impressions during GC sniffing are indicated by arrows (positions 1 to 4). As apparently no significant FID signals appear at these positions, the olfactory active components must be present at a very low concentration. In contrast, the nitrogen-specific detection in the lower part provides signals which correlate quite well with olfactory perception. This implies nitrogen-containing components with rather low odour thresholds like pyrazines or related compounds in the sample – an assumption which was verified by further data (Buchbauer et al. 1994b), as shown in Section 2.3 (handling multivariate analysis) later on. Thus, decisive hints for a solution may be given by a comparatively simple and cheap device.

Generally speaking, GC detectors like FID or related types, deliver for each point of the chromatogram only a scalar value, i.e. the recorded signal is essentially a chromatogram again. If for each point of the chromatogram instead of a scalar value, a spectrum of values is recorded, then more complex data arrays, like GC-

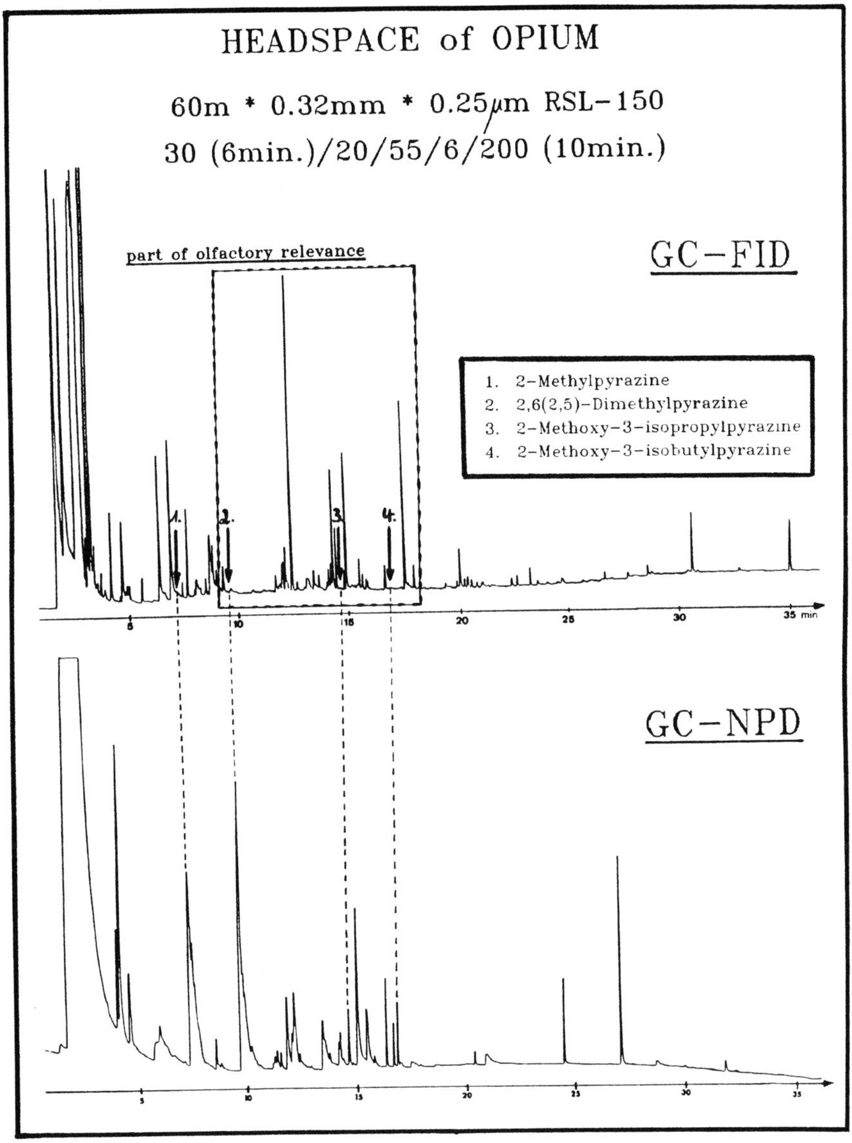

Fig. 1. Comparison of FID and N-specific detection on example of opium headspace

MS or GC-FTIR, will result. The former are sometimes referred to as first-order data, the latter as second-order data sets. Most of the analytical methods discussed in this chapter may be applied in either of these ways. In particular, these are the combinations of GC with one of the following methods: atomic emission spectroscopy (AES), mass spectrometry (MS), Fourier-transformed infrared spectroscopy (FTIR), and "sniffing".

In this respect, e.g. depending on whether in GC sniffing only olfactory intensities or a list of attributes of the individual components in the chromatogram are assigned, GC sniffing produces a simple chromatogram or a second-order data set, as shown below (Sect. 2.2.3).

2.2 Coupled Methods and Multivariate Data Analysis

2.1.1 GC-FTIR-MS

The combination of GC separation with Fourier-transformed infrared spectroscopy (FTIR) and mass spectrometry (MS) is the most effective method for analysis of aroma compounds (Jirovetz et al. 1994b; Woidich et al. 1994). On the one hand, the GC-FTIR measurements (light pipe technique or cryo-technique) furnish chemical information on functional groups, molecular geometry and stereochemical arrangements of the already separated odour components. On the other hand, the GC-MS data (electron impact or chemical ionization) supply molecular information via molecular weight, number of carbons, appearance of heteroatoms and their numbers, and roughly the total structure (by fragmentation of the compound). The combination of GC-FTIR with MS is generally possible in three ways: both the GC-FTIR and the GC-MS are two completely separated systems; GC-FTIR and GC-MS detectors use a column splitter and two columns; GC-FTIR and GC-MS detectors use a single column with sequential connection of FTIR and an MS-detecting device.

The last version was used in all investigations described in this chapter. The chemical information for each separate compound in a complex aroma system discussed above leads to an enormous amount of data. Although different libraries (e.g. MS library of Wiley, NIST, NIH and EPA or IR libraries of Sadtler) render the identification of single constituents somewhat easier and faster, the comparison of a higher number of similiar samples is still too time-consuming to be effective. In this case, the combined evaluation of the data sets by multivariate data analysis (MDA) is the best alternative (Brereton 1992; McLachlan 1992; Nikiforov et al. 1992; Graham 1993; Meier and Zünd 1993; Nikiforov et al. 1994; Wasicky et al. 1995). The basic principles, as well as applications, of MDA are given in the following.

2.2.2 GC-AES

GC-atomic emission spectroscopy coupling is a powerful tool in investigating fragrances with an oxygen, sulphur or nitrogen functional group as trace or target compound in a complex odour system. The registration of a selective oxygen, sulphur or nitrogen trace using GC-AES is in the sensitivity range of GC-ECD and therefore an additional dimension in the analysis of, e.g. essential oils with gas chromatographic-spectroscopic systems. The combined information of the various data is also of basic interest for the development of classifiers for multivariate data analysis (Nikiforov et al. 1992, 1994). An example of the qualitative and quantitative determination of monoterpenic constituents with an oxygen functional group in complex essential oils (lavender and lavandin oils) is given.

2.2.3 GC Sniffing Technique

The GC sniffing technique is the most powerful medium for the aroma analyst to evaluate the odour impression of an eluted compound during GC separation. In this technique, a splitter is used at the end of the GC column, where one part of the eluate flows towards the detector (normally FID) and the other part to the nose (Nikiforov et al. 1989; Jirovetz et al. 1990b; Guichard et al. 1995). Thus, a new dimension, namely the odour description of the single constituents of a complex system, can be obtained in addition to the chromatographic and spectroscopic data of the detected compounds. The quantification of single odour impressions in a complex mixture is possible via aroma extract dilution (AED) analysis (Grosch 1990). This method also allows identical reproduction of a natural sample, including all compounds responsible for its typical odour and its correlation with the genuine odour of the fragrance sample (Buchbauer et al. 1995a,b). This analytical procedure, as well as the combined use of achiral and chiral GC phases, will be discussed in the following.

2.3 Multivariate Data Analysis (General)

The underlying principle of multivariate data analysis of instrumental data is based on the notion of instrumental data as entities corresponding to scalars, vectors and higher-order entities. In this respect, the chromatograms or spectra are considered to be vectors (though) of different types and more complex data structures like GC-MS data are counterparts for second-order tensors etc. This means that the chromatogram may be alternatively considered to be a one-row matrix, the spectrum to be a one-column matrix and a single GC-MS data set with n spectra of m masses to be a [m * n] matrix (set of n vectors or set of m chromatograms). For visualisation, comparison, or related operations with spectral vectors, the latter are considered to be elements of hypothetical m-dimensional hyperspace, where the criteria of identity or similarity are based simply on the geometric relation between the spectral vectors involved (e.g. the criterion for similarity may be simply the Euclidian distance between two spectral vectors involved; the spectral vectors are identical when the Euclidian distance between them is zero). The m-dimension coordinates correspond to the intensities of particular masses (wave numbers etc.) in the particular spectra. Of course, every spectral vector of similar compounds is near each other, so that separable clusters of similar compounds exist in the hypothetical hyperspace.

To visualise the geometric displacement of the spectral vectors in the n-dimensional space, a projection of the end points of spectral vectors in a plane (less common by a projection on one axis) is used. In order to find the optimal plane for projection of spectral vectors, usually a plane is used which is defined by the first two eigenvectors resulting from principal component analysis (PCA) of the original [m * n] data matrix. Once a suitable optimised projection is found, it is easy to use it for semiautomatised classification of large quantities of mass spectra, or spectra in general. The important tools to perform this task are the classifiers and training data sets of the spectra.

The classifier (Fig. 2) is a two-step procedure which computes the similarity between reference spectrum and a given spectrum in the first step and compares it with a threshold value (or with a range) to decide if a given spectrum has a certain property. This procedure, in turn, may be repeated with all spectra of the set, thus performing a classification. Figure 3 shows the working principle of planar classifier (i.e. classifier based on the projection of spectra in a plane) compared with the less effective option of linear classifier.

Essential for an effective classifier is the definition of the threshold value (or of the relevant similarity range). For this purpose spectral libraries with substructure search option are indispensable. First, the substructure search option in a spectral library enables the arbitrary spectrum selection of similar substructure components to built a training data set, as indicated in Fig. 3. A simple combination of spectral library like the NIST/EPA/NIH with an MDL ltd. ISIS/BASE (database operating on a defined MOL format) together with MOL format structures of an NIST/EPA/NIH library file (Chemical Concepts GmbH, Weinheim, Germany) solves this problem.

The following example shows the efficiency of a planar classifier in performing the task of classifying the set of GC-MS spectra to find possible pyrazine compo-

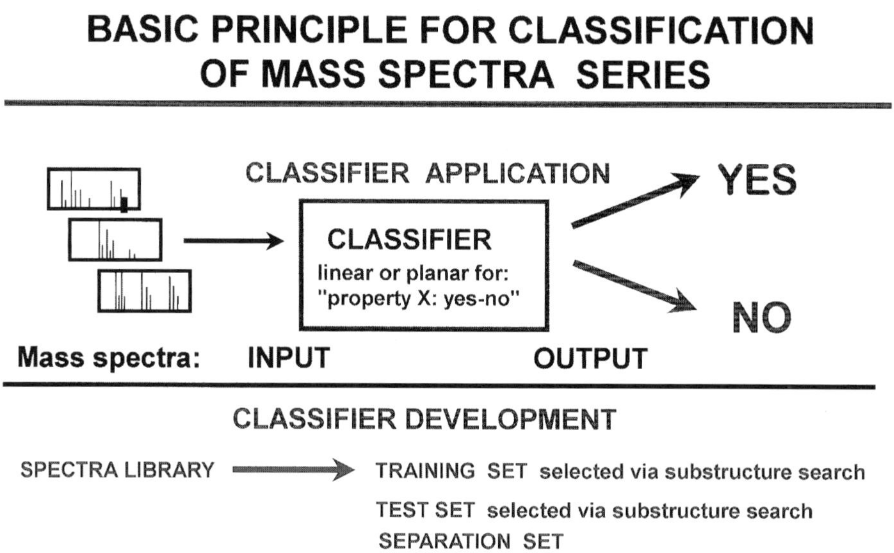

Fig. 2. Basic principle for classification of mass spectra series

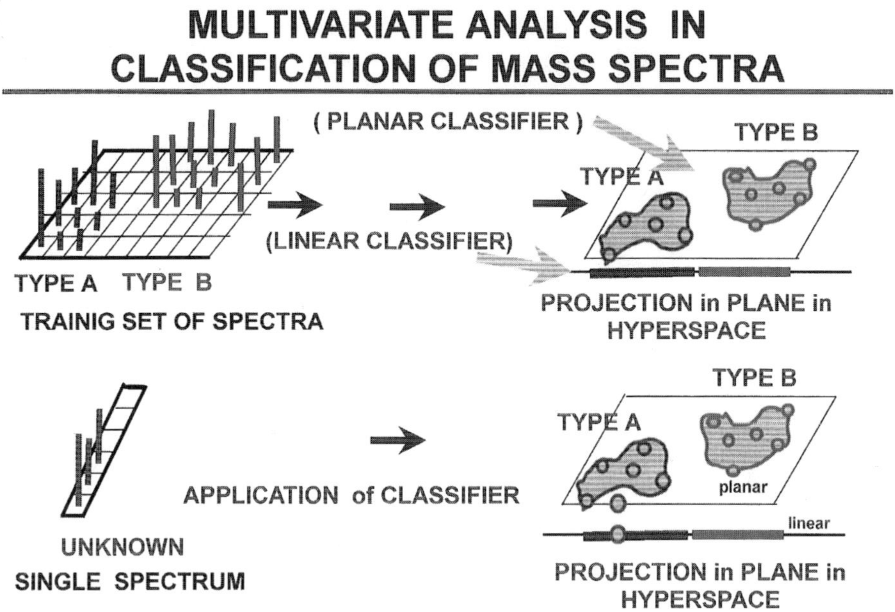

Fig. 3. Multivariate analysis in classification of mass spectra

nents. The data set originates from the same sample as used for comparison of GC-FID and GC-N(PD)-specific detection in Fig. 1. Using a training set of mass spectra of 97 heterocyclic compounds (pyrazines, alkoxypyrazines, thiazoles, pyridines and related heterocycles), a planar classifier (in this case, a polygon area in Fig. 4) for pyrazines was defined. After subsequent projection of all spectra from the GC-MS data set in the same plane (each spectrum is just a dot), it can be seen which spectra of the GC-MS data set are confined to being possible pyrazines. The program indicates their numbers, so that a reduced data set (reduced chromatogram) for any arbitrary further processing (e.g. library search) is produced. Library search in the reduced data led to the identification of components responsible for opium smell indicated in Fig. 1.

In this way, the use of multivariate analysis brings the benefits of fast data reduction accompanied by efficient target analysis for classes of compounds.

3 Experimental Details

3.1 Sample Preparation

The genuine plant material, headspace samples (after Knudsen et al. 1993), essential oils and extracts were olfactorily evaluated by professional perfumers and

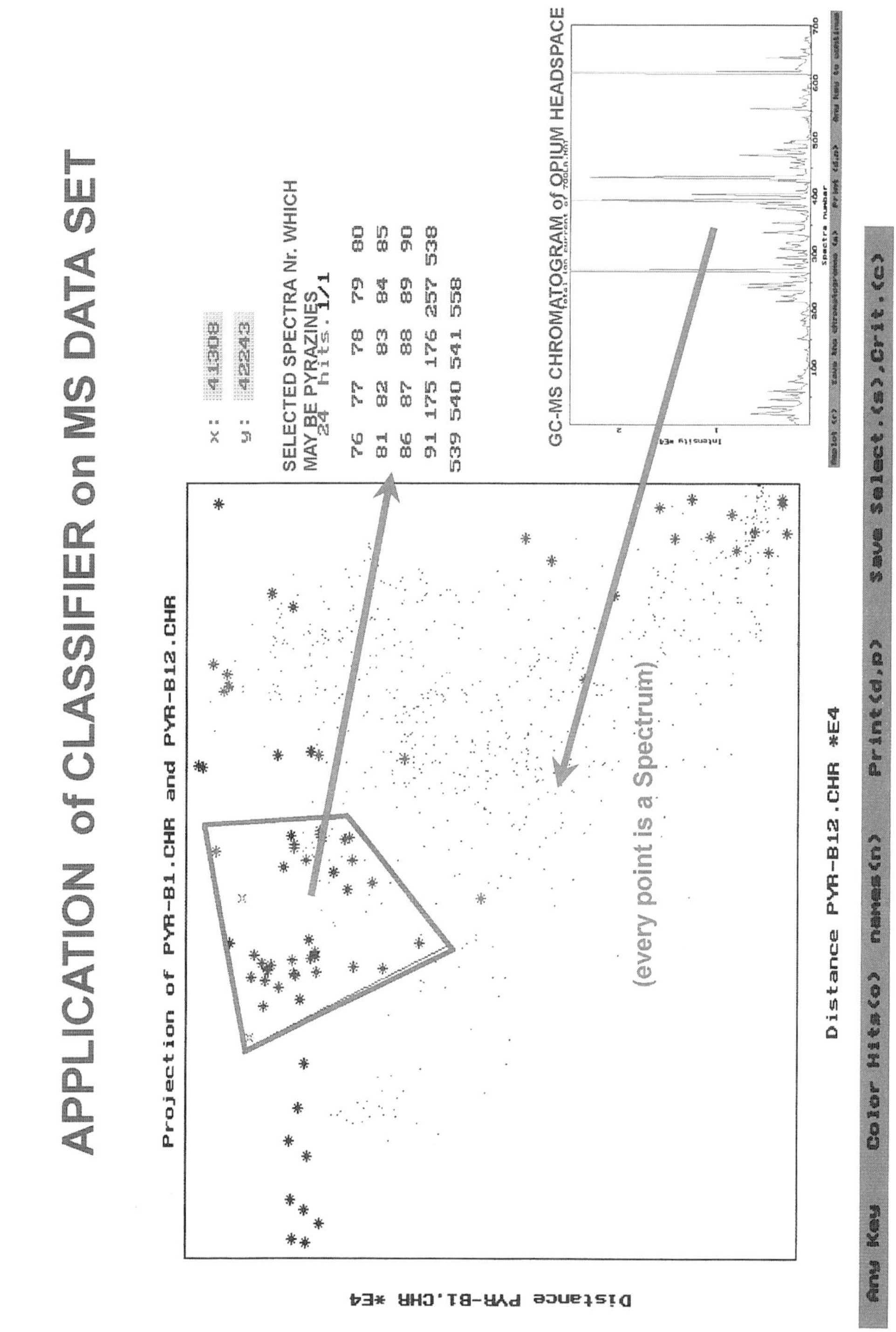

Fig. 4. Application of classifier on mass spectra set

aroma chemists. For the procedure of each single sample preparation, see the references cited according to the presented applications.

3.2 GC-FID

GC-FID analyses were performed with a GC-14A combined with a C-R6A integrator (Shimadzu Co.) or a GC-8160-00 (Fisons Instruments) with an SP-4270 (Spectra Physics Co.). For other parameters, see the references cited.

3.3 GC-NPD

GC-NPD: nitrogen-phosphorus detector NPD-40 (Fisons Instruments) was operated with FDP-500 control module (Fisons Instruments), carrier: hydrogen 80 kPa (30 m DB wax), air: 11 kPa, hydrogen 50 kPa, make up gas: helium 110 kPa, temperature program: 45 °C/2 min – 8 °C/min – 200 °C.

3.4 GC-MS

For GC-MS analyses, GC-HP-5890A gas chromatograph in connection with an HP-5970C MSD and the data system HP-9000/300 data system as well as the HP-GCD system, further [GC(HR)MS]: GC-HP-5890A (all Hewlett-Packard Co.) in connection with a MAT-95 mass spectrometer (Finnigan Co.) under control of the data system DEC5000 (Digital Co.) or a GC-17A with a QP-5000 mass spectrometer (Shimadzu Co.) and a ProLinea 4/50 data system (Crompaq Co.) were used. For other parameters, see the references cited.

3.5 GC-FTIR-MS

The GC-FTIR-MS analyses were done by a GC-HP-5890A with an HP-5965 IRD (MCT detector) and an HP-5970C MSD (Hewlett-Packard Co.). For other parameters, see the references cited.

3.6 GC-AES

The GC-AES analyses were carried out with an HP-5890A gas chromatograph and an HP-5971A atom emission detector (AED) in combination with an HP-59970C ChemStation (all Hewlett-Packard products). For other parameters, see the references cited.

3.7 GC Sniffing Technique

The GC Fractovap-Series-2101 with the LT-Programmer-210 and the Electrometer-160 (Carlo Erba Co.) in connection with a BBC-Servogor-S-Recorder (Brown Boveri Co.) were used for GC sniffing technique investigations. For other parameters, see the references cited.

3.8 Multivariate Data Analysis, File Transfer

The MDA was performed on an AT-486DX-50 computer (Highscreen Co.) with our own programs (Turbo-Pascal 6.0) in combination with commercial NIST/EPA/NIH mass spectra library, the MDL Information Systems AG ISIS/BASE data base structures of NIST/EPA/NIH in ISIS/BASE format data file from Chemical Concepts GmbH, Weinheim, Germany. The data transfer was performed using LAN-Ethernet. For other parameters, see the literature cited.

4 Applications

4.1 Application 1: Analysis of *Melissa officinalis* and *Cymbopogon* Fragrance Compounds

The first application using the combined information of gas chromatographic-spectroscopic and olfactory methods shows the determination of the dominant odour compounds of balm and cymbopogon samples. Various investigations were done (Kreis and Mosandl 1994a,b; Hener et al. 1995) to verify the authenticity of these olfactorily and therapeutically valuable oils. The analysis of odorous constitutents of these oils and a correlation between the single samples were performed to find out which compounds were responsible for the typical odour of the aroma samples. The production of synthetic mixtures was enclosed to obtain a simple odour system comparable to the complex genuine one.

Different oils of balm (*Melissa officinalis* L.) and cymbopogon [*Cymbopogon nardus* Rendle (Ceylon type) and *Cymbopogon winterianus* Jowitt (Java type)] samples as well as the headspace and hydrosteam distillates of the oils were investigated (after olfactory evaluation by perfumers) in the first step by GC-FID to obtain information about their qualitative (retention time correlations and Kovats indices) and quantitative distribution (peak area calculations). More than 20 main compounds were detected in the different samples. The headspace compounds (trapped by a dynamic headspace method) possess the highest relevance for the total odour impression of the single oils [see Fig. 5: GC-FID of headspace (HS) sample from Java] and therefore were of primary interest for this investigation.

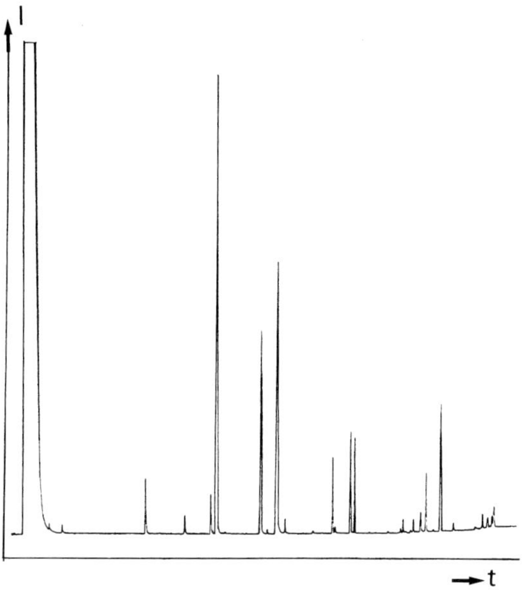

Fig. 5. GC-FID chromatogram (*I* intensity; *t* retention time) of the headspace from a Java *Cymbopogon winterianus* Jowitt oil

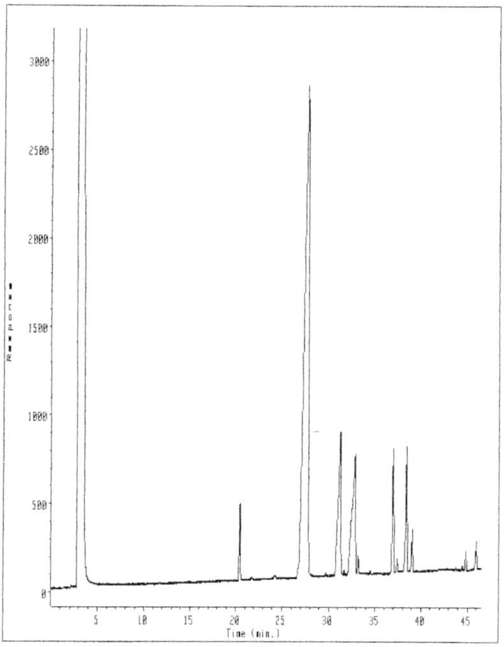

Fig. 6. GC-FTIR chromatogram of the essential oil from a sample of Java *Cymbopogon winterianus*

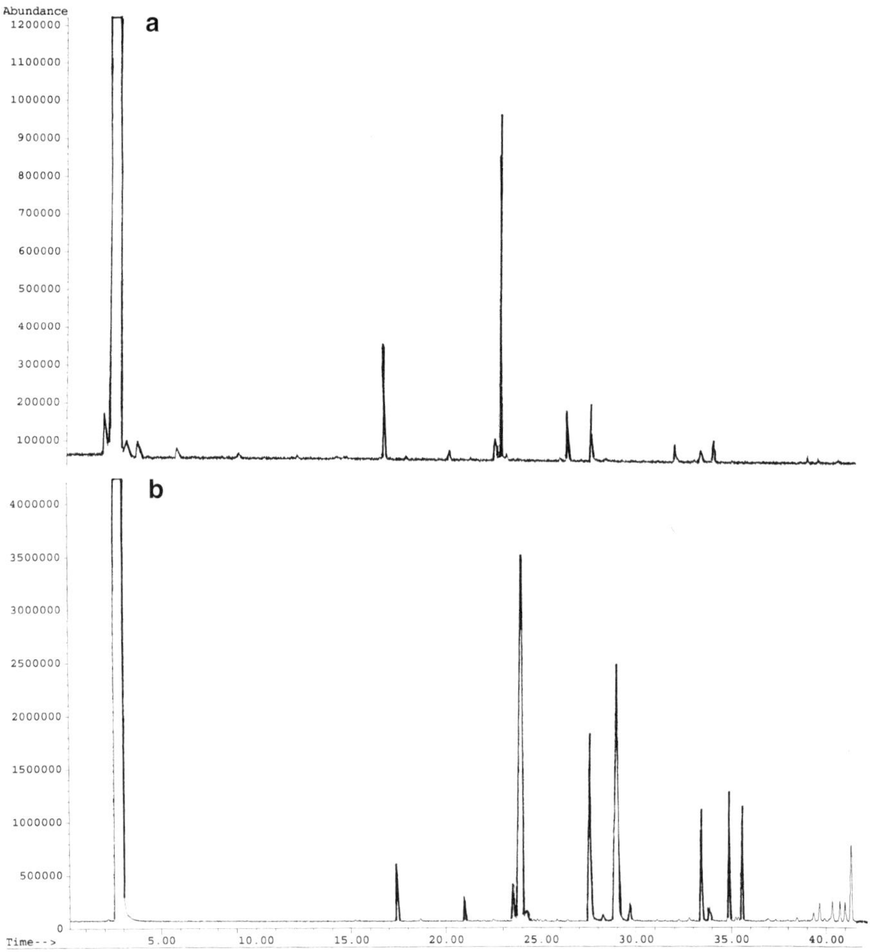

Fig. 7a,b. GC-MS chromatogram of the headspace (**a**) and of the essential oil (**b**) from a sample of Java *Cymbopogon winterianus*

The next step was to perform GC-FTIR-MS analyses to identify the single fragrance compounds in each sample (see Fig. 6: GC-FTIR of the Java oil and Fig. 7a,b: GC-MS of the Java oil).

More than 98% of all detected compounds (GC-FID) were identified (IR and mass spectra correlations with library spectra on- and off-line as well as coinjection of pure compounds) as genuine constitutents of *Melissa* or *Cymbopogon* oils (but also from relevant headspace and hydrosteam distillation samples). As main constituents of all investigated samples, citronellal, limonene, citronellol, nerol, geraniol, caryophyllene and some others (mainly monoterpenes) were found (see Table 1: identified compounds).

The determination of the odour value of the single constituents in correlation to the total odour impression of the samples was possible using the above-

Table 1. Identified compounds of the headspace from a Java *Cymbopogon winterianus* Jowitt oil

Compound	Name	MW[a]	Oil[b]	hs[c]	ls[d]
01	Limonene	136	2.31	20.3	93
02	Linalool	154	0.81	–	91
03	Beta-iso-pulegol	154	1.64	Trace	99
04	Citronellal	154	32.5	61.3	97
05	Citronellol	156	12.1	9.2	98
06	Geraniol	154	22.9	9.2	93
07	Geranial	152	0.75	–	97
08	Citronellyl acetate	198	4.36	–	91
09	Eugenol	164	5.3	–	98
10	Geranyl acetate	196	4.6	–	91
11	Elemene	204	0.62	–	95
12	Germacrene D	204	0.98	–	80
13	Gamma-cadinene	204	0.96	–	93
14	Delta-cadinene	204	0.80	–	94

[a] Molecular weight.
[b] Concentration in essential oil (peak area percentage of GC-FID).
[c] Concentration in headspace (peak area percentage of GC-FID).
[d] Correlation factor (percentage) of mass spectra (GC-MS) library search.

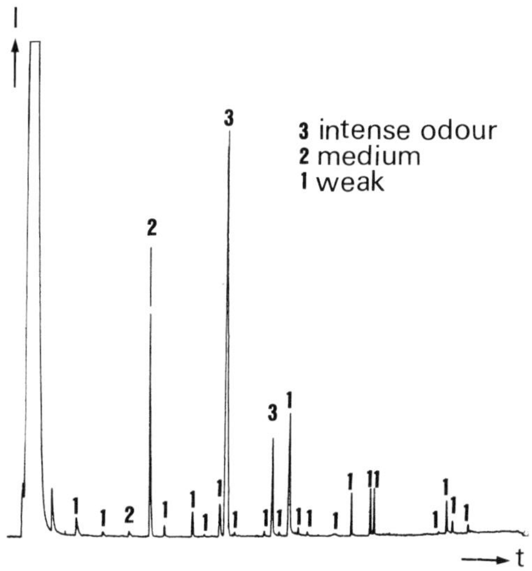

Fig. 8. GC sniffing technique profile of samples of the essential oil of Ceylon *Cymbopogon nardus* Rendle

mentioned GC sniffing technique. As a result of these analyses, the compounds citronellal, alpha-pinene, beta-pinene, 3-carene, camphene, limonene and citronellol were found to be essential for the typical (high quality) odour of these natural products (see Fig. 8: GC sniffing technique of Ceylon oil).

For quantitative calculation, the aroma extract dilution analysis was used: each sample was diluted (always factor 10) and the dilutions checked by the GC sniffing technique. This investigation showed that the typical odour of the four basic oil samples (*Cymbopogon* oils from Jaya and Ceylon respectively *Melissa* oils from commercial trade of pharmacy and industrial origin) can be combined in systems of pure compounds:

1. Java oil: limonene : citronellal = 1 : 3.
2. Ceylon oil: beta-carene : alpha-pinene : camphene : limonene = 1 : 2 : 5 : 3.
3. Pharmacy oil: limonene : citronellal = 1 : 1.
4. Industry oil: beta-pinene : limonene : citronellal = 2.5 : 3 : 1.

Each sample – headspace, essential oils, steam distillates and synthetic compositions – was olfactorily evaluated by professional perfumers. Surprisingly, the combination of the pure compounds limonene and citronellal 1 : 3 (Java oil type) was recognized by the perfumers as the fragrance sample with the best olfactory quality. The headspace sample, the essential oil and the steam distillate of *Cymbopogon* from Java, the combination of limonene and citronellal 1 : 1, the headspace sample, the essential oil and the steam distillates of *Melissa officinalis* of pharmaceutical origin were also found to be characterised by a medium to high olfactory quality suited to be used in fine perfumery. The other samples investigated showed only poor odour values and therefore were of less importance for a possible perfumery product.

The enclosed multivariate correlations (Brereton 1992; McLachlin 1992; Nikiforov et al. 1992, 1994; Graham 1993; Meier and Zünd 1993; Wasicky et al. 1995) of the oils and extracts, on the basis of these seven olfactorily important compounds, demonstrate the great similarity in headspace and in essential oil of the pharmaceutical grade *Melissa officinalis* and of the Java *Cymbopogon* samples (Fig. 9).

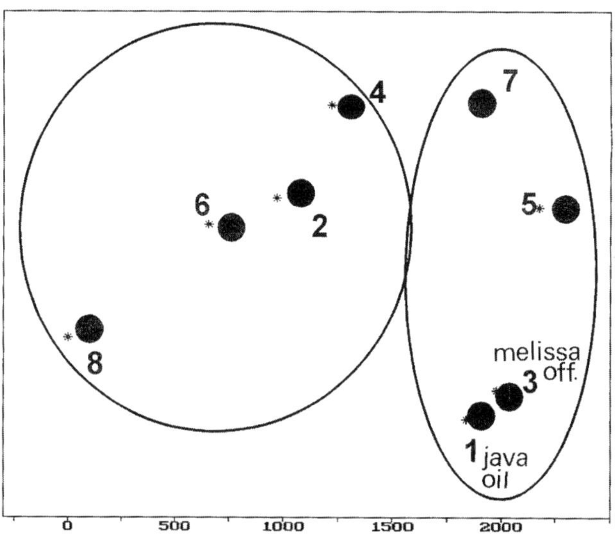

Fig. 9. Multiavariate data analysis (MDA) of samples of *Melissa* and *Cymbopogon*

4.2 Application 2: Analysis of Hay Blossom Fragrance Compounds

Hay blossoms possess a long tradition in pharmaceutical and cosmetic applications. The complex composition (mainly grasses) of this plant mixture renders it very difficult to identify the compounds responsible for the typical odour impression (fresh, green, coumarin-like, aromatic and spicy).

The total analysis of the headspace, essential oil and pentane extract of a commercial product using TLC, HPLC-UV, HPLC-MS, GC-FID, GC-FTIR-MS and the GC sniffing technique resulted in the identification of more than 100 constitutents (Buchbauer et al. 1990; Jirovetz et al. 1991). The correlation of chromatographic-spectroscopic and olfactory data determined more than 20 potent aroma compounds. The odour impression of each single fragrance compound contributes to the typical odour of hay blossoms, as mentioned above. Only the identification of free coumarin (normally bonded with glycosides) was not possible by routine techniques. Although in the appropriate retention time range, where coumarin normally elutes, a significant odour was recognised by the GC sniffing technique, but no peak could be found. Only in the GC-FTIR measurement was a trace of coumarin hidden beneath another peak detected (without the characteristic coumarin IR spectra).

Therefore, the OCI technique (on-column injection) was used for all gas chromatographic-spectroscopic methods. The GC-FTIR with SWC (selected wavelength chromatogram) and GC-MS with MID (multiple ion detection: Fig. 10) showed all significant structure information of the coumarin molecule. Thus, the correlation of the coumarin-like odour impression with free coumarin was pos-

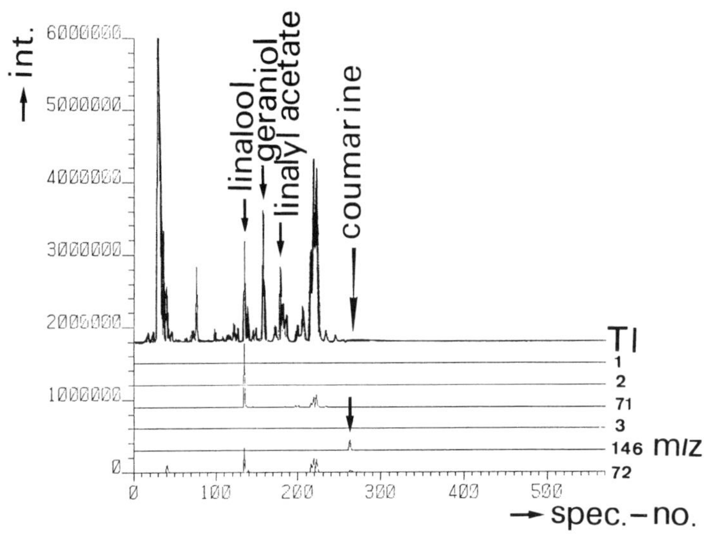

Fig. 10. GC-MS chromatogram of hay blossom sample using multiple ion detection for the identification of coumarin

sible and the combination of all odours of the single components was found to be essential for the fresh, green, coumarin-like, aromatic and spicy odour of hay blossoms.

4.3 Application 3: Analysis of the Content of (–)- and (+)-Carvone in Dill Oils

The principal component of dill oils, (+)-carvone, a monoterpene ketone with a caraway-like odour, is a good indicator for the quality of dill oil samples (dill oil, dill herb oil and dill seed oil). A content higher than 5% of (–)-carvone (spearmint-like odour) in genuine dill oils may be an adulteration by a synthetic, racemic carvone.

Some oil samples from Kazachstan were investigated by GC (GC-FID, GC-FTIR-MS and GC sniffing technique) with achiral and chiral phases (polypeptide and cyclodextrin; König 1992) to control these samples for stereoisomeric carvone content.

The use of achiral phases furnished no clear answer, why the oils possess a non-characteristic side-note (evaluated by professional perfumers); but analyses (GC-FID and GC-MS) with chiral GC phases (the best results were obtained by trifluoracetyl-cyclodextrin phases: Fig. 11) showed a rate of (–)-carvon to (+)-carvone of 10:90% (correlations of high quality dill oils from Romania have a rate of 4:96%). This indicates an adulteration of the dill oils by the cheaper, synthetic, racemic carvone to increase the price of this valuable product for the food industry (Jirovetz et al. 1994a). This result was verified by GC sniffing technique investiga-

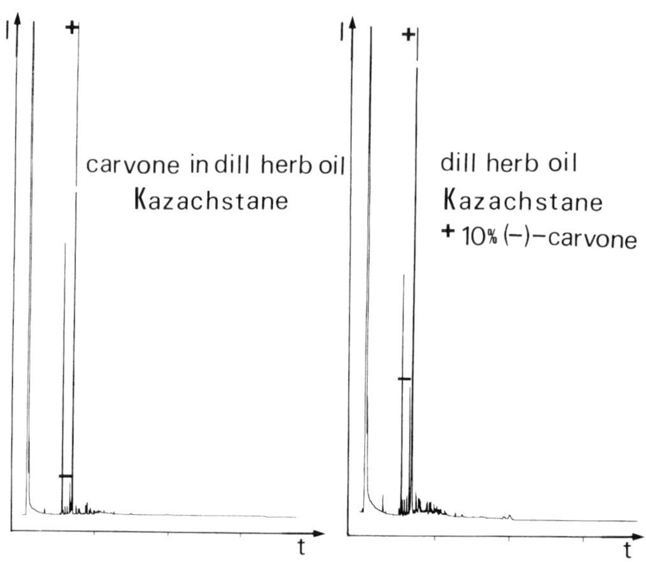

Fig. 11. GC-FID chromatogramme of (+)/(–) carvone in dill herb oil from Kazachstan

tions, where (–)- and (+)-carvone (two peaks) could be olfactorily and separately determined.

4.4 Application 4: Aroma of Red and White Wine Blossoms – Correlation of Sensory Data with Headspace Constituents

By combining gas chromatographic-spectroscopic systems (GC-FID and GC-FTIR-MS) with the GC sniffing technique in the headspace analysis of different wine blossoms, it was possible to correlate different wine varieties in terms of the composition of their corresponding wine blossom fragrance compounds. Red wine blossoms possess an odour similar to white wine blossoms (Buchbauer et al. 1994a) with a beautiful floral and fruity-fresh note, as well as a mignonette-like, acrid-dusty and green side-note. The blossoms differ, depending on the wine variety, only in the intensity of single odour impressions, but not in the direction of the total odour. The analysis of the headspace constituents of different samples of blossoms using GC-FID, GC-FTIR-MS and in combination with the GC sniffing technique identifies odour-dominating sesquiterpenes and odour-assisting monoterpenes (Buchbauer et al. 1994a; Buchbauer et al. 1995). Various correlations (odour profiles, qualitative and quantitative statistics and multivariate data analysis) give additional information to characterise the red and white wine species investigated and show a clear ampelographic correlation between single wine varieties. A clear separation of the red from the white wine species via multivariate data analysis of the headspace constituents of single blossoms is also possible (see Fig. 12).

4.5 Application 5: Use of GC-AES and GC-FTIR-MS to Identify the Characteristic Fragrance Compounds with an Oxygen Functional Group in Lavender and Lavandin Oils

The quality of lavender and lavandin oils is especially determined by monoterprenic compounds with an oxygen functional group like linalool, linalyl acetate, camphor, 1,8-cineole, lavandulol, lavandulyl acetate, geraniol and geranyl acetate. The identification of these constituents using GC-FID and GC-FTIR-MS is more efficient in combination with GC-AES analyses. Because some of these are often trace compounds in a high number of lavender or lavandin samples, registration of the oxygen trace, with a principal sensitivity in the range of GC-ECD, is a very fast and powerful tool to detect these monoterpenes (see Fig. 13). Correlation with the simultaneously registered carbon trace, as well as comparison with GC-FID and GC-FTIR-MS chromatograms allows the selective separation of the oxygen functional group constituents in odour systems from other compounds.

An implementation of this method for the development of oxygen selective classifiers to investigate natural products (essential oils, extracts, headspace and

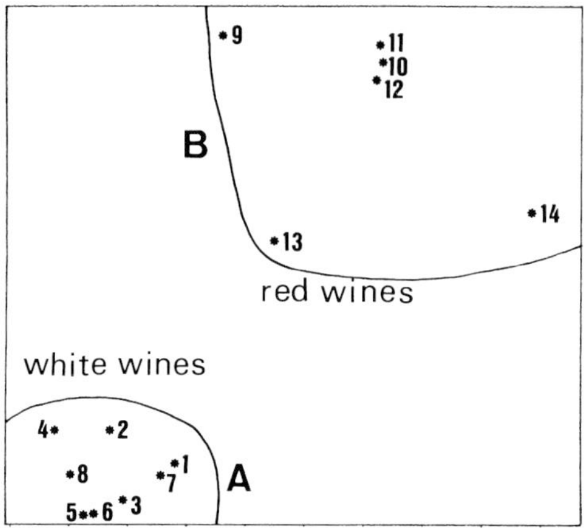

Fig. 12. Separation of white and red wine varieties using multivariate analysis of headspace data of single wine blossoms

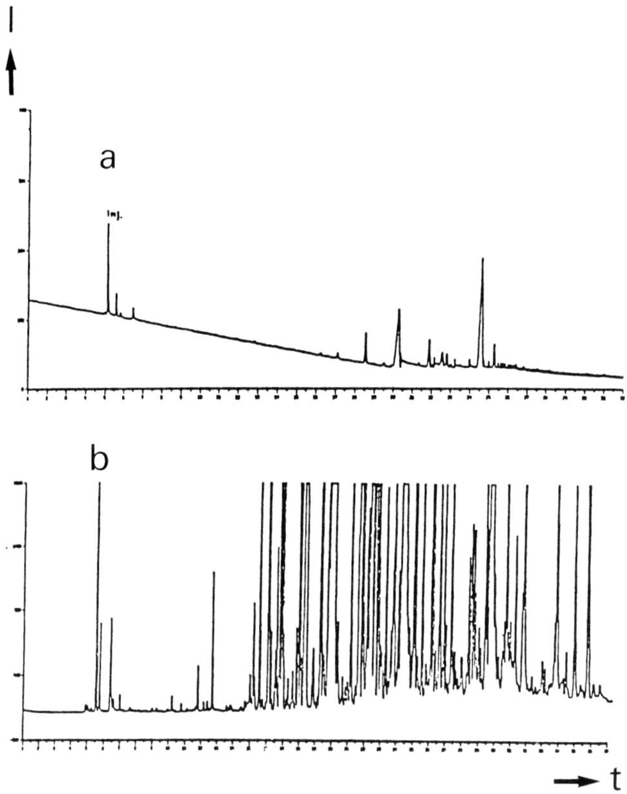

Fig. 13a,b. GC-AES chromatogram of French lavender oil using oxygen (**a**) trace and carbon (**b**) trace to identify odorous monoterpenes with an oxygen functional group

genuine samples in food and fragrance chemistry) by multivariate analysis of gas chromatographic-spectroscopic data has been demonstrated using a semiautomatic interpretation (Nikiforov et al. 1992, 1994).

GC-AES coupling is also of essential importance in the analysis of target compounds in a complex biological matrix (blood/serum). Therefore, this method was used to identify fragrances in blood samples after inhalation experiments with single-odour compounds (e.g. linalool and linalyl acetate) or essential oils (lavender and sandalwood oil) to find out which derivatives were responsible for the pharmacological effects (e.g. mild sedation or stimulation) of aromatherapeutical applications (Jirovetz et al. 1990a; Jirovetz et al. 1992).

5 Summary

The combined use of gas chromatographic-spectroscopic (GC-FID, GC-NPD, GC-AES, GC-FTIR and GC-MS) and gas chromatographic-olfactory (GC sniffing technique) methods is the most powerful tool for aroma analysis in the field of perfumery, fragrance and food chemistry as well as health science (e.g. aromatherapy). Main, minor and trace components of olfactory importance for the odour impression of complex mixtures (headspace, essential oil and extract samples) can be detected, identified and their olfactory values determined. Additional information on stereochemical aspects is possible by means of chiral phase (especially cyclodextrin phases) gas chromatography with various detectors (FID, NPD, AES, MS, FTIR and nose). These applications include all these points in a short and compact manner to discuss the single problems in aroma analysis and give possible answers to some analytical questions. Detailed data are mentioned in the references cited corresponding to each single application.

Acknowledgements. We acknowledge with gratitude the sample preparations, extractions and chromatographic-spectroscopic measurements by Dr. Michael Wasicky, Dr. Barbara Remberg and Mag. A. Shafii-Tabatabai. We are also very grateful to Mr. W. Höppner, V. Hausmann and A. Obrocki (chief perfumers of Dragoco Co., Vienna, Austria), who performed the olfactory evaluations, as well as to their company for their interest in our investigations. We are indebted to the *Austrian Fonds zur Förderung der wissenschaftlichen Forschung* (Projects P7587CHE, P8299CHE and P9568CHE) and the Hochschuljubiläumsstiftung of the City of Vienna (H-78/94, H-185/95) for the financial support. For help in multivariate data analysis and library search software programming, we thank Mr. Josef Plangger and Dr. Alexander Woidich.

References

Blum W (1991) Hochtemperatur Gas-Chromatographie. Hüthig, Heidelberg

Brereton RG (ed) (1992) Multivariate pattern recognition in chemometrics. Elsevier, Amsterdam

Buchbauer G, Jirovetz L, Nikiforov A, Remberg G, Raverdino V (1990) Headspace analysis and aroma compounds of Austrian hay-blossoms (Flores Graminis, Graminis Flos) used in aromatherapy. J Essent Oil Res 2:185–191

Buchbauer G, Jirovetz L, Wasicky M, Hertlitschka A, Nikiforov A (1994a) Aroma von Weissweinblüten: Korrelation sensorischer Daten mit Headspace-Inhaltsstoffen. Z Lebensm Unters Forsch 199:1–4

Buchbauer G, Nikiforov A, Remberg B (1994b) Headspace constituents of opium. Planta Med 60:181–183

Buchbauer G, Jirovetz L, Wasicky M, Nikiforov A (1995a) Aroma von Rotweinblüten: Korrelation sensorischer Daten mit Headspace-Inhaltsstoffen. Z Lebensm Unters Forsch 200:443–446

Buchbauer G, Jirovetz L, Wasicky M, Nikiforov A (1995b) On the odour of old books. J Pulp Pap Sci 21:398–400

Graham RC (1993) Data analysis for the chemical science. VCH, Weinheim

Grosch W (1990) Analyse von Aromastoffen. Chem unserer Zeit 24:82–84

Guichard H, Guichard E, Langlois D, Issanchou S, Abbot N (1995) GC sniffing analysis: olfactive intensity measurement by two methods. Z Lebensm Unters Forsch 201:344–350

Hener U, Faulhaber S, Kreis P, Mosandl A (1995) On the authenticity evaluation of balm oil (*Melissa officinalis* L.). Pharmazie 50:60–62

Jirovetz L, Buchbauer G, Jäger W, Raverdino V, Nikiforov A (1990a) Determination of lavender oil fragrance compounds in blood samples. Fresenius J Anal Chem 338:922–923

Jirovetz L, Nikiforov A, Woidich A, Braun D, Buchbauer G (1990b) Olfaktometrie und Analytik von Vetiverölen und Vetiverduftstoffen. Seifen-Öle-Fette-Wachse 63:593–597

Jirovetz L, Buchbauer G, Jäger W, Remberg G, Raverdino V, Nikiforov A (1991) Zur Untersuchung der Inhaltsstoffe und Wirkungen von Heublumen. Seifen-Öle-Fette-Wachse 117:252–256

Jirovetz L, Buchbauer G, Jäger W, Woidich A, Nikiforov A (1992) Analysis of fragrance compounds in blood samples of mice by gas chromatography, mass spectrometry, GC/FTIR and GC/AES after inhalation of sandalwood oil. Biomed Chromatogr 6:133–134

Jirovetz L, Buchbauer G, Nikiforov A (1994a) Vergleichende Inhaltsstoffanalyse verschiedener Dillkraut- und Dillsamenöle mittels GC/FID und GC/MS. Nutrition 18:534–536

Jirovetz L, Buchbauer G, Nikiforov A (1994b) Anwendungen der GC-FTIR-MS in der Analytik von Naturstoffen. GIT Fachz Lab 38:449–453

Knudsen JT, Tollsten L, Bergström LG (1993) Floral scents – a checklist of volatile compounds isolated by head-space techniques. Phytochemistry 33:253–280

König WA (1992) Gas chromatographic enantiomer separation with modified cyclodextrins. Hüthig, Heidelberg

Kreis P, Mosandl A (1994a) Chiral compounds of essential oils. Part XVI. Enantioselective multidimensional gas chromatography in authenticity control of balm oil (*Melissa officinalis* L.). Flav Fragr J 9:249–256

Kreis P, Mosandl A (1994b) Chiral compounds of essential oils. Part XVII. Simultaneous stereoanalysis of *Cymbopogon* oil constituents. Flav Fragr J 9:257–260

Maas B, Dietrich A, Mosandl A (1994) Collection of enantiomer separation factors obtained by capillary gas chromatography on chiral stationary phases. J High Resolut Chromatogr Commun 17:109–115; 169–173

McLachlin GJ (1992) Discrimination analysis and statistical pattern recognition. J Wiley, New York

Meier P, Zünd RE (1993) Statistical methods in analytical chemistry. J Wiley, New York

Mosandl A (1992) Capillary gas chromatography in quality assessment of flavours and fragrances. J Chromatogr 624:267–292

Nikiforov A, Stanek W, Woidich A, Jirovetz L, Buchbauer G (1989) Eine einfache und effiziente Methode der mikropräparativen Gas-Chromatographie mit Quarzkapillarsäulen und Splitsystem. Fresenius Z Anal Chem 333:46–47

Nikiforov A, Jirovetz L, Woidich A (1992) Multivariate correlations of GC/MS/FTIR data of strawberry aroma. In: Woidich H, Buchbauer G (eds) Proc 12th Int Conf Flav Fragr Essent Oils, Austrian Assoc Flav Fragr Ind, Vienna

Nikiforov A, Jirovetz L, Woidich A (1994) Evaluation of combined GC/MC/FTIR data sets of strawberry aroma. Food Qual Pref 5:135–137

Wasicky M, Buchbauer G, Jirovetz L, Nikiforov A (1995) Zur Korrelation olfaktorischer Eigenschaften mit Headspace-Inhaltsstoffen in komplexen Geruchssystemen. EUROCOSMETICS 3:20–26

Woidich A, Nikiforov A, Jirovetz L (1994) Kombination der achiralen und chiralen GC mit FTIR- und Massenspektrometrie. Nutrition 18:221–224

Chemometric Analysis of Data from Essential Oils

D.B. Hibbert

1 Introduction

Chemometrics is "the use of computational and mathematical methods to extract information from analytical data" (Hibbert and James 1987). Advances in a chemist's ability to make measurements of the nature and composition of essential oils has led to a potential bottleneck in obtaining the maximum information from the plethora of high-quality data that arises. It is now perfectly possible to measure the concentrations of hundreds of essential oils from a plant, and how these data are analysed is the concern of this chapter.

Thirty years ago, McKern (1965) stated that although many promising leads had appeared, it was considered that both chemical and botanical data were still too insufficient to enable the chemical compositions of volatile leaf oils of trees and plants to be used for taxonomic purposes. The quality of the chemical data now allows us to use statistical methods to make taxonomic judgements with some confidence (Canigueral et al. 1994; Dunlop et al. 1995).

Other uses of chemometric data may be in the processing of essential oils, to monitor the progress of mixtures of oils in foods, cosmetics or pharmaceuticals (Cotroneo et al. 1990). Quality control by spectroscopy regularly involves chemometrics (Schulz and Losing 1995). Authentication of oils may also require chemometric analysis (Cotroneo et al. 1988).

Chemometrics is an established branch of analytical chemistry, with two international journals (Journal of Chemometrics, Wiley, and Chemometrics and Intelligent Laboratory Systems, Elsevier) and a number of textbooks (see, for example, Massart et al. 1988; Meloun et al. 1992, 1994). There are a great number of techniques that may fall under the heading of chemometrics, and many may be found in general or specialist works and embodied in chemometric software packages.

This chapter will survey, with examples, the chemometric methods that have been used to investigate essential oils. Papers in the literature often are obscure when reporting chemometric manipulations. This chapter, therefore, also has a goal of helping researchers to write up their work so that it can be properly judged by journal readers and reviewers.

Finally, it cannot be overemphasised that conclusions arrived at are only as good as the original data. No amount of chemometrics can compensate for poor science at an earlier stage.

Modern Methods of Plant Analysis, Vol. 19
Plant Volatile Analysis
Edited by H.-F. Linskens and J.F. Jackson

2 Before Chemometrics

2.1 Defining the Problem

If data are to be treated by the methods given in this chapter, planning at an early stage will improve the chance of a successful analysis. A proper "experimental design" (Morgan 1991) will lead to adequate data to distinguish between the possible sources of variability in the system. An investigation is usually undertaken with particular questions in mind: what is the seasonal variation in oil from a species?; does a species exist in a single chemotype?. It is therefore necessary to choose samples for chemical analysis that will answer the required questions and will not cause unwanted factors to influence the outcome. There is a limit on the number of analyses that may be performed. The limit may be of time, cost or amount of sample. Designed experiments can be shown to give the maximum information for the minimum number of analyses.

2.2 Data Collection

There are many possible sources of variation in the essential oil content of a plant. In addition, changes may occur because of sampling and storage, and there is an underlying uncertainty in the experimental results. Replicate analyses of a sample allow estimation of the method uncertainty. The word error, with connotations of wrongdoing, is no longer fashionable.

By pooling samples, for example old and new leaves from a given tree, will *confound* the effects that arise from these differences (here, age of leaf). Contrariwise, if all the old leaves from a number of trees were pooled and all the young leaves were also pooled (analysed together), the variations between trees would have been confounded and the variation between old and new leaves investigated. The variance associated with a particular effect may be determined in an ANOVA (analysis of variance).

A typical initial strategy when several factors may be important is to employ a two-level *factorial design* in which samples are analysed that have every combination of two values of each factor, e.g. old/new leaf, summer/winter, fresh/dried sample, species 1/species 2. For k factors there are 2^k analyses plus any replicates (Fig. 1a). The number of analyses rises quickly with the number of factors rendering six factors (64 experiments plus replicates) a practical upper limit. Data analysis gives each main effect – the change in the measured property that is associated with the low to high change in each variable – plus interaction effects. For example, the effect of day length on the production of an essential oil may be more pronounced in a young leaf than an old leaf. In this case, the interaction effect between day length and leaf age would be significant. It is the interaction effects that are not found by simple analysis of a series of experiments in which one factor is changed while the others are kept constant. To investigate numbers of factors that are greater than the experimenter's capacity to perform the corresponding full factorial design, it is possible to sacrifice information of high order interactions by performing a subset of experiments in a *fractional factorial design* of 2^{k-n} experiments. Highly fractionated designs that yield only main effects have also been published. More sophisticated designs that can investigate non-linear effects are *star* and *central composite* designs (Fig. 1b).

Whether the intention is to submit the data to multivariate analysis or other forms of statistical treatment, the use of designed experiments is strongly recommended. However, it is recognised that plants are not industrial assembly lines (where the methods of experimental design originated), and it may not always be possible to fit the samples that are analysed into a comfortable high/low pattern of factor values.

Although experimental designs per se have not been used extensively in essential oil analysis, the importance of proper sampling is well known. Adams (1991) explains a sampling strategy for the collection of juniper leaves. By collecting three leaf samples from each ordinal direction from old and new foliage, the within-tree variation may be estimated. Further samples then may investigate different locations, ages, species, and diurnal, seasonal and ontogenic variation. Butcher et al. (1994) also discuss the need to confound effects due to season and physiological age of leaves of *Melaleuca alternifolia*.

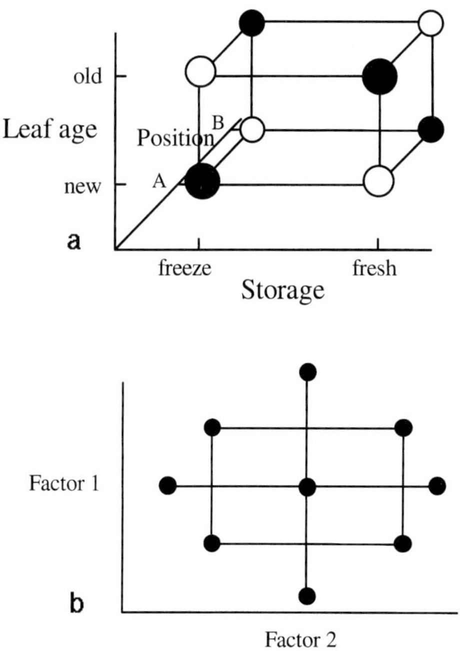

Fig. 1. **a** Experimental design for investigating three factors: leaf age, position of the plant, method of sample storage. For a full factorial design, experiments should be done at *each circle in the factor space* (eight experiments). A fractional design may be accomplished by experiments at either the *filled circles* or *open circles* (four experiments). **b** Central composite design for investigating two factors. Note that five different levels of each factor are specified in the design

2.3 Data

Data that are obtained in experiments on essential oils are usually *multivariate*, that is, more than one variable is measured per sample. It is possible to arrange the data as a matrix in which each row has measured values from a particular sample (or *object*) and each column corresponds to a variable. Examples of what is measured are GC peak heights (or areas), spectroscopic absorbances and the signal from a mass spectrometer. For each of these examples, the variables would be elution time (or possible chemical species), wavelength and mass, respectively (see Table 1 for more examples). The samples may have been chosen to represent variation over a number of factors, as discussed above, or may be a number of opportunistic samples. The data matrix may therefore be long and thin (many samples, few variables) or short and fat (many variables, few samples), and the methods of investigating the data may be constrained by this.

Experimenters should be careful when taking peak areas or heights that are divided by the summed area or height. This is a useful method of accounting for changes in gain of an instrument over time. However, it removes a degree of freedom and the data set is known as *closed*. When performing principal components analysis, it has been shown to lead to spurious correlations between variables, often seen as a negative correlation between compounds present in high concentration. To avoid this, a selective normalisation is recommended. Vaterova et al. (1995) discuss this problem in a partial least-squares analysis of monoterpenes in xylem oleoresins from species of *Pinus*.

Some methods use information from the known attributes of the samples, such as may have been determined in setting up an experimental design. Extra columns may be added to accommodate this information, which may be used, or not, as the method dictates.

It is also worth knowing if there is likely to be a high degree of correlation in the data, that is, do the variables or objects tell much the same story? Some methods such as multilinear regression do not tolerate high degrees of correlation.

Table 1. Examples of data used for chemometric analysis of essential oils

Variable	Measured property
Wavelength	Absorbance, transmittance, fluorescence
Elution time	Chromatographic peak height or peak area
Compound identified from GC	Calibrated concentration
Mass (amu)	Mass spectrometer detector response
Element from ICP analysis (trace elements in plants)	Element concentration
Human member of sensory panel	Score based on perceived odour
Gas sensor array (electronic nose)	Voltage developed by detector

3 Chemometric Methods

The methods described below will be under three headings according to the type of information required. Unsupervised pattern recognition is used to glean any groupings within the samples based on the data presented, but will no additional information. If groups are already known for some or all of the samples, then classification methods may be used that build models using this information. How well a model "fits" may be used as a test of the a priori groupings, or the model may be applied to data for which the group is not known in order to effect a classification. Finally, calibration is used when a model is required that links one or more measured properties of the data (dependent variables) to one or more other properties of the data (independent variables). Once the model is established this, too, may be used to predict the property (independent) of an unknown sample for which the dependent property is measured. The trivial example of this is linear regression used in univariate calibration.

3.1 Data Preprocessing

Preprocessing data may be vital in order to obtain sensible results. Data may be transformed to cause data to conform to the model to be used. Typically, the logarithm of a peak height may be taken if it is thought that the signal varies in this way with the dependent variable. If transmittance were measured as a function of concentration of a compound, it would be reasonable to transform the data logarithmically so that the absorbances scale linearly through Beer's Law. Taking logarithms is also an excellent way of compressing measurements that may range over several orders of magnitude.

Multivariate data, especially when the variables are qualitatively different, may also need scaling to remove variability inherent in the nature of the variables (Siebert and Stenroos 1989). Column centring (mean centring), in which the average value of a column of data is subtracted from each value in the column, is applied to nearly all data before processing. This brings all the variables to a mean of zero. If there is a large variation in values, standardisation, the division of each value in a column by the column standard deviation, reduces the variance of each column to 1. Autoscaling combines both column centring and standardisation. Range scaling causes the values to fall between 0 and 1. It is possible to apply these scaling methods to rows (i.e. samples) or globally by using the mean and standard deviation of all data. Formulae to effect these transformations are given in Table 2.

Whether or not any transformation should be applied is a question answered by general experience, and often trial and error. In a recent study of *Eucalyptus* species, Dunlop et al. (1995) concluded that scaling of GC peak heights at 180, 6-s time intervals was not appropriate before a principal components analysis or linear discriminant analysis. This means that the compounds that were in greatest

Table 2. Commonly used methods for preprocessing data

Method	Function to transform original variable x_{im} to scaled variable y_{im} [a]
Column centring	$y_{im} = x_{im} - \bar{x}_m$, where $\bar{x}_m$ is the mean of the column
Column standardisation	$y_{im} = \frac{x_{im}}{s_m}$, where s_m is the sample standard deviation of the column
Autoscaling	$y_{im} = \frac{(x_{im} - \bar{x}_m)}{s_m}$
Column range scaling	$y_{im} = \frac{x_{im} - \min_m(x_{im})}{\max_m(x_{im}) - \min_m(x_{im})}$

[a] x_{im} refers to a datum in row i and column m.

abundance (and therefore contributed most to an unscaled data set) really did give most information about the species. Standardisation compresses the difference between variables, thus increasing the importance of compounds that have a small concentration.

3.2 Classical Statistics

Before a great deal of chemometrics is done, it is always advisable to perform standard statistical tests for normality, outliers etc. For deciding if differences are significant between two or more groups of data, paired t-tests or analysis of variance (ANOVA) are recommended if the data are normally distributed. This is an important caveat, as ANOVA is sensitive to non-normal distributions, and unless very large samples are taken in a genuinely random way, it is not necessary that essential oil data sets be normally distributed. As a rare example of this being checked, Butcher et al. (1994) concluded that ANOVA was not appropriate to be used on ratios of terpenes from 109 samples of *Melaleuca alternifolia*.

3.3 Unsupervised Pattern Recognition

This may be as prosaic as a simple plot of one variable against another followed by visual inspection to reveal groups, or it may involve more complex processing. If groupings are observed, it may then be possible to determine which variables are important for the classification.

3.3.1 Principal Components Analysis (PCA)

At the heart of many multivariate methods is the use of principal components analysis (PCA). The essence of PCA is that the data are transformed to combinations of a number of orthogonal principal components or factors. The data matrix is reduced to two matrices, the scores matrix, which gives information about the objects, and the loadings matrix, which is concerned with variables. Each principal component may be thought of as a spectrum, or chromatogram, or whatever the set of the variables is, that represents some underlying feature of the data, and the loadings are the absorbances or peak heights. The first principal component accounts for the maximum amount of variance in the set, the second accounts for most of the remaining variance, and so on. A bonus of a PCA is that noise in the set which conveys no information is relegated to the later principal components, and thus the method may be used to clean up data.

The scores matrix gives the contribution of each object to each principal component. The use of a PCA in grouping objects is by inspection of the scores matrix. Typically, the score of the first principal component is plotted against the objects, or against the score of the second principal component with each point being identified by its associated object. Figure 2 shows such a plot for gas chromatographic data for oils distilled from two species of eucalyptus, *E. woodwardii* and *E. torquata*. The data set consisted of gas chromatographic peak heights for 37 samples at 40 selected retention times. Some of these times corresponded to known compounds (as identified by GC mass spectrometry), and some were unknown compounds. The oils of these species are expected to have different chemical compositions that can be revealed by a PCA. Indeed, two distinct groups may be seen based on the first principal component alone, which, in this case, accounts for 83% of the variance of the set. A PCA of these data to which has been added a number of crosses (*E. torwoodii*) is given by Dunlop et al. (1995).

The contributions of each variable to the principal components are given by the loadings matrix. A plot of the loading of the first principal component against

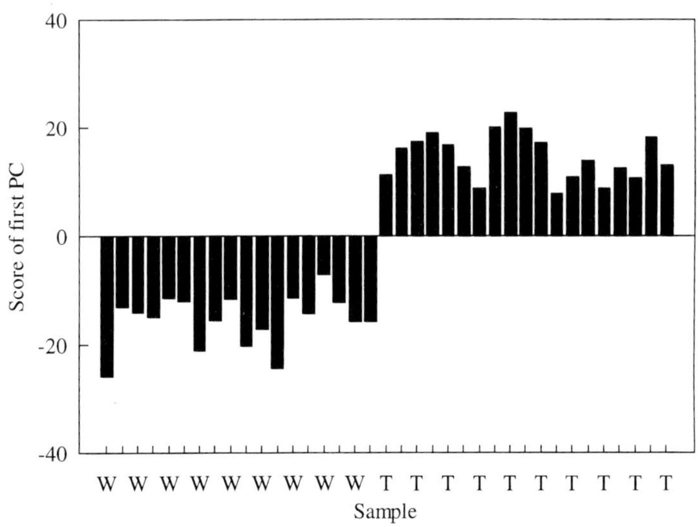

Fig. 2. Plot of the scores of the first principal component of a data set comprising 40 GC peak heights of samples of *E. woodwardii* and *E. torquata*

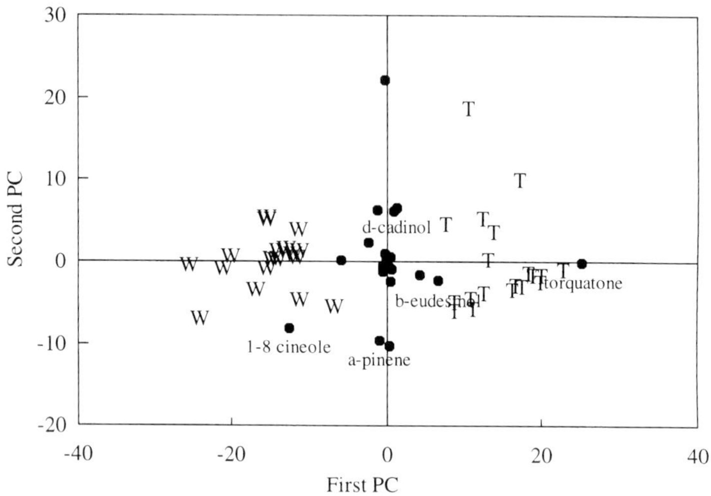

Fig. 3. Biplot of the scores and loadings of the first and second principal components of a data set comprising 40 GC peak heights of samples of *E. woodwardii* and *E. torquata*. *W* Score of sample of *E. woodwardii*; *T* score of sample of *E. torquata*. *Points* are loadings of peaks, with chemical identification where shown

the samples, or against the loading of the second principal component, indicates which variables are the most important for which principal components. If the variables represent chemical species, as they do with many gas chromatographic or mass spectroscopic analyses, useful information may be obtained about the relationships between compounds and the plant species. A biplot superimposes suitably scaled scores and loadings plots which can show these relationships in one graph. In Fig. 3, we show a biplot for the above example. It may be concluded that torquatone is a marker for *E. torquata* and that 1,8-cineole is associated with *E. woodwardii*. Alpha pinene, although present in both species, is not a good discriminator between them.

3.3.2 Hierarchical Methods of Cluster Analysis

The vector of data points for a given sample may be considered as a point in an n-dimensional hyperspace, where n is the number of variables. Cluster analysis tries to group objects by their proximity in this variable space. A typical aggregation method would have the following steps: find the two objects closest together and replace them by a cluster with coordinates midway between them. Repeat this step, treating clusters as points in the variable space until the entire set is clustered. The *similarity* ($s_{i,j}$) between two points i and j in space may be simply defined as

$$s_{i,j} = 1 - \frac{d_{i,j}}{d_{max}},$$

where $d_{i,j}$ is the distance between them and d_{max} is the maximum distance between any two points. An alternative approach is to divide the whole set into two clusters with the greatest dissimilarity, and then to continue splitting until single objects remain. The groupings that are built up may be displayed as a graph, called a dendrogram, showing the groups as a function of similarity. To accomplish this clustering, two procedures must be defined; what is the definition of "distance", and what criterion is to be used for linking clusters? Table 3 gives the methods that are most often found in statistical packages. In the essential oil literature, where hierarchical clustering is one of the most commonly used pattern recognition methods, authors rarely defend their choice of linkage method or clustering rules and sometimes do not mention the methods at all (e.g. Weber et al. 1994). The *Eucalyptus* data of Figs. 2 and 3 were clustered by an unweighted pair group average linkage on Euclidean distances leading to the dendrogram of Fig. 4. The computer time and memory requirements of hierarchical clustering increases as the square of the number of objects, which, together with the display requirements, puts a limit of one or two hundred on the number of objects treated by these methods. It is also possible to cluster the variables to show which give similar information. This has been done by Harch et al. (1995) for fatty acids in Australian groundnut germplasm.

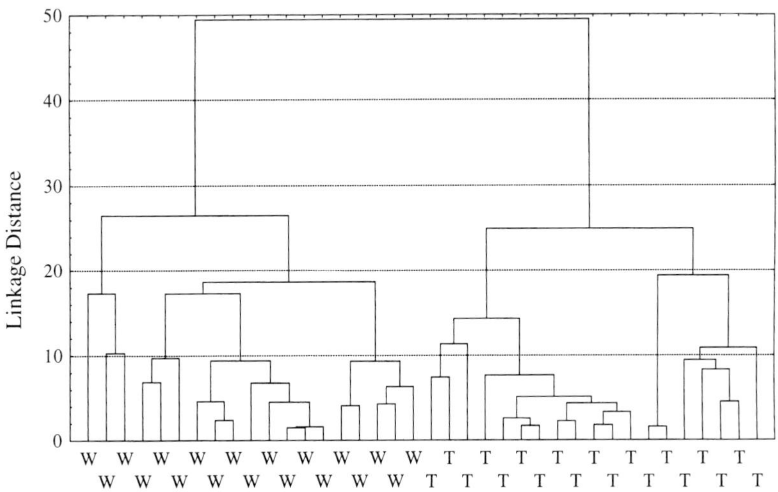

Fig. 4. Dendrogram of a data set comprising 40 GC peak heights of samples of *E. woodwaraii* and *E. torquata*. The method of complete linkage was used, with Euclidean distances

Table 3. Hierarchical clustering. Distance formulae and linkage rules

Distance metric	Formula[a]	Comments (references to use)
Euclidean distance	$d_{ij} = \sqrt{\sum_{k=1}^{k=m}\left(x_{i,k} - x_{j,k}\right)^2}$	The geometric distance in the multidimensional space (Maffei et al. 1993; Maffei 1994; Saez 1995)
Squared Euclidean distance	$d_{ij} = \sum_{k=1}^{k=m}\left(x_{i,k} - x_{j,k}\right)^2$	Emphasises points far apart (Harch et al. 1995)
City-block (Manhattan) distance	$d_{ij} = \sum_{k=1}^{k=m}\left\|x_{i,k} - x_{j,k}\right\|$	Similar to the Euclidean distance, but the effect of outliers is less (Li et al. 1994, 1995)
Chebychev distance	$d_{ij} = \max\left\|x_{i,k} - x_{j,k}\right\|$	Takes maximum difference between any variables
Power distance	$d_{ij} = r\sqrt{\sum_{k=1}^{k=m}\left\|x_{i,k} - x_{j,k}\right\|^p}$ p and r are user-defined	Parameter p controls the progressive weight that is placed on differences on individual dimensions and r controls the overall difference
Percent disagreement	$d_{ij} = 100 \times \frac{\text{number}\left\|x_{i,k} \neq x_{j,k}\right\|}{n}$	Useful for categorical data (Spring et al. 1994). See also Tanimoto index (Dunlop et al. 1995)
Linkage rules		
Single linkage (nearest neighbour)	The distance of the two closest objects (nearest neighbours) in the different clusters	Forms long, stringy clusters (Siebert and Stenroos 1989; Maffei et al. 1993; Maffei 1994; Saez 1995)
Complete linkage (furthest neighbour)	The greatest distance between any two objects in the different clusters	Forms clumpy clusters
Unweighted pair-group average	The average distance between all pairs of objects in the two clusters	Efficient for both clumped and stringy clusters (Li et al. 1994, 1995)
Weighted pair-group average	As the unweighted pair-group average method weighted by the size of the respective clusters	Used if sizes of clusters are very disparate
Unweighted pair-group centroid	The difference between centroids (average point in the variable space)	Links centres of gravity of clusters
Weighted pair-group centroid	As unweighted pair-group centroid weighted by the size of the respective clusters	Used if sizes of clusters are very disparate
Ward's method	Finds clusters which have a minimum sum of squares (cf. ANOVA)	Very efficient, but tends to create clusters of small size

[a] $d_{i,j}$ is the distance between clusters (objects) i and j in n dimensional variable space. $x_{i,k}$ is the coordinate (data value) for cluster i and variable k.

3.3.3 Icon Plots

A useful method of providing a quick overview of data is to plot the values of the chosen variables for each object as a geometric figure, and then visually compare the plots. In a *star plot* the length of lines emanating radially from a central point are proportional to the data values. The lines are equally spaced and the ends are joined. The lines in a *sun ray plot* are of equal length and are proportional to four standard deviations of the values at each variable. For a given object joins are made between the lines at points proportional to the values. Figure 5 is a star plot for the *Eucalyptus* data, including *E.woodwardii* and *E. torquata* crosses, using only seven variables. Because of the nature of these plots, only a small number of variables may be accommodated. In the example, seven variables having the greatest variance across the sets were chosen. These types of plot have been used in displaying the results from arrays of gas detectors (electronic noses) when distinguishing among types of soy sauce, beers and other comestibles.

A more bizarre method of plotting data is to represent the data as features on a cartoon face. The method of *Chernoff faces* relies on the particular acuity with which humans can observe differences in physiognomy. Twenty features (e.g. size of nose, spacing of eyes, length of face) may be used. Figure 6 is such a plot for the *Eucalyptus* data using 20 peaks in the GC. The reader is left to decide if these plots allow any useful discrimination between the groups.

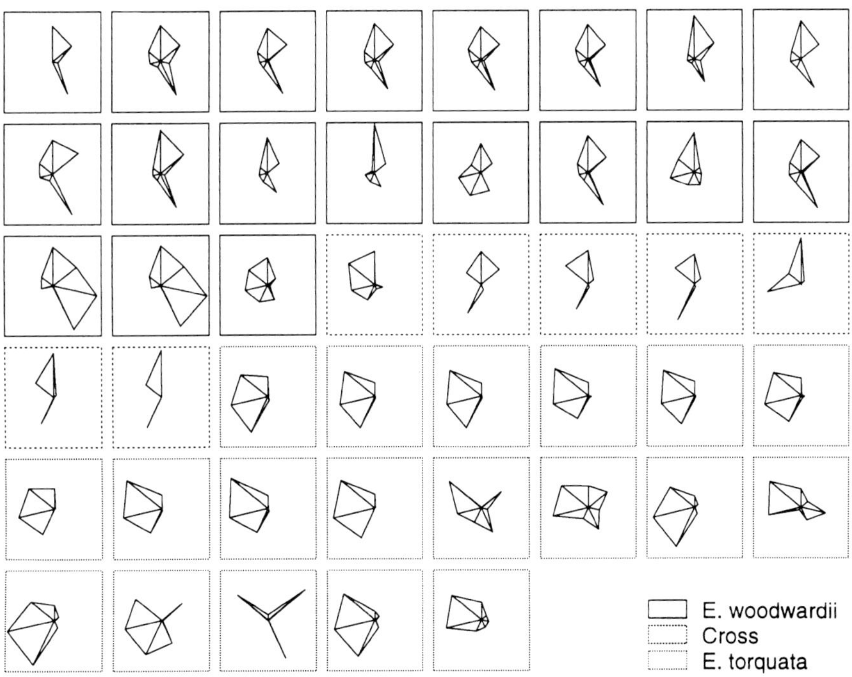

Fig. 5. Star plot of seven GC peak heights of samples of *E. woodwardii* and *E. torquata*

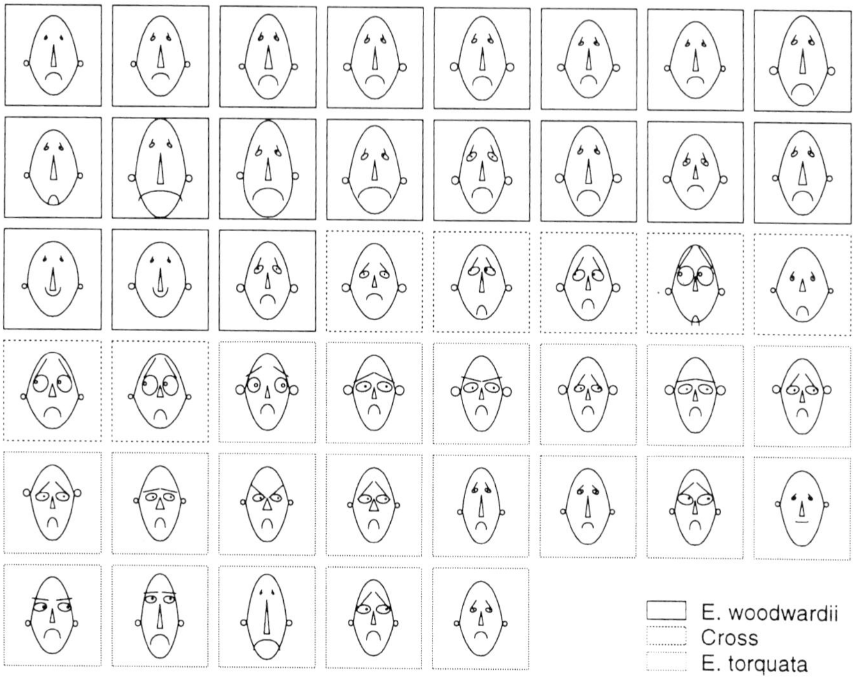

Fig. 6. Chernoff faces plot of 20 GC peak heights of samples of *E. woodwardii* and *E. torquata*

3.4 Supervised Classification

In supervised classification some information is given about the membership of classes, either the number of classes, or the membership of a given set of objects (*training set*). One example is given of each of these methods; *k* means clustering and linear discriminant analysis.

3.4.1 *k*-Means Clustering

If it is believed that the data fall naturally into *k* groups, this may be tested by *k* means clustering. The method creates *k* random clusters, and then objects are moved between those clusters to minimise variability within each cluster and maximise variability between clusters. In some senses, this approach is analysis of variance (ANOVA) in reverse. The output of the method gives the means for each cluster on each variable, which should be significantly different if the analysis has been successful. The F value from an ANOVA is used to determine how well the respective variable discriminates between clusters.

3.4.2 Linear Discriminant Analysis (LDA)

Linear discriminant analysis (Fisher discriminant analysis for two classes) has been shown to perform well in classifying essential oils from chromatographic data (Gaydou et al. 1988) and the quality of soy sauce (Aishima et al. 1979). For two groups the method builds a linear model of a discriminant function from the data, given by

$$G = a + \sum_{i=1}^{i=n} b_i x_i,$$

where G is the group membership and a and b_i are coefficients of the variables x_i. The membership of an unknown is determined by computing G for each group and choosing the group having the greatest value. The success of discrimination is seen in a plot of the Mahalanobis distance of each object from each of the group centroids. An object should belong to the group to which it is closest, that is, where the Mahalanobis distance is smallest. (The Mahalanobis distance is a statistical distance taking into account the variance of each variable and the correlation between variables). For more than two groups, a set of canonical discriminant functions are computed, with classification functions being generated to classify unknowns. A plot of the first canonical root against the second is a good indicator of the discrimination: as with PCA scores, the groups should be distinct. When there are many variables in a data set, it is useful to be able to choose those variables that contribute most to the discrimination. Forward and backward stepwise LDA accomplish this. At each step in a forward stepwise LDA, all variables are reviewed, and the one that will contribute most to the discrimination between groups is chosen to add to the discriminant function. This process is repeated until the added variables make no further significant addition to the discrimination, as measured by an F-test. In backward stepwise LDA, all variables are included initially and the least useful variables are removed, again until a suitable limit is reached.

Figure 7 is a plot of the Mahalanobis distances from each group centroid of the *Eucalyptus* data set from a two-group forward stepwise LDA. Excellent discrimination is achieved with 15 variables. The crosses have been presented to the model and plotted accordingly. It is seen that, while *E. torquata* and *E. woodwardii* group respectively together, the crosses fall more or less between these groups, slightly favouring *E. woodwardii*.

It must be noted, however, that because information of group membership is given a priori, it is expected that the supervised methods would inherently outperform unsupervised methods.

3.5 Regression (Calibration)

For many purposes in the analysis of essential oils, the goal is classification or discrimination between groups of samples. However, there may be the need to

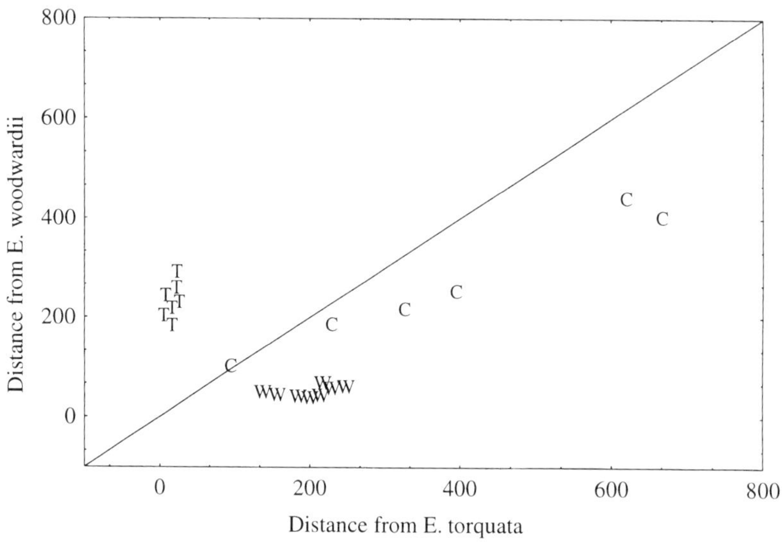

Fig. 7. Plot of the Mahalanobis distances of samples of *E. woodwardii* (*W*), *E. touquata* (*T*) and their crosses *E. torwoodii* (*C*) from the group centroids of the *E. woodwardii* and *E. torquata* sets from a two-group forward stepwise LDA. Fifteen variables were used for the LDA

predict the value of a variable given measurements of a sample. Usually, multilinear regression, in which the variable to be predicted is regressed against a number of independent variables (for example the peak heights at different retention times), cannot be employed because of the high degree of correlation of the variables.

We have already seen that principal components analysis produces orthogonal variables, and therefore the principal components would be ideal candidates for the independent variables in a regression. In principal components regression (PCR), the dependent variables (Y values) are regressed against the scores of the principal components. It is necessary to decide how many principal components to include in the model (a PCA produces as many principal components as there are objects). One method is to inspect how much extra variance is contributed with each successive principal component. When the change is no longer significant, as measured by an F test, no further principal components should be added. To test the predictive ability of the model, the mean square error is calculated by cross-validation (MSECV). For the first principal component the PCR model is established for n-1 data points, with the datum that has been left out being predicted by the model. This is repeated for the whole data set (n points), leaving out a different data point each time and calculating the difference between the predicted and actual value for this point. The sum of the squares of these residuals is thus determined for one, then two, then three etc. principal components. A graph of MSECV against the principal component falls to a minimum at the optimum number of principal components.

Partial least squares (PLS) is an extension of PCR in which the data to be predicted (the Y-block) are used in extracting the principal components of X so that more information that is relevant to the Y variables appears in the first few factors. PLS has been shown to outperform PCA for a number of data sets and has the added advantage that it can treat multiple Y sets (so-called PLS2), i.e. it is possible to predict a number of variables in a single calibration. However, PLS is more complex and gives intercorrelated scores. Also noisy Y data tend to lead to overfitting. A thorough description and critique of these methods is given by Martens and Naes (1989).

In an example by J.J. Brophy et al. (in prep.), an investigation of the essential oils of *Callistemon viminalis*, from the eastern coast of Australia, by GC and GC mass spectrometry, determined that the chemical evidence did not support distinct northern and southern species, as had been suggested by Byrnes (1986), but that there was a continuum of chemotypes. This was shown by regressing the latitude of a given specimen against 54 peaks in the GC analysis. It was possible to calibrate the latitude between 29° S and 12° S with a standard error of 0.9°.

4 Examples

4.1 Eucalypts and Other Australian Trees

Dunlop et al. (1995), using PCA and LDA on GC peak heights, investigated the relationships among a number of *Eucalyptus* species (see above). By adding published data to their analysis of essential oils from 62 samples of native *E. camaldulensis* (red river gum), the authors also demonstrated that the two chemotypes proposed by Boland et al. (1991) appear to represent the extremes of a continuum of types that may be identified.

Brophy et al. (1994) reported two entities in Australian populations of *Murraya paniculata* (L.) Jack. The large and small leaf forms are quite different in their chemical composition, being distinguished by the first principal component. The difference may be traced to differences in the percentages of e-nerolidol and germacrene-d.

Brophy and Clarkson (1992) also investigated the essential oils from three species of *Neofabricia* using hierarchical clustering and a method called ordination. Some 60 compounds were quantified by GC, and the numerical analysis supported the taxonomic revision of the genus and the separation of *Neofabricia* from *Leptospermum*. In this paper a statistical package is cited, but it was locally produced and not widely available and is probably now out of circulation. Although the software used should be cited, it is also very important to give references to the methods used.

A clear result which supported the separation of *E. nitens* and *E. denticulata* into two separate species was found by Li et al. (1994). They collected leaves from two trees of each of five replicate plots for seed collected from 20 localities. Principal component analysis, t-tests and cluster analysis (Manhattan distance, average cluster linkage) all showed two chemotypes. The *E. nitens* chemotype was characterised by high levels of 1,8-cineole and α-pinene, and *E. denticulata* had high levels of *p*-cymene and γ-terpinene. The statistical t-tests were used to distinguish between juvenile and mature leaves and showed significantly higher yields of oils from juvenile *E. nitens*. A similar study was conducted by Li et al. (1995) on 70 populations of Tasmanian eucalypts of the subgenus *Monocalyptus*, which included five species from the series Obliquae and seven from the series Piperitae. Series Obliquae exhibited little intraspecific variation in percentage oil composition, whereas Piperitae showed great intraspecific variation with poor differentiation. From loadings plots of the PCA the oils grouped as A = all the monoterpenoids, B = 1,8-cineole, α-terpineol, α-terpinylacetate, limonene and tasmanone, and C = all the sesquiterpenoids. No data transformation was recorded, but the variance explained by the first two principal components was only 28.7 and 16.9%, respectively. Li and Madden (1995) also used ANOVA to determine the significance between subgenera, species and provenances of *Eucalyptus* species. In a PCA the monoterpenoids and sesquiterpenoids were used to separate a number of species by a scores plot. Choosing the compounds on which to base an analysis may be a good approach and should be considered.

Butcher et al. (1994) analysed leaves from 109 trees in 11 populations of *Melaleuca alternifolia*. Five chemotypes were identified and the conclusion from the study was that the existence of chemical forms suggests a selective advantage for increased variablity in specific populations. This also had implications for selection of *M. alternifolia* in breeding programs to maximise the yield of commercial oil. The latitude, longitude and elevation of each population is given, so it would be of interest to see if any of these factors could be regressed against the data set of essential oils.

4.2 *Thymus*

The essential oils of *Thymus* show a variation across species and localities of production. The compositions of 11 populations of *T. carnosus* were investigated by GC, GC-MS and C^{13} NMR. Eighty five samples were analysed for 20 compounds, and the data were treated by cluster analysis and principal components analysis (Salgueiro et al. 1995). Three main clusters were observed and a sample from the region Estremadura in Portugal was shown to be high in linalool. The same group (Canigueral et al. 1994) studied *T. moroderi* and *T. antoninae*. Forty to fifty plants were chosen from each group and a PCA was performed on GC data from a number of samples and chemical compounds (84 samples and 48 variables for *T. moroderi* and 93 samples and 39 variables for *T. antoninae*). Samples of *T. moroderi* from different locations were easily distinguished by differences in the chemical composition, but there was no real difference for *T. antoninae*. Ninety plants of *T. willkomii* were analysed for 42 chemical compounds (Adzet et al.

1991). Three principal components accounted for only 47% variance, but three groups were identified, of which one half showed high linalool content and one third high α-terpinylacetate.

Thymus hyemalis from SE Spain was analysed for 37 essential oils at 24 sites (55 samples; Sáez 1995). A cluster analysis (Euclidean distance, single linkage), principal components analysis and multidimensional scaling analysis were used to analyse the variability of essential oils. A correlation was found, using Spearman correlation analysis, between the amount of camphor and the altitude of a sample. Again, a PLS calibration between the altitude and the chemical analysis may have been of interest.

4.3 *Pinus*

Vaterova et al. (1995) analysed the xylem oleoresins from four species of *Pinus* growing in Cuba to try and determine if attack by *Dioryctria horneana* caused differences in the composition of monoterpene hydrocarbons. They chose 16 trees with about five samples from each, and performed a PCA, PLS and PLS with discriminant analysis on the area of the GC peaks normalised by the total area. As discussed above, closure is a problem and a selective normalisation procedure was employed. In the PLS analysis, the Y variable was a dummy variable taking the values 0 for no attack and 1 for attack by the insect. The first principal component accounted for only 28% of the variance. While it has been established that this approach has been successful for a variety of problems, other methods to discriminate between attacked and unattacked trees may have been tried. The results of the analysis suggested that there was no difference between attacked and non-attacked trees of *P. cubensis*, while *P. caribaea* did reveal differences.

The attack by leaf-cutter ants (*Atta laevigata*) on *P. caribaea* was the subject of an investigation by Barnola et al. (1994). Discriminant analysis on 60 trees (30 foliated and 30 defoliated, split between January and May) and 19 compounds was used to classify the trees. Plots of frequency versus canonical variable showed differences between foliated and defoliated trees, but no difference between January and May harvests.

In a departure from the usual GC, GC-MS approach, Ochocka et al. (1992) performed differential thermal analysis, and thermal gravimetric analysis on samples of *P. sylvestris*: 12 samples, *Abies alba* (silver fir): 4 samples and *Coriandrum sativum* (coriander): 9 samples. For TGA the temperature at 5, 30, 70, 80, 90 and 95% loss of weight, and for DTA the height and temperature of maximum height were used as variables. In a PCA, 80% of the variance was explained by the first principal component and this was linear with the time of distillation. However, only three points were used in this analysis. The large variance in the first principal component presumably stems from the somewhat disparate samples, and the separation achieved is not really surprising. This study is a good example of the use of indirect variables, i.e. temperatures at which weight loss occurs rather than the concentrations of compounds found. If there is a correlation between the measured variables and the properties under investigation, then this approach is likely to work. However, any information from loading plots in a PCA will yield less information if the variables are indirect than if they are chemical species.

4.4 *Juniperus*

Gas chromatographic analysis of steam-distilled (an extraction method criticised by Milner et al. (this Vol.)) juniper samples was subjected to ANOVA and principal coordinate analysis (Adams 1994). *Juniperus monospermum* had the least amount of populational variability with a slight east-west cline. *J. osteospermum* showed a more typical infraspecific variability with a north-south trend.

4.5 *Mentha*

High resolution gas chromatography of the oils from *Mentha piperita* gave 39 peaks having an area greater than 0.1% of the largest peak. The data were subjected to PCA, SIMCA, and PLS analysis (Chialva et al. 1993). Of interest were the zone of origin, the distillery, the period of distillation, the age of the tree, the soil and the type of still used. In the PLS regression the Y values were the date of distillation and the age of the plant. From 99 samples, 6 were rejected as outliers as having too high leverage. Seven principal components were identified as being significant. The scores plots were not useful, showing a grouping only for the distillery. The conclusion was that it is not possible to find a fully certain classification model for Piedmontese mint in the various areas of origin and producing distilleries.

Maffei et al. (1993) used PCA and cluster analysis on an unspecified number of a population of *M.* × *verticillata* L. Ten chemotypes are quoted and a comparison was made of clustering by univariate (selected single compounds) and multivariate data. Published data on a wide range of mint species allowed discrimination between *M. piperita* from many different countries by PCA.

4.6 Beer

The analysis of odour from hops and the final product, beer, has been a topic of research for some years. Siebert and his group have employed chemometric techniques to a number of problems associated with brewing (Peppard et al. 1989; Siebert and Stenroos 1989). A number of samples of each of four American beer brands produced at six different plants were taken at 4-weekly intervals. Eighteen beer aroma compounds were extracted by Freon 12 and analysed by capillary GC. The GC peak areas divided by the area of an added internal standard were used in the data analysis. These were autoscaled (see Table 2) and further weighted by the ratio of interclass variance to intraclass variance for some of the analyses. This study is interesting to the chemometrician in that the effects of different weighting schemes on scores plots of the PCA are investigated. A number of classification techniques, including SIMCA, LDA and Bayesian analysis were compared. It was easier to distinguish the brand than the plant of manufacture. Siebert and Stenroos note that differences in chemicals found in the beer do not necessarily indicate differences in taste.

Peppard et al. (1989) investigated the relationships among the variety of hop, essential oil composition of the hop and flavour of the resulting beer. ANOVA and factor analysis were used to discriminate between four imported European hop varieties and four American hop varieties, using the chemical analysis of the hop and sensory evaluation data. A PLS regression was performed with sensory data of the beers from a taste panel (nine flavour attributes) being regressed against the compounds in hops analysed by GC (36 compounds). The taste test concentrated on the "hoppy" flavour of the beers. A small number of test sets (16) precluded a proper validation, but the authors report that 75% of the taste test scores could be predicted by the PLS model to within 0.5 (out of 10). The major distinction between imported and local hops was easily made by these methods, but subdivision of each group was not complete, even though the human testers could accomplish this task. Canonical correlation analysis failed because of the high degree of correlation between some of the independent variables (i.e. the chemical constituents). In a careful discussion, the authors suggest that PLS would be well suited for predicting taste panel scores from instrumental data.

A direct attempt to mimic the human nose is embodied in an array of sensors, the output of which is presented to a multivariate program, and a judgement on what has been "smelled" by the electronic nose, is given. There are several commercial devices on the market. One, the Aromascan, is based on resistance changes in semiconducting polymers and was used by Tomlinson et al. (1995) to analyse the quality of a number of beers. The normalised resistance and fractional resistance change of 20 sensors was measured for four lagers and five ale brands together with one lager spiked with off-flavour compounds. The difference in scale of the two measurements necessitated weighting the data. A non-linear mapping technique from the computer engineering literature (Sammon 1969) was used to reduce the number of dimensions in the data. It is not stated whether this was used because other methods failed, or if it was part of the software supplied with the instrument. The method was mostly successful in distinguishing between ales and lagers, but not so among the brands. Hop and malt aroma detection were also included in this study.

4.7 Mandarin Oil

The commercial treatment of mandarin oil, whether the fruit, green or ripe, was pressed or peeled, has been the subject of a number of studies by Cotroneo et al. (1988, 1990). The chemometric method used was PCA on the GC peak heights of 13 or 17 compounds for 38 and 59 samples respectively. The data were autoscaled (one of the few papers to report what data transformation was used) and four principal components explained 89% of the variance. Rotation of the principal components, a technique used in factor analysis, was found to improve the discrimination between the ripeness of fruits and their method of extraction.

Continuing the study, Campisi et al. (1995) showed that linear discriminant analysis gave complete separation between mandarin oils extracted by peeling or pressing, whereas principal factor analysis could not. As discussed above, the extra information given to LDA concerning the grouping of the data set usually allows an apparently better result. To be sure of the model, it should be validated by presenting a test set of cases that have not been used in creating the discriminant functions.

4.8 Others

Correspondence analysis of the concentrations of 17 flavanoids measured by headspace GC was used to show that for a sample of 30 ancient roses, the present biochemical communities clustered around the founder species of modern roses, *Rosa chinensis*, *R. gallica* and *R. moschata* (Raymond et al. 1995).

A similarity index, referred to as Sørenson's index was employed by Knudsen and Ståhl (1994) to determine similarities in the head space GC-MS analysis of floral odours from seven species of Theophrastacae. This was defined as $\frac{2M_w}{M_a + M_b} \times 100$, where M_w was the sum of smaller values (the authors may mean similar values) and M_a and M_b the sum of all compounds. It is not clear if peaks were counted (cf. Tanimoto index) or if the heights were summed.

Spring et al. (1994) investigated the chemistry of sesquiterpene lactones from 45 taxa of *Hymenoxys*. Fifty nine compounds were analysed by HPLC and the data subjected to an unpublished program that created a phenogram based on the distance metric defined as the difference between two chromatograms divided by the total characteristics present.

Harch et al. (1995) used a variety of methods including hierarchical cluster analysis and principal components analysis to study the diversity in fatty acid composition of Australian groundnut germplasm.

Acknowledgments. The author gratefully acknowledges botanical help and advice from Dr. J. Brophy (University of New South Wales, Sydney, Australia) and Dr. P. Dunlop (University of Adelaide, Australia).

References

Adams RP (1991) Analysis of juniper and other forest tree oil. In: Linskens HF, Jackson JF (eds) Modern methods of plant analysis, vol 12, Essential oils and waxes. Springer, Berlin Heidelberg New York, pp 131–157

Adams RP (1994) Geographic variation in the volatile terpenoids of *Juniperus monosperma* and *J. osteosperma*. Biochem Syst Ecol 22:65–71

Adzet T, Canigueral S, Gabalda N, Ibanez C, Tomas X, Vila R (1991) Composition and variability of the essential oil of *Thymus willkomii*. Phytochemistry 30:2289–2293

Aishima T, Nagasawa M, Fukushima D (1979) Differentiation of the aroma quality of soy sauce by statistical evaluation of gas chromatographic profiles. J Food Sci 44:1723–1727

Barnola LF, Hasegawa M, Cedano A (1994) Mono and sesqiterpene variation in *Pinus caribaea* needles and its relationship to *Atta laevigata* herbivory. Biochem Syst Ecol 22:437–445

Boland DJ, Brophy JJ, House APN (1991) *Eucalyptus* leaf oils. Inkata Press, Melbourne

Brophy JJ, Clarkson JR (1992) The essential oils of *Neofabricia* (Myrtacae). Biochem Syst Ecol 20:689–696

Brophy JJ, Forster PI, Goldsack RJ (1994) Diversity in Australian populations of *Murraya paniculata* (Rutacae): new evidence from volatile leaf oils. Aust Syst Bot 7:409–418

Butcher PA, Doran JC, Slee MU (1994) Intraspecific variation in leaf oils of *Melaleuca alterifolia* (Myrtaceae). Biochem Syst Ecol 22:419–430

Byrnes NE (1986) A revision of *Melaleuca* L. (Myrtacae) in northern and eastern Australia, 3. Austrobaileya 2:254–273

Campisi B, Dugo G, Cotroneo A, Favretto L (1995) Chemometric analysis and extraction processes of mandarin essential oils. Anal Chim Acta 312:199–205

Canigueral S, Vila R, Vicario G, Tomas X, Adzet T (1994) Chemometrics and essential oils analysis – chemical polymorphism in two *Thymus* species. Biochem Syst Ecol 22:307–315

Chialva F, Ariozzi A, Decastri D, Manitto P, Clementi S, Bonelli D (1993) Chemometric investigation on Italian peppermint oils. J Agric Food Chem 41:2028–2033

Cotroneo A, Dugo G, Favretto L, Gabrielli Favretto L (1988) On the genuineness of citrus essential oils. 23. Gas chromatographic separation and chemometric analysis of mandarin essential oils. J Chromatogr 449:183–190

Cotroneo A, Dugo G, Favretto L, Gabrielli Favretto L (1990) Variations in essential oils of mandarine related to processing and ripening. J Chemom 4:379–385

Dugo G, Gabrielli Favretto L, Favretto L, Cotroneo A (1991) Chemometric methods in the investigation of Sicilian citrus essential oils. Riv Soc Ital Sci Aliment 20:321–326

Dunlop P, Bignell CM, Jackson JF, Hibbert DB (1995) Chemometric analysis of gas chromatographic data of oils from *Eucalyptus* species. Chemom Intell Lab Syst 30:59–68

Gaydou EM, Randriamiharisoa RP, Bianchini JP, Llinas JR (1988) Multidimensional data analysis of essential oils. Application to ylang-ylang (*Cananga odorata* Hook Fil. et Thomson, Forma genuina) grades classification. J Agric Food Chem 36:574–579

Harch BD, Basford KE, DeLacy IH, Lawrence PK, Cruickshank A (1995) Patterns of diversity in fatty acid composition in the Australian groundnut germplasm collection. Genet Resour Crop Evol 42:243–256

Hibbert DB, James AM (1987) Macmillan dictionary of chemistry. Macmillan Press, London

Knudsen JT, Ståhl B (1994) Floral odours in the Theophrastacae. Biochem Syst Ecol 22:259–268

Li H, Madden JL (1995) Analysis of leaf oils from a *Eucalyptus* species trial. Biochem Syst Ecol 23:167–177

Li H, Madden JL, Davies NW (1994) Variation in leaf oil of *Eucalyptus nitens* and *E. denticulata*. Biochem Syst Ecol 22:631–640

Li H, Madden JL, Potts BM (1995) Variation in volatile leaf oils of the Tasmanian *Eucalyptus* species. I. Subgenus *Monocalyptus*. Biochem Syst Ecol 23:299–318

Maffei M (1994) Discriminant analysis of leaf wax alkanes in the Lamiaceae and four other plant families. Biochem Syst Ecol 22:711–728

Maffei M, Peracino V, Sacco T (1993) Multivariate methods for aromatic plants: an application to mint essential oils. Acta Hortic 330:159–170

Martens H, Naes T (1989) Multivariate calibration. Wiley, Chichester

Massart DL, Vandeginste BGM, Deming SN, Michotte Y, Kaufman L (1988) Chemometrics: a textbook. Elsevier, Amsterdam

McKern HHG (1965) Volatile oils and plant taxonomy (Presidential Address). J R Soc N S W 98: 1–10

Meloun M, Militký J, Forina M (1992, 1994) Chemometrics for analytical chemistry, vols 1,2. Ellis Horwood, New York

Morgan E (1991) Chemometrics: experimental design. Wiley, Chichester

Ochocka RJ, Wesolowski M, Lamparczyk H (1992) Thermoanalysis supported by principal component analysis (PCA) in quality assessment of essential oil samples. Thermochim Acta 210:151–162

Peppard TL, Ramus SA, Witt CA, Siebert KJ (1989) Correlation of sensory and instrumental data in elucidating the effect of varietal differences on hop flavour in beer. Am Soc Brew Chem 47:18–26

Raymond O, Biolley J-P, Jay M (1995) Fingerprinting the selection process of ancient roses by means of floral phenolic metabolism. Biochem Syst Ecol 23:555–565

Sáez F (1995) Essential oil variability of *Thymus hyemalis* growing wild in S.E. Spain. Biochem Syst Ecol 23:431–438

Salgueiro L, Vila R, Tomas X, Canigueral S, Casanova J, Dacunha AP, Adzet T (1995) Chemical polymorphism of the essential oil of *Thymus carnosus* from Portugal. Phytochemistry 38:391–396

Sammon JW Jr (1969) A non-linear mapping for data structure analysis. IEEE Trans Comput C 18:401–409

Schulz H, Losing G (1995) Quality control of essential oils by NIR spectroscopy. Dtsch Lebensm-Rundsch 91:239–242

Siebert KJ, Stenroos LE (1989) The use of multivariate analysis of beer aroma volatile compound patterns to discern brand-to-brand and plant-to-plant differences. Am Soc Brew Chem 47:93–101

Spring O, Zitterell-Haid B, Bierner MW, Mabry TJ (1994) Chemistry of glandular trichomes in *Hymenoxys* and related genera. Biochem Syst Ecol 22:171–195

Tomlinson JB, Ormrod IHL, Sharpe FR (1995) Electronic aroma detection in the brewery. J Am Soc Brew Chem 53:167–173

Vaterova I, Sjodkin K, Norin T, Urkoc J (1995) Contents and enantomeric compositions of monoterpene hydrocarbons from four *Pinus* species growing in Cuba. Comparison of trees unattacked and attacked by *Dioryctria horneana*. Biochem Syst Ecol 23:1–15

Weber DJ, Gang DR, Halls SC, Smith BN, McArthur ED (1994) Inheritance of hydrocarbons in subspecific big sagebrush (*Artemisia tridenta*) hybrids. Biochem Syst Ecol 22:689–697

Supercritical CO_2 Extraction of the Essential Oils of Eucalypts: A Comparison with Other Methods

C.P. Milner, R.D. Trengove, C.M. Bignell, and P.J. Dunlop

1 Introduction

The genus *Eucalyptus* (L'Heritier) is the most widely studied of all the Australian flora. Whilst Australia is the original source of nearly all eucalypts, Spain, Portugal, the Republic of South Africa, Swaziland, Chile and China all have substantial plantations that are used for the production of eucalyptus oil for export (Rothschild 1991).

The medicinal potential of eucalyptus oil was noted almost immediately by the European settlers in Australia in 1788 (Boland et al. 1991); however, McKern (1968) reports that the establishment of a eucalyptus essential oil industry took over 60 years. The industrial and perfumery or flavouring uses also started around this period (Boland et al. 1991). The primary component of oils used for medicinal purposes in 1,8-cineole, and these medicinal oils are required by the *British Pharmacopoeia* to contain at least 70% 1,8-cineole and to be practically free of α- and β-phellandrene. Industrial oils have piperitone and α-phellandrene as their major constituents. The phellandrene-rich oils are used for scenting inexpensive disinfectants and industrial liquid soaps, and the piperitone-rich oils are used for the production of synthetic menthol, as a flavouring agent and an additive for various medicinal preparations.

The exploitation of eucalypts for use in the perfumery/flavouring industry has been very limited and only a handful of species have been used directly as perfumery or flavouring additives. Hay and Svoboda (1993) state: "The diversity of the family Myrtaceae in Australia offers considerable scope for commercial exploitation, both culinary and medicinal, but to date, only the *Melaleuca* (tea tree) and a few *Eucalyptus* species have been cropped for pharmaceutical, perfumery and cosmetic uses."

Boland et al. (1991) edited a monograph on the historical and current commercial potential for eucalyptus leaf oils. This compilation reports essential oil yields and compositions for 111 species of northern and eastern Australian eucalypts extracted using a modified Dean and Stark steam distillation apparatus. The three-volume compilation by Brooker and Kleinig (1990a,b, 1994) gives the botanical information and geographical habitat of many Australian eucalypts.

Dunlop and coworkers have recently initiated a systematic study of essential oils of eucalypts from Southwestern and Southern Australia using the vacuum distillation technique (Bignell et al. 1994a,b,c, 1995a,b,c, 1996a,b,c). They have reported significant differences in the volatile oil composition extracted with the vacuum distillation technique when compared with the composition of oil extracted using hydrodistillation for species in which the major component of the oil extracted with the vacuum distillation technique is bicyclogermacrene (Bignell et al. 1995c, 1996a).

The aim of the present chapter is to compare the essential oils obtained by supercritical fluid extraction and vacuum distillation of the two eucalyptus species with that obtained by hydrodistillation. The purpose is to determine which of

Modern Methods of Plant Analysis, Vol. 19
Plant Volatile Analysis
Edited by H.-F. Linskens and J.F. Jackson

the techniques provides essential oils which most closely reflect the in vivo oil composition.

2 Traditional Extraction Methods for Essential Oils

The traditional term essential oil is still widely used today even though the "essence" of the plant is an ill-defined concept dating from mediaeval pharmaceutical techniques and the preferred term is volatile oil (Hay and Waterman 1993). This refers to the components of the oils that are stored in the extracellular spaces in the epidermis or mesophylla that can be extracted using steam distillation of the plant tissues.

The most commonly used techniques for extraction of essential oils from leaf material are steam or hydrodistillation and solvent extraction. The essential or volatile oils that are commercially available are generally obtained using steam distillation (Koedam 1987). Since the 1980s, this situation has changed considerably and by 1992 there were in excess of 50 oils and extracts commercially available that have been prepared using supercritical and subcritical carbon dioxide technologies (Moyler 1986, 1988, 1989, 1991, 1993a,b; Moyler and Heath 1988).

2.1 Steam Distillation

Waterman (1993) states that: "Volatile oils are, as the name implies, volatile, and most have traditionally been obtained by the process of steam distillation."

This technique is usually carried out in two slightly different ways. In the first, leaves are covered with water in a suitable vessel fitted with a condenser, and while the mixture is gently boiled, a condensate is collected (hydrodistillation) and the oil then separated from the water. In the second, steam is passed through a vessel containing a leaf-water mixture (steam distillation) to yield a similar condensate. Many different apparatus designs have been discussed (Guenther 1948; Likens and Nickerson 1964; Guenther et al. 1975; Boland et al. 1991). In both of these techniques, the pH of the liquid phase is usually not controlled, with the result that the composition of the oil in the condensate may be pH-dependent (see Nishimura et al. 1969; Koedam and Looman 1980; Garnero and Tabacchi 1987). Adams (1991b) has described a Clevenger-type apparatus, based on the work of von Rudloff (1967), in which the leaves are suspended above boiling water and the oil is collected in an ether trap; it is still possible that small amounts of water in contact with the leaves might not have a neutral pH.

For all classes of experiment carried out under the heading of steam distillation, the plant material is subjected to temperatures up to 100 °C, but the degree of liquid or condensate that contacts the plant varies substantially between the extremes of steam distillation.

2.2 Solvent Extraction

As with steam distillation, the term solvent extraction covers a class of techniques ranging from simply soaking the plant material in organic solvents at ambient temperatures or below, to extraction by boiling solvent in a Soxhlet-type apparatus (Koedam 1987; Adams 1991a). To recover the volatile oils requires evaporation of the solvent, and care must be taken that the more volatile compounds are not lost.

Solvent extraction has the advantage over steam distillation that lower temperatures are used during most of the extraction. The boiling point of solvents used in Soxhlet-type extractions is typically <60 °C, and when the plant material is simply soaked in the organic solvent the temperature is typically from 5 to 25 °C.

As the volatile oils extracted using organic solvents are kept at relatively low temperatures during most of the extraction process, they are claimed to have a more "natural" composition that is far superior to steam-distilled oils, in which thermal and pH-induced alterations may have occurred.

3 Recent Extraction Methods

In contrast to the more traditional extraction techniques, supercritical fluid extraction is a gentler extraction method and causes little degradation of the essential or volatile oil components during the extraction (Stahl et al. 1988; Deans 1991). Supercritical fluid extraction of plant materials was initially developed on a large scale (Williams 1981; Stahl et al. 1988) and only relatively recently on an analytical scale (Dean 1993; Smith 1995; King 1990).

A major application area for the use of supercritical fluids has been in food science, with the extraction of caffeine from coffee beans being the most commonly quoted commercial-scale application of supercritical fluids (Dean 1993). This particular application has been the subject of many patents in the USA and Europe (McHugh and Krukonis 1986, 1994; Bruno and Ely 1991). The extraction of caffeine from tea has also received much attention, and a review is provided by Lack and Seidlitz (1993). Several reports appear in the literature of conditions for the analytical scale extraction of caffeine from coffee and tea (Sandra et al. 1990; Hills et al. 1991; Jenkins et al. 1991).

Moyler and coworkers (Moyler 1986, 1988, 1991, 1993a,b) extracted botanicals with liquid CO_2 over the range 0 and 10 °C and at pressures between 50 and 80 bar. They report that the essential oil plus the light fractions of the resins are selectively extracted, producing an extract with the characteristics of the "absolute". This character results from the extraction of components with molecular weight greater than 250 that are not obtained in steam distillation.

Liquid CO_2 extracts are claimed to be superior to steam-distilled equivalents because they have none of the "off-notes" produced in steam distillation due to decomposition of plant matrix to produce trace levels of nitrogen and sulfur containing compounds; more "top notes" from the low temperature capture of the volatiles; and more "back notes" due to the extraction of components in the 200–400 molecular weight range not extracted in steam distillation (Moyler 1986, 1988).

The "back notes" obtained with liquid CO_2 can be vital (Moyler 1988), as found with ginger extracts. The gingerols and shogaols responsible for the pungent taste of ginger are totally absent in the steam-distilled oil but are present in the oil extracted with liquid CO_2. Chen and Ho (1988) reported that extraction with liquid carbon dioxide gave extracts that contained a lot of previously unknown volatile sesquiterpene esters not found in ginger oils from steam distillation and, in addition, they found very low concentrations of the thermal degradation products of non-volatile gingerols, Although van Beek (1991) states that liquid carbon dioxide extraction has no commercial importance for preparation of ginger oil, from a plant biochemistry perspective this technique offers many advantages. Moyler et al. (Moyler 1993a) report that commercial ginger extract produced with liquid CO_2 is now available.

Inman et al. (1991) developed a vacuum distillation apparatus to obtain the pure *leaf oils* from eucalypts uncontaminated by non-volatile material. They demonstrated that this technique gives *quantitative* recovery from mixtures of α-pinene, *d*-limonene and *d,l*-citronellol, that have a boiling point range of 150–250 °C. They achieved almost quantitative extraction in 240 min. Dunlop and coworkers have used this technique to extract the essential oil of several eucalypt species (Bignell et al. 1994a,b,c, 1995a,b,c, 1996a,b,c) and they have reported significant differences in the composition of the essential oil extracted using steam distillation and vacuum distillation (Bignell et al. 1995c, 1996a).

3.1 Supercritical Fluid Extraction

Supercritical fluid extraction (SFE) is a solvent extraction process that uses a supercritical fluid as the extraction solvent. A substance is said to be supercritical when it is above its critical temperature (T_c) and critical pressure (P_c), as indicated in the generic P vs. T plot in Fig. 1.

A supercritical fluid is intermediate in its thermophysical properties between those of a liquid and those of a gas, and the term dense gas extraction was used in the 1960s and 1970s (Stahl et al. 1988). Typically, supercritical fluids have solvent strengths that approach those of liquid solvents and thus will dissolve many substances. Supercritical fluids also have diffusion coefficients that are close to those of gases and are therefore able to rapidly transfer dissolved solutes through materials. The low viscosities of supercritical fluids, combined with the high diffusion rates of liquids, makes them ideal for the extraction of diffusion-controlled matrices such as plant tissues (Smith 1995). Supercritical fluid extraction is faster than liquid extraction and the supercritical fluid solvents are easily removed by reducing the pressure to remove the extracted analytes.

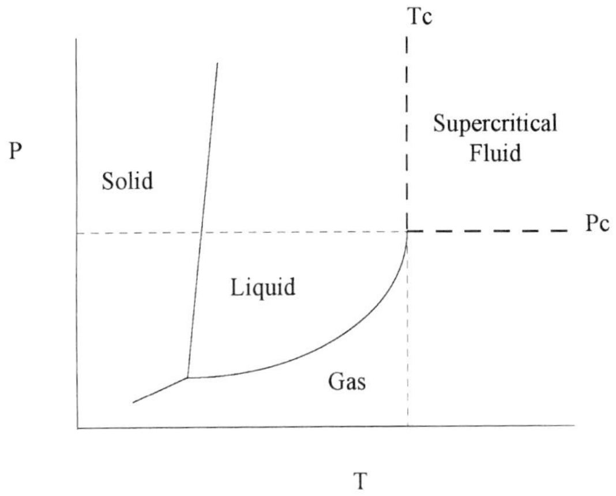

Fig. 1. Generic phase diagram

The physical properties of supercritical fluids such as density, viscosity and diffusion coefficients vary considerably with temperature and pressure and, in fact, typically cover a range rather than being assigned a single value as indicated in Table 1 (Clifford 1993). The variations of these properties are important, as they lead to a single supercritical solvent having a range of solvating powers. The ability to vary the solvating power of supercritical fluids has the major advantage that it is possible to extract classes of compounds using the same physical solvent by simply varying the extraction conditions, i.e. temperature and pressure. Supercritical fluids make possible separation of multicomponent mixtures by capitalising on differences in component volatilities, which is the prominent feature of distillation, and differences in the specific interactions between the mixture components and the supercritical fluid solvent, or the solubility, the prominent feature of liquid extraction (McHugh and Krukonis 1994).

Table 2 shows the critical points for several pure components that have been used as supercritical fluid solvents (Brunner 1994). Inspection of Table 2 shows that critical temperatures of gases and liquids can differ by hundreds of degrees. For processing of heat-sensitive flavours and pharmaceutical it would appear that ethylene, xenon, sulfur hexafluoride, ethane, nitrous oxide, triflouromethane and carbon dioxide would be attractive solvents. However, for these classes of compounds, carbon dioxide is the solvent of choice because it is readily available in high purity and offers low cost, operator safety and minimal environmental impact. The other possible supercritical solvents indicated above would be ruled out because of cost and safety considerations (Hitchen and Dean 1993; Raynie 1993).

The solvating power of supercritical carbon dioxide is generally equated with that of hexane. This should be regarded only as a guide, and varying the temperature and pressure (and therefore density) changes the solvating power of pure carbon dioxide from a gas through hexane equivalence to solvating power approaching that of 50/50 hexane/ether (Knipe et al. 1993). The solvating power can

Table 1. Typical physical property values for gases, supercritical fluids and liquids. (Clifford 1993)

	Density ($kg\,m^{-3}$)	Viscosity ($mPa\,s$)	Self-diffusion coefficient ($10^4\,m^2\,s^{-1}$)
Gas (30 °C, 1 atm)	0.6–2	$(1–3) \times 10^{-2}$	0.1–0.4
Supercritical fluid			
Near T_c, P_c	200–500	$(1–3) \times 10^{-2}$	0.7×10^{-3}
Near T_c, $4P_c$	400–900	$(3–9) \times 10^{-2}$	0.2×10^{-3}
Liquid (30 °C, 1 atm)	600–1600	0.2–3	$(0.2–2) \times 10^{-5}$

Table 2. Critical conditions for pure components. (Brunner 1994)

Compound	Critical temperature T_c (°C)	Critical pressure P_c (bar)
Carbon dioxide	31.0	73.8
Ethylene	9.3	50.4
Xenon	16.6	58.4
Trifluoromethane (CHF_3)	26.2	48.6
Ethane	32.2	48.8
Nitrous oxide (N_2O)	36.5	72.4
Sulphur hexafluoride (SF_6)	45.6	37.6
Propane	96.7	42.5
n-Butane	152.1	38.0
n-Pentane	196.6	33.7
n-Hexane	234.4	30.1
Methanol	239.5	80.9
Ethanol	240.8	61.4
Benzene	289.1	48.9
Toluene	318.7	41.0
Water	374.2	221.2

then be further increased by the addition of a more polar modifier such as methanol, but this is not generally necessary for the extraction of essential or volatile oils.

There are many reports of supercritical fluid extraction of plant materials (Stahl et al. 1988; King and Bott 1993; Bevan and Marshal 1994; Brunner 1994; Smith 1995). A number of papers have focussed on the extraction and separation of essential oils using supercritical fluids (Matos et al. 1989; Reverchon 1992; Gomes de Azevedo et al. 1993).

The use of supercritical CO_2 to extract the volatile oils and oleoresins from aromatic plants has the desirable property that it behaves like a solvent; it can be manipulated to obtain differential or sequential fractions; all traces of gas are removed from the volatile oil; and it is particularly useful for heat-labile compounds (Deans 1991). Supercritical fluid extraction can produce an essential oil that has organoleptic properties that more closely resemble those of the plant from which they were extracted (Stahl et al. 1988; Reverchon and Senatore 1992). Hirvi et al. (1986) extracted coriander using steam distillation, Soxhlet extraction with diethyl ether and supercritical carbon dioxide. They reported that carbon dioxide extracts were more "coriander-like" than those of steam-distilled and Soxhlet extracts.

Stahl et al. (1988) report that the volatile oils can be totally extracted from plant matter at densities between 0.4 and 0.6 $g\,cm^{-3}$ or 400 to 600 $kg\,m^{-3}$, which corresponds to approximately 87 to 97 bar at 40 °C. Extractions at densities of 0.6 $g\,cm^{-3}$ and above will also extract some of the non-volatile materials present in the plant matrix such as C25–C35 hydrocarbons and other waxy materials (Stahl et al. 1988; Hawthorne et al. 1993; Reverchon and Senatore 1992). For industrial-scale extractions, where rapid extraction times and maximum solubility conditions are desirable, this requires that the extract be fractionated to remove the waxy materials and this is generally carried out using a two-stage separator, with the first separator recovering the waxes and the second stage recovering the desired volatile oils (Stahl et al. 1988; Reverchon and Senatore 1992, 1994). In some cases, where only the oxygenated fraction of the volatile oils is required, the fraction responsible for taste in foods, a further third-stage separation is carried out to deterpenate the volatile oil, removing hydrocarbon terpenes such as limonene which are somewhat unstable and can cause the volatile oil to "go off" when stored. This fractionation process or deterpenation of volatile oils is an active area of research (Stahl et al. 1988; Matos et al. 1989; Reverchon 1992; Barth et al. 1994; Chouchi et al. 1995; Dugo et al. 1995).

In general, the essential oil yields from liquid and supercritical carbon dioxide extractions are higher than the corresponding extraction with steam distillation (Moyler 1993a,b; Piggott et al. 1996; Raghuram Rao et al. 1992). Some of this higher yield may be attributed to the concentration of waxy material, and in particular C25–C35 hydrocarbons, that is extracted with liquid CO_2 and supercritical CO_2 (Stahl et al. 1988; Reverchon and Senatore 1992; Hawthorne et al. 1993). The degree of this in the extract is dependent on the extremity of the conditions used for the extraction.

Milner and Trengove (1996) have used supercritical carbon dioxide to investigate the differences in composition of the volatile or essential oils from Western Australian sandalwood trees, *Santalum spicatum*, that were both living and dead when harvested. The variation in the composition of the major components of trees from several locations was also investigated. Initially, they investigated the extraction kinetics at 40 °C over the density range of 0.25 $g\,cm^{-3}$ to 0.90 $g\,cm^{-3}$, and found that the higher density led to much more rapid extractions, but that extended extraction times led to extraction of significant amounts of non-volatile material.

Piggott et al. (1996) have recently studied the extraction of the essential or volatile oils from Western Australian sandalwood powder, and in particular the material extracted at 40 °C and densities from 0.50 $g\,cm^{-3}$ to 0.90 $g\,cm^{-3}$ over 80 min.

The total yield extracted was determined and the volatile oil fraction yield at each density was determined by calculating the fraction of the total extract that was eluted from a polyethylene glycol GC column (HP Innowax). At the lowest density (0.5 $g cm^{-3}$), the volatile oil fraction was 93% of the total extract yield. The total yield and the total volatile oil yield increased with density, and for the highest extraction density the volatile oil fraction was only 56%.

No systematic study of the application of supercritical fluid extraction to eucalypts has been carried out to date. Only exploratory extraction studies and essential oil fractionation studies have been reported in the literature (Hawthorne et al. 1989; Matos et al. 1989; Gomes de Azevedo et al. 1993).

3.1.1 Analytical Scale Supercritical Fluid Extraction Equipment

A generic analytical supercritical fluid extraction apparatus consists of a solvent cylinder, a fluid pump, a heater, an extraction chamber, a pressure reduction device and an analyte collection device as shown in Fig. 2. Commercially available analytical SFE systems use either piston pumps or reciprocating pumps with pulse dampeners, which are generally modified HPLC pumps. After compression above its critical pressure, the fluid is heated above its critical temperature so that it is supercritical before entering the extraction chamber. The extraction chamber for analytical systems is typically 10 ml or less in volume. The pressure reduction device is most commonly a linear restrictor for home-made systems and for more sophisticated systems a modified back pressure regulator. The linear restrictors are prone to *plugging* (or blockage by formation of solids) and most instrument manufacturers have now designed a heated linear restrictor pressure-reduction system. The collection systems are either a solid phase bed or a liquid trap consisting of either a pure liquid or liquid mixture. Systems that have solid phase bed for collection of the extracted analytes need to have some way of reconstituting the trapped analyte at the end of the extraction process, and this is generally done by rinsing the trap with solvent.

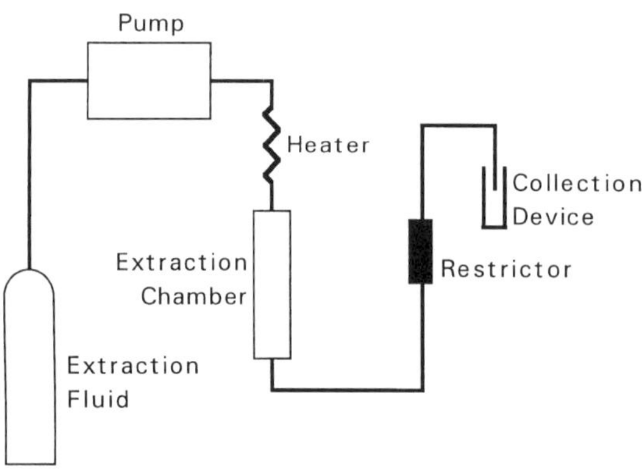

Fig. 2. Generic analytical supercritical fluid extraction apparatus

3.1.2 SFE Equipment Used for this Work

The SFE equipment used for this work is an HP 7680T SFE apparatus and a schematic is shown in Fig. 3. Liquid carbon dioxide is compressed above the critical pressure with a modified HPLC pump cooled to 4 °C to give reproducible volume flow rates, which can be varied between 0.5 and 4 ml min^{-1} at the pump head. The compressed liquid is heated above the critical temperature prior to entering the 7-ml extraction vessel, and then the pressure is reduced in the flow restrictor or nozzle prior to analyte collection on a solid phase trap. At the end of the extraction the pressure isolation valve switches position and liquid solvent rinses the collected analytes from the solid trapping material into a vial, which is then ready for analysis by an appropriate technique.

The important feature of this instrument is the flow restrictor or nozzle, a back-pressure regulator, which allows independent control of the flow and the pressure. This is not possible when linear restrictors are used. The analyte collection system is on a solid phase trap rather than in liquid solvent or mixture of liquid solvents. The solid phase trap can be any solid material that will trap the desired analytes as the pressure drops from supercritical conditions to near atmospheric pressure in the nozzle because the analytes are then no longer soluble. Typically, the solid phase trapping material is solid phase extraction material or GC column packing material. During extraction, the solid phase bed is acting like a GC column and during the sample reconstitution (rinse) step the solid phase trap is acting like an LC column (Milner et al. 1995, 1996). The rinsing step is a frontal displacement mechanism.

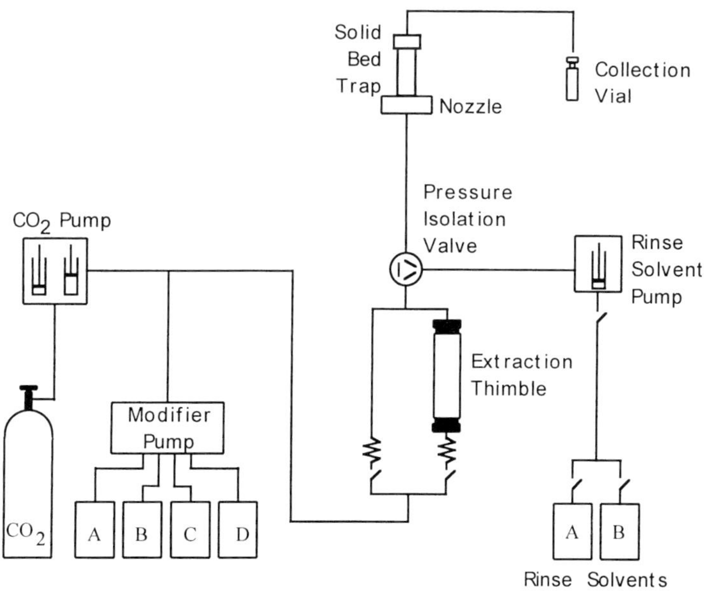

Fig. 3. HP7680T supercritical fluid extraction apparatus

The equipment is controlled by HP Chemstation software. The extraction parameters set with the software are temperature and density, with the associated pressure given at each density. The PTρ conditions for extraction are calculated using the equation of state developed by Reynolds (1979). Supercritical extraction temperatures can be varied between 32 and 150 °C and the extraction pressure can be varied between 79 and 383 bar. These conditions correspond to a density range of 0.25 to 0.95 g ml^{-1} at 40 °C, the extraction temperature used for this study, reducing to a density range of 0.15 to 0.59 g ml^{-1} at 150 °C. The solid phase analyte collection trap can be varied from −30 to +150 °C. The pressure restrictor or nozzle can be temperature-controlled to avoid any ice formation associated with the pressure reduction at this point. Liquid CO_2 extraction can also be performed with a temperature range of 25 to 30 °C and pressure range of 90 to 351 bar (0.80 to 0.99 g ml^{-1}).

As with liquid trapping, the solid trap packing material to be used is one of the optimisation conditions that must be determined as part of the overall extraction optimisation. Milner et al. (1995, 1996) carried out an extensive study of different solid packing materials for essential oil components and the results suggest that a mixture of solid phase trapping materials may be necessary to retain all the different functionalities contained in essential oils.

4 Extraction Conditions

4.1 Sample Preparation for Supercritical Fluid Extraction

Prior to extraction, the leaf material was ground with dry ice in a stainless steel Waring blender to produce a powder and then the dry ice was allowed to sublime. A known amount (approximately 1 g) of the ground leaf material was loaded into the extraction thimbles between filter flocs (Macherey Nagel) and then the leaves were extracted using the conditions outlined in Table 3.

4.2 Supercritical Fluid Extraction Conditions

The supercritical fluid extraction conditions used for this work are presented in Table 3; however, the table needs some explanation. Our aim was to remove the essential oils only and leave intact, if possible, the high degree of waxy material found in eucalypt leaves. The ground leaf material was extracted at 77 bar for 10 min, 81 bar for 10 min and then 84 bar for 10 min, and the extracted essential oil components collected with a single rinse of the trap with 1.5 ml of hexane into a precapped vial. These three steps constituted stage 1 of the extraction process. The ground leaf material in the extraction thimble was then extracted at 87 bar for 10 min and 89 bar for 10 min, followed by rinsing with 1.5 ml of hexane into a second precapped vial. This constituted stage 2 of the extraction process. As expected, the composition of the two vials was different and only the concentration

Table 3. SFE extraction conditions used for *E. sparsa* and *E. varia*

Extraction conditions	Extraction step 1	Extraction step 2	Extraction step 3	Extraction step 4	Extraction step 5	Extraction step 6
CO_2 pressure/bar	77	81	84	87	89	281
CO_2 density (calculated) g/ml	0.25	0.3	0.35	0.4	0.45	0.9
Extraction chamber temperature/°C	40	40	40	40	40	40
CO_2 flow rate/ml (measured at 4 °C)	1.0	1.0	1.0	1.0	1.0	1.0
Equilibration time/min (static extraction)	0.0	0.0	0.0	0.0	0.0	0.0
Extraction time/min (dynamic extraction)	10.0	10.0	10.0	10.0	10.0	15.0
Thimble volumes swept	5.3	4.4	3.8	3.3	0.9	0.9
Nozzle extraction temperature/°C	45	45	45	45	45	45
Trap extraction temperature/°C	0	0	0	0	0	0
Trap packing	ODS[a]	ODS	ODS	ODS	ODS	ODS
Rinse conditions						
Nozzle rinse temperature/°C	45	45	45	45	45	45
Trap rinse temperature/°C	40	40	40	40	40	40
Rinse solvent 1	Hexane	Hexane	Hexane	Hexane	Hexane	Hexane
Rinse solvent flow rate $ml\,min^{-1}$	1.0	1.0	1.0	1.0	1.0	1.0
Rinse solvent volume ml^{-1}	1.5	1.5	1.5	1.5	1.5	1.5
Void volume compensation	0.9	0.9	0.9	0.9	0.9	0.9

[a] Octadecylsilane.

for vial 1 is presented here for comparison in Table 3. The second vial contained much smaller concentrations of bicyclogermacrene (approximately 11%), indicating that most of this compound had been extracted in the first three short extraction steps.

4.3 Vacuum Distillation

In this method, developed by Inman (1991), leaves are frozen with liquid nitrogen and then macerated in a stainless steel Waring blender to release the oil from the glands in which it is encapsulated. The dry leaf powder is then cooled to –75 °C and the temperature is slowly raised to 80 °C over a 4-h period: the oil and a small amount of water condense onto a gold-plated copper rod maintained at –75 °C. When the extraction is complete, the temperature of the cold finger is raised and the oil aspirated from the water. Complete details of this technique have been given by Inman et al. (1991).

4.4 Hydrodistillation

The leaf samples for hydrodistillation were prepared in the same manner as for vacuum distillation. The leaf powder was then hydrodistilled using a modified Dean and Stark apparatus (Boland et al. 1991).

5 Essential Oil Analysis

All analysis for this work used a HP5890 Series II Gas Chromatograph coupled to a HP3396 Series II integrator. On-column injection was used with a SGE BP20 capillary column (25 m × 0.33 mm × 0.5 µm) and hydrogen carrier gas flow rate of 2.0 ml min^{-1} (inlet pressure of 25 kPa). The inlet temperature was 83 °C with a temperature program of 80 to 220 °C at 5 °C min^{-1}.

The SGE BP20 column had been calibrated using GC-MS results (Bignell et al. 1994a,b,c, 1995a,b,c, 1996a,b) to identify the various components in the extracted oil. This calibration procedure has been described by Bignell et al. (1996a,b).

6 Effects of Extraction Methods on Modification and Composition of Essential Oils

To compare some of the methods described above, essential oils from two species of Australian eucalyptus (1) *Eucalyptus varia* subsp. *varia* Brooker and Hopper and (2) *Eucalyptus sparsa* Boomsma were extracted. Approximately 100-g samples

Table 4. Comparison of essential oils extracted from leaves of *E. varia* subsp. *varia* and *E. sparsa* using the VD, SFE and HD techniques

Compound[a]	*E. varia*			*E. sparsa*		
	SFE	VD	HD	SFE	VD	HD
Aromadendrene	0.1	0.2	1.2	3.9	1.8	7.4
Allo-aromadendrene	1.2	1.1	0.5	2.0	1.4	1.8
Viridiflorene	0.8	0.8	6.0	1.2	0.8	8.5
Bicyclogermacrene	65.0	71.3	3.0	63.6	62.6	17.6
Globulol	0.3	0.3	23.0	0.3	0.5	7.3
Viridiflorol	0.3	0.3	9.2	0.3	0.4	4.0
Spathulenol	7.2	2.8	5.4	3.1	3.5	4.3
α-Eudesmol	0.1	0.1	3.1	0.1	0.2	0.2
β-Eudesmol	0.1	0.1	4.2	0.1	0.1	0.2
Unidentified sesquiterpene alcohols	0.5	0.1	24.2	–	–	8.6

[a] The values in each column are percentage concentrations.

of leaves were picked from one tree of each species and dried in an oven at 38 °C. The essential oils were then extracted from the dry leaves using supercritical fluid extraction (SFE), vacuum distillation (VD) and hydrodistillation (HD) techniques and analysed using GC-MS and GC (Bignell et al. 1994a,b,c, 1995a,b,c, 1996a,b,c). Only the percentage concentrations of the important sesquiterpene hydrocarbons and alcohols are listed in Table 4; the other oil components are not relevant to the following discussion.

The major point indicated in these results is the extremely high concentration of bicyclogermacrene listed in the SFE and VD columns of Table 4 and the much lower corresponding yield obtained by HD in which the pH of the water/leaf mass was ~4; we have reported similar problems with *E. rudis* Endl. It seems clear that HD is inappropriate if an oil sample close to the in vivo composition is necessary (in some cases this may not be important).

There have been many previous reports in the literature describing the isomerisation and chemical reactions of bicyclogermacrene in aqueous solution at non-neutral pH. Nishimura et al. (1969) isolated bicyclogermacrene for the first time from *Citrus pinus* and reported several rearrangements. Tressl et al. (1983) studied oil obtained by distillation of Hops (*Humulus lupulus* var *Hersbrucker Spät*) and indicated possible reaction schemes for the transformation of bicyclogermacrene into tricyclic sesquiterpenes. Inspection of Table 4 indicates that such transformations have indeed occurred for *E. varia* and *E. sparsa*. Garnero and Tabacchi (1987) have also recently commented on the effects of heat and pH on the reactions of bicyclogermacrene.

Koedam and Looman (1980) in an excellent study have shown that problems also arise during the extraction of monoterpenes. Their experiments indicated that the sabinene concentration in oil obtained from *Juniperus sabina* depended markedly on the pH of the liquid in the distillation vessel. At high pH, about 40% of sabinene was present in the oil collected, while at pH ~2, the concentration of sabinene fell to approximately 7%; the concentrations of α-terpinene, γ-terpinene, terpinolene and terpinen-4-ol, formed by the decomposition of sabinene (Koedam et al. 1981), rose markedly. In another study, Dellacassa et al. (1990) reported a benzaldehyde concentration of 34% in oil obtained by steam distillation from leaves of *E. cladocalyx*. In a recent study, Bignell and Dunlop (unpubl. data 1995) were unable to detect any benzaldehyde in the oil obtained by using the VD technique, undoubtedly due to the fact that large amounts of a glucoside of the cyanohydrin of benzaldehyde occur in the leaves of *E. cladocalyx* (Finnemore et al. 1935) and that this compound hydrolyses in solution to a degree which may depend on pH.

From the previous discussion it would seem that either the SFE or VD techniques should be used if essential oils of *Eucalyptus* with compositions close to their in vivo values are to be obtained. This opinion has also been expressed by Koedam and Looman (1980). As a test of this conclusion, we have removed oils directly from the glands of *E. torquata* Luehm., *E. woodwardii* Maiden. and *E. sparsa* and found their compositions to be quite close to those found by SFE and VD techniques (C.M. Bignell and P.J. Dunlop, unpubl.). The leaves were mounted on a gauge block with double-sided adhesive tape and the glands exposed by slicing off a thin layer of the outer surface of the leaf. Then, using a dissecting microscope, the oil glands were observed as "large puddles" from which the oil was removed with a small stylus and placed directly into diethyl ether for GC analysis. In addition, we used the VD technique to extract oils and used their GC profiles for taxonomic purposes with reasonable success (Dunlop et al. 1995, 1996).

"Steam distillation is a relatively 'hard' extraction method and will often result in a modification of the original constituents by heat-induced reactions. This does not matter as long as the product obtained has proved acceptable to users. Indeed, if one were now to change to gentler methods, then the change in the product would probably render it unacceptable!" (Waterman 1993). The extraction technique used should be based on the final use for the essential oil, whether it is for chemotaxonomical purposes or for industrial use as flavours, fragrances etc. "If the intention of the study is to determine the true composition of the oil as it exists in the plant material, then the isolation procedure should not exert any harmful effects on the oils constituents" (Koedam 1987). It should be mentioned that oils may also be extracted with ethyl alcohol at room temperature (Ammon et al. 1985). However, while the compositions of these oils are quite close to those obtained by vacuum distillation, the additional material extracted causes the GC columns to eventually degrade as shown by very large drifts in the retention times.

7 Future Studies

This work is part of an ongoing study to determine the in vivo essential oil composition of members of the Australian flora. We are currently using manual harvesting from the oil glands as a baseline for the oil composition, but this is a very time-consuming, labour-intensive process and serves merely to provide a baseline for comparison. We are developing the methodology for quantitative extraction of the eucalypt essential oils using SFE and this will be published in the near future.

Acknowledgments. The authors would like to thank Associate Professor E.L. Ghisalberti, Dr. G.E. Gream and Dr. J.J. Brophy for many useful and enlightening discussions and suggestions in the preparation of the manuscript.

References

Adams RP (1991a) Analysis of juniper and other forest tree oil. In: Linskens HF, Jackson JF (eds) Modern methods of plant analysis new series, vol 12. Springer, Berlin Heidelberg New York, pp 131–157

Adams RP (1991b) Cedar wood oil – analysis and properties. In: Linskens HF, Jackson JF (eds) Modern methods of plant analysis new series, vol 12. Springer, Berlin Heidelberg New York, pp 152–173

Ammon DG, Barton AFM, Clarke DA, Tjandra J (1985) Rapid and accurate determination of terpenes in the leaves of *Eucalyptus* species. Analyst 110:921–924

Barth D, Chouchi D, Della Porta G, Reverchon E, Perrut M (1994) Desorption of lemon peel oil by supercritical carbon dioxide: deterpenation and psotalens elimination. J Supercrit Fluids 7:177–183

Bevan CD, Marshal SP (1994) The use of supercritical fluids in the isolation of natural products. Nat Prod Rep 11:451–466

Bignell CM, Dunlop PJ, Brophy JJ, Jackson JF (1994a) Volatile leaf oils from some south-western and southern Australian species of the genus *Eucalyptus*. Part I. Subgenus *Symphyomyrtus*, Section Dumaria, Series Incrassatae. Flav Fragr J 9:113–117

Bignell CM, Dunlop PJ, Brophy JJ, Jackson JF (1994b) Volatile leaf oils from some South-western and Southern Australian species of the genus *Eucalyptus*. Part II. Subgenus *Symphyomyrtus*, Section Dumaria, Series Torquatae. Flav Fragr J 9:167–171

Bignell CM, Dunlop PJ, Brophy JJ, Jackson JF (1994c) Volatile leaf oils from some South-western and Southern Australian species of the genus *Eucalyptus*. Part III. Subgenus *Symphyomyrtus*, Section Bisectaria, Series Macrocarpae. Flav Fragr J 9:309–313

Bignell CM, Dunlop PJ, Brophy JJ, Jackson JF (1995a) Volatile leaf oils from some South-western and Southern Australian species of the genus *Eucalyptus*. Part IV. Subgenus *Symphyomyrtus*, Section Dumaria, Series Dumosae. Flav Fragr J 10:85–91

Bignell CM, Dunlop PJ, Brophy JJ, Jackson JF (1995b) Volatile leaf oils from some south-western and southern Australian species of the genus *Eucalyptus*. Part V. Subgenus *Symphyomyrtus*, Section Bisectaria, Series Oleosae. Flav Fragr J 10:313–317

Bignell CM, Dunlop PJ, Brophy JJ, Jackson JF (1995c) Volatile leaf oils of some South-western and Southern Australian species of the genus *Eucalyptus*. Part VI. Subgenus *Symphyomyrtus*, Section Adnataria. Flav Fragr J 10:395–364

Bignell CM, Dunlop PJ, Brophy JJ, Jackson JF (1996a) Volatile leaf oils of some south-western and southern Australian species of the genus *Eucalyptus*. Part VII. Subgenus *Symphyomyrtus*, Section Exsertaria. Flav Fragr J 11:35–41

Bignell CM, Dunlop PJ, Brophy JJ, Jackson JF (1996b) Volatile leaf oils from some south-western and southern Australian species of the genus *Eucalyptus*. Part VIII. Series subgenus *Symphyomyrtus* (a) Section Bisectaria, Series Cornutae and Series Bakeranae, and (b) Section *Dumaria*, unpublished series Furfuracae group. Flav Fragr J 11:43–7

Bignell CM, Dunlop PJ, Brophy JJ, Jackson JF (1996c) Volatile leaf oils from some south-western and southern Australian species of the genus *Eucalyptus*. Part XIII. Series (a) Subulatae (b) Curviptera (c) Contortae (d) Incognitae (e) Terminaliptera (f) Inclusae (g) Microcorythae and (h) Cornutae. Flav Frag J 11:••–••
Boland DJ, Brophy JJ, House APN (1991) Eucalyptus leaf oils: use, chemistry, distillation and marketing. Inkata Press, Melbourne
Brooker MIH, Kleinig DA (1990a) Field guide to the eucalypts, vol 1, 2nd edn. Inkata Press, Melbourne
Brooker MIH, Kleinig DA (1990b) Field guide to the eucalypts, vol 2. Inkata Press, Melbourne
Brooker MIH, Kleinig DA (1994) Field guide to the eucalypts, vol 3. Inkata Press, Melbourne
Brunner G (1994) Gas extraction: an introduction to fundamentals of supercritical fluids and the application to separation processes. Steinkipf, Darmstadt, Springer, Berlin Heidelberg New York
Bruno TJ, Ely JF (eds) (1991) Supercritical fluid technology: reviews in modern theory and applications. CRC Press, Boca Raton
Chen C-C, Ho C-T (1988) Gas chromatographic analysis of volatile components of ginger (*Zingiber officinale* Roscoe) extracted with liquid carbon dioxide. J Agric Food Chem 36:322–328
Chouchi D, Barth D, Reverchon E, Della Porta G (1995) Supercritical CO_2 desorption of bergamot peel oil. Ind Eng Chem Res 34:4508–4513
Clifford AA (1993) Introduction to supercritical fluid extraction in analytical science. In: Westwood SA (ed) Supercritical fluid extraction and its use in chromatographic sample preparation. Blackie Academic & Professional, Glasgow, pp 1–38
Dean JR (1993) Applications of supercritical fluids in industrial analysis: applications of supercritical fluids in food science. Chapman and Hall, London
Deans SG (1991) Evaluation of antimicrobial activity of essential (volatile) oils. In: Linskens HF, Jackson JF (eds) Modern methods of plant analysis new series, vol 12. Springer, Berlin Heidelberg New York, pp 309–320
Dellacassa E, Menendez P, Moyna P, Soler E (1990) Chemical composition of *Eucalyptus* essential oils grown in Uruguay. Flav Fragr J 5:91–95
Dugo P, Mondello L, Bartle KD, Clifford AA, Breen GPA (1995) Deterpenation of sweet orange and lemon grass essential oils with supercritical carbon dioxide using silica gel as an adsorbent. Flav Frag J 10:51–58
Dunlop PJ, Bignell CM, Jackson JF, Hibbert DB (1995) Chemometric analysis of gas chromatographic data of oils from *Eucalyptus* species. Chemometr Intelligent Lab Syst 30:59–67
Dunlop PJ, Bignell CM, Hibbert DB (1996) Aust J Bot (submitted)
Finnemore H, Reichard SK, Large DK (1935) Cyanogenetic glucasides in Australian plants – part 3 *E. cladocalyx*. J Proc R Soc NSW 69:209–214
Garnero J, Tabacchi R (1987) Examples of artefact formation by chromatographic techniques. In: Sandra P, Bicchi C (eds) Capillary chromatography in essential oil analysis. Heutig, New York, pp 359–366
Gomes de Azevedo E, Matos HA, Nunes da Ponte M (1993) Phase equilibria of ethene + limonene and ethene + cineole from 285 K to 308 K and pressures to 8 Mpa. Fluid Phase Equilib 83:193–202
Guenther E (1948) The essential oils, vol 1. van Nostrand, New York
Guenther E, Gilbertson G, Koenig RT (1975) Essential oil and related products. Anal Chem 47:139R–157R
Hawthorne SB, Miller DJ, Kreiger MS (1989) Coupled SFE-GC: a rapid and simple technique for extracting, identifying, and quantitating organic analytes from solids and sorbent resins. J Chromatogr Sci 27:347–354
Hawthorne SB, Riekkola M-L, Serenius K, Holm Y, Hiltunen R, Hartonen K (1993) Comparison of hydrodistillation and supercritical fluid extraction for the determination of essential oils in aromatic plants. J Chromatogr 634:297–308
Hay RKM, Svoboda KP (1993) Botany. In: Hay RKM, Waterman PG (eds) Volatile oil crops. Logman Scientific & Technical, Harlow, UK, pp 5–22
Hay RKM, Waterman PG (1993) Introduction. In: Hay RKM, Waterman PG (eds) Volatile oil crops. Logman Scientific & Technical, Harlow, UK, pp 1–4

Hills JW, Hill HH Jr, Maeda T (1991) Simultaneous supercritical fluid derivatization and extraction. Anal Chem 63:2152–2155

Hirvi T, Salovaara I, Oksanen H, Honkanen E (1996) Volatile constituents of coriander fruit cultivated at different localities and isolated by different methods. In: Brunke EJ (ed) Progress in essential oil research. Walter de Gruyter, Berlin, pp 111–116

Hitchen SM, Dean JR (1993) Properties of supercritical fluids. In: Dean JR (ed) Applications of supercritical fluids in industrial analysis. Blackie Academic & Professional, Glasgow, pp 1–11

Inman RB, Dunlop PJ, Jackson JF (1991) Oils and waxes of eucalypts: vacuum distillation method for essential oils. In: Linskens HF, Jackson JF (eds) Modern methods of plant analysis new series, vol 12. Springer, Berlin Heidelberg New York, pp 195–204

Jenkins TJ, Kaplan M, Simmonds MR, Davidson G, Healy MA, Poliakoff M (1991) Novel methods of instrumental analysis using supercritical fluid-based systems. An overview. Analyst 116:1305–1311

King JW (1990) Applications of capillary supercritical fluid chromatography-supercritical fluid extraction to natural products. J Chromatogr Sci 28:9–14

King MB, Bott TR (eds) (1993) Near-critical extraction of natural products. Chapman & Hall, London

Knipe CR, Miles WS, Rowland F, Randal LG (1993) Designing a sample preparation method that employs supercritical fluid extraction. Hewlett-Packard Co, USA

Koedam A (1987) Some aspects of essential oil preparation. In: Sandra P, Bicchi C (eds) Capillary gas chromatography in essential oil analysis. Huethig, Heidelberg, pp 13–27

Koedam A, Looman A (1980) Effect of pH during distillation on the composition of the volatile oil from *Juniperus sabina*. Planta Med Suppl:22–28

Koedam A, Scheffer JJC, Svendsen AB (1981) Comparison of isolation for essential oils IV. Leyland cypress. Perf Flav 5:56–65

Lack E, Seidlitz H (1993) Commercial scale decaffeination of coffee and tea using supercritical CO_2. In: King MB, Bott TR (eds) Extraction of natural products using near-critical solvents. Chapman and Hall, London, pp 101–139

Likens ST, Nickerson GB (1964) Detection of certain hop oil constituents in brewing products. Proc Am Soc Brew Chem •:5–13

Matos HA, Gomes de Azevedo E, Simoes PC, Carrondo MT, Nunes da Ponte M (1989) Phase equilibria of natural flavours and supercritical solvents. Fluid Phase Equilib 52:357–364

McHugh M, Krukonis V (eds) (1986) Supercritical fluid extraction. Butterworth, Boston

McHugh M, Krukonis V (eds) (1994) Supercritical fluid extraction, 2nd edn. Butterworth, Boston

McKern HHG (1968) Research into the volatile oils of the Australian flora, 1788–1967. In: Centenary of scientific progress, centenary volume. Royal Society of NSW, Science Society House, Sydney, pp 310–331

Milner CP, Trengove RD (1996) Supercritical fluid extraction of Western Australian Sandalwood (*Santalum spicatum*) (submitted)

Milner CP, Minett RD, Edwards TJ, Gere DR, Trengove RD (1995) Solid trapping processes in off-line supercritical fluid extraction: normal phase traps for essential oil components. Pittcon '95, March 5–10, New Orleans

Milner CP, Trengove RD, Gere DR (1996) Solid phase trapping in SFE of essential oil components (submitted)

Moyler DA (1986) Liquid carbon dioxide extraction of flavour materials. In: Birch GG, Lindly MG (eds) Developments in food and flavours. Elsevier, London, pp 119–129

Moyler DA (1988) Liquid CO_2 extraction in the flavour and fragrance industries. Chem Ind 18:660–662

Moyler DA (1989) Supercritical fluids for extract preparation. In: Piggott JR, Paterson A (eds) Distilled beverage flavour. Ellis Horwood, Chichester, pp 319–328

Moyler DA (1991) Natural ingredients for flavours and fragrances. Chem Ind 21:11–14

Moyler DA (1993a) Extraction of flavours and fragrances with compressed CO_2. In: King MB, Bott TR (eds) Near-critical extraction of natural products. Chapman and Hall, London, pp 140–183

Moyler DA (1993b) Extraction of essential oils with carbon dioxide. Flav Frag J 8:235–297

Moyler DA, Heath HB (1988) Liquid carbon dioxide extraction of essential oils. In: Lawrence BM, Mookherjee BD, Willis BJ (eds) Development in food science 18: flavours and fragrances, a world perspective. Proc 10th Int Congr Essential oils, fragrances and flavors, Washington, DC, 16–20 Nov 1986. Elsevier, Amsterdam, pp 41–63

Nishimura K, Shinoda N, Hirose Y (1969) A new sesquiterpene, bicyclogermacrene. Tetrahedron Lett 36:3097–3100

Piggott MJ, Ghisalberti E, Trengove RD (1996) Western Australian Sandalwood oil: extraction by different techniques and variations of the major components in different sections of a single tree. (in press)

Raghuram Rao GV, Srinivas P, Sastry SVGK, Mukhopadhyay M (1992) Modeling solute-co-solvent interactions for supercritical-fluid extraction of fragrances. J Supercrit Fluids 5:19–23

Raynie DE (1993) Warning concerning the use of nitrous oxide in supercritical fluid extractions. Anal Chem 65:3127–3128

Reverchon E (1992) Fractional separation of SCF extracts from marjoram leaves: mass transfer and optimizing. J Supercrit Fluids 5:256–261

Reverchon E, Senatore F (1992) Isolation of rosemary oil: comparison between hydrodistillation and supercritical CO_2 extraction. Flav Frag J 7:227–230

Reverchon E, Senatore F (1994) Supercritical carbon dioxide extraction of chamomile essential oil and its analysis by gas chromatography-mass spectrometry. J Agric Food Chem 42:154–158

Reynolds WC (1979) Thermodynamic properties in SI. Department of Mechanical Engineering, Stanford University, Stanford

Rothschild GHL (1991) Foreword in: Eucalyptus leaf oils: use, chemistry, distillation and marketing. Inkata Press, Melbourne

Sandra P, David F, Stottmeister E (1990) Recovery studies by offline SFE. J High Resolut Chromatogr Chromatogr Commun 13:284–286

Smith RM (1995) Supercritical fluid extraction of natural products. LC-GC 13:930–939

Stahl E, Quirin K-W, Gerard D (1988) Dense gases for extraction and refining. Springer, Berlin, Heidelberg New York

Tressl R, Engel K-H, Kassa M, Köppler H (1983) Characterization of the tricyclic sesquiterpenes in hop (*Humulus lupulus*, var. Hersbrucker Spät). J Agric Food Chem 31:892–897

van Beek TA (1991) Special methods for the essential oil of ginger. In: Linskens HF, Jackson JF (eds) Modern methods of plant analysis new series, vol 12. Springer, Berlin Heidelberg New York, pp 79–98

von Rudloff E (1967) Chemosystematic studies in the genus *Picea* (Pinaceae). Can J Bot 45:891–901

Waterman PG (1993) The chemistry of volatile oils. In: Hay RKM, Waterman PG (eds) Volatile oil crops. Logman Scientific & Technical, Harlow, UK, pp 47–62

Williams DF (1981) Extraction with supercritical gases. Chem Eng Sci 36:1769–1788

Principal Components Analysis of Essential Oils

M.L. Arrebola

1 Introduction

The importance of aromatic plants is considerable because many of the essentials oils obtained from them have very useful properties and can be used in many ways. From the earliest times there was great interest in the essential oil of *Thymus*.

The genus *Thymus* L. belongs to the family Lamiaceae and occurs in all temperate regions of the northern hemisphere, being especially common in the Mediterranean area. Because of the large number and diversity of species comprised in this genus, it has been taxonomically subdivided into sections. One of the most widely accepted taxonomic classifications is that proposed by Morales (1986), based on the criteria set down by Jalas and Kaleva (1970) and Jalas (1971), who in turn based their work on Velenovsky (1906). Morales (1986) divided the genus *Thymus* into seven sections, some of which are in turn subdivided into subsections. In these categories (section or subsection) the type species is mentioned, as well as those taxa found in the Iberian Peninsula included in them.

Different phytochemical groups occur in *Thymus* species, including polyphenol derivatives and, especially, essential oils. These essential oils are quite complex. Moreover, their quantitative composition may be affected by several factors: site where plants are collected, stage of the growth cycle at the moment of collection, environmental factors (climate, soil type, density of plantation, intensity and manner of irrigation, etc.), part of the plant used, method of extraction of essential oil, way of storing the essential oil, etc. (Stahl 1982; Trease and Evans 1986; Bruneton 1987; Janssen et al. 1987; Muñoz 1987; Arrebola et al. 1994b, 1995, and others). In general, it is observed that phenols occur when plants grow at high temperatures and dry climate; linalol and α-terpineol content increase in plants growing under humid environmental conditions, and geraniol contents is higher in a wet and cold climate (Morales 1986).

Because of this complexity, before conducting an experiment, time should be spent in planning an experimental design and analysis of the results, in order to resolve the questions or objectives proposed. Statistical analyses are, therefore, an essential part of essential oils research.

One of the main problems of biological research is that it focuses on complex phenomena, in which many factors or elements coexist and interact, and together cause the final effect that we observe. One major difficulty is how to identify the component elements, and how to understand the mutual interactions and influ-

Modern Methods of Plant Analysis, Vol. 19
Plant Volatile Analysis
Edited by H.-F. Linskens and J.F. Jackson

ences. The variate techniques belonging to the family of factorial analyses (principal components analysis, factor analysis, correspondence analysis, canonical analysis, etc.) are particularly suited to help in understanding complex phenomena (Benigni and Giuliani 1991).

Several statistical methods have been used for the analysis of essential oils (canonical analysis, Tukey test, variance analysis, etc.). This chapter deals with a statistical method suitable for the analysis of essential oil composition data, called principal components analysis. The results obtained in the case of *Thymus serpylloides* Bory subsp. *serpylloides* essential oil will be used as an example. Special emphasis will be put on the methodical aspects concerning statistical analysis.

2 Definition of Some Statistical Terms

2.1 Correlation Matrix

A correlation matrix is a variance-covariance matrix for standardized variables (see below). In the diagonal of this matrix, we therefore find values equal to unity. In the other cells we find correlations. In the correlation matrix the relationships considered are only pairwise (van de Geer 1971; Takeuchi et al. 1982). Sometimes, sample correlation matrices are used instead of sample covariance matrices (by standardizing the variables relative to the sample); the problem of units is thereby avoided, since the "principal components" will not be affected because standardized variables are dimensionless (Press 1972). An example of a correlation matrix could be as follows:

	x_1	x_2	x_3
x_1	1.000		
x_2	0.556	1.000	
x_3	0.628	0.584	1.000

2.2 Data Matrix

A data matrix is nothing but a systematic collection of observed data. Generally, in multivariate analysis, we have observations on a certain number of variables. If there are n objects, $o_1, \ldots, o_n$, and p variables, $x_1, \ldots, x_p$, the data contains np pieces of information. They may be conveniently arranged using an $(n \times p)$ data matrix, in which each row corresponds to an object and each column corresponds to a variable (van de Geer 1971; Mardia et al. 1979).

2.3 Eigenvalue, Latent Root, or Characteristic Root

When an orthogonal rotation of factors is performed, the variance explained by each factor is the eigenvalue for that factor. Total variance is defined as the sum of the positive eigenvalues of the correlation matrix. In principal components analysis, the first eigenvalue sometimes accounts for a considerable proportion of the total variability (Seal 1964; Dixon et al. 1990).

2.4 Linear Combination

Taking linear combinations of the variables is one of the most important tools in multivariate analysis. A few suitably chosen combinations may provide more information than a multiplicity of the original variables, often because the dimension of the data is reduced. Linear transformations can also simplify the structure of the covariance matrix, making interpretation of the data more straightforward (Mardia et al. 1979). As an example of linear combination we could consider:

$$y_1 = a_1x_{r1} + \ldots + a_px_{rp}, \text{ where } r = 1, \ldots, n \text{ and } a_1, \ldots, a_p \text{ are given.}$$

2.5 Model

A model is a mathematical formulation of assumptions that are taken to be valid. According to van de Geer (1971), the researcher has to make decisions in two stages. The first decision stage is to choose a model, the second stage is to determine whether the data are consistent with a solution of the model.

2.6 Variables

Variables can be classified according to different criteria (Takeuchi et al. 1982). This chapter will deal only with those mentioned in the text. *Independent* or *explanatory* variables are variables ($x_1, \ldots x_p$) whose variation may be considered to be the cause of variation of other variables called *dependent* or *criterion variables* ($y_1, \ldots, y_p$). *Observable variables* are those variables that are directly observable in some situations. *Unobservable* or *latent variables* are variables for which we have no direct observations; they may influence the dependent variables.

3 What is Principal Components Analysis?

Principal components analysis (PCA) is one of the methods for analyzing data included in *multivariate analysis*. In 1957, Kendall defined multivariate analysis as "the branch of statistical analysis which is concerned with the relationships of sets of dependent variables". The main object of various methods of multivariate analysis lies in transforming a mass of observations into a smaller number of composite scores such that they may reflect as much information as possible contained in raw data. Mathematically, this procedure may be interpreted as forming a linear composite vector in a vector subspace, which can be represented in terms of the projection of a vector onto certain specified subspaces (Takeuchi et al. 1982). Multivariate analysis comes down to operations on columns or rows of the data matrix. Operations to be performed depend on the specific model that inspires the analysis. The analysis depends on the nature of the specific questions asked about the variables and their interrelations (van de Geer 1971). Multivariate statistical techniques have been strongly applied in fields such as biology, medicine, agriculture, pedagogy, sociology, economics, psychology, and anthropology, in which it is common that many variables are supposed to be highly correlated with each other.

Four methods of initial factor extraction are available: principal components analysis (PCA); iterated principal factor analysis (PFA); maximum likelihood factor analysis (MLFA); and Kaiser's Second Generation Little Jiffy (LJIFFY). For a first analysis, PCA is recommended (Dixon et al. 1990).

PCA was initially described by Pearson (1901) and further developed by Hotelling (1933) in order to analyze the structural interdependence of a set of variables. Data vectors taken from multivariate populations involve independent observations on p possibly correlated random variables. It is sometimes the case that $p = 50$, 100, or even 1000. For this reason, it is natural to seek ways of rearranging or summarizing the data with as little loss of information as possible (in many problems, there is much more to gain than to lose by such approach). PCA looks for a few linear combinations of the original variables which can be used to summarize the data. The set of dependent original variables $x_1, \ldots, x_p$ (which are correlated with each other) are linearly transformed into a set of new variables $y_1, \ldots, y_p$ called *principal components* of the x's, which are mutually uncorrelated (independent) and can thus be considered separately. The transformation is performed by assigning weights to each variable so that the resulting composite variable as a set may have maximum variance under the restriction that different linear composites are orthogonal to each other. The order of variance associated with the principal components is decreasing. The first eigenvalue usually accounts for a considerable proportion of the total variability. Sometimes, however, it is necessary to use other eigenvalues to summarize the whole of the variability and covariability of the original variables. This attempt to reduce the dimension of the problem can be described as parsimonious summarization of a mass of observations. PCA may also be said to be a method of projecting a set of vectors onto subspaces which are pairwise orthogonal (Seal 1964; Press 1972; Mardia et al. 1979; Takeuchi et al. 1982; Dixon et al. 1990).

PCA is often used in applied research as merely a first step in studying the reasonableness of various hypotheses about the phenomenon under investigation. Typically, the data will first be subjected to a PCA to expose linear combinations of variables with the greatest variance. Those with relatively small variance are dropped and the remaining subset of transformed variables is then used as the set of regressors in a regression, in conjunction with some dependent variables; or they may be further rotated according to some criterion in a factor analysis model; or they may be used in a clustering analysis; or in a multivariate analysis of variance (Press 1972).

4 Factor Analysis

Factor analysis has been described (Burt and Banks 1947) as "a statistical technique for reducing a large number of correlated variables to a small number of uncorrelated variables. The correlated variables consist usually of measurements for observable traits; the uncorrelated variables (called 'factors') are abstract hypothetical components". The main aim of factor analysis is, as Holzinger and Harman (1941) put it, "to attain scientific parsimony or economy of description, by determining the minimum number of theoretical variables (factors) which can be used to describe and 'explain' the interrelations of a large set of observable variables".

These definitions may sound suspiciously like PCA because the overall goals are quite similar, but factor analysis is distinguished from PCA by four characteristics (Seal 1964; Mardia et al. 1979; Takeuchi et al. 1982):

1. Each of the p original variables is supposed analyzable into $m < p$ mutually uncorrelated "common factors" with an uncorrelated residual ("unique") component which is not correlated with any of the remaining $p - 1$ variables.
2. The number of common factors obtained in factor analysis is usually less than the number of components in PCA.
3. The m orthogonal axes of "common factors" may be rotated to new orthogonal or oblique axes to conform with theoretical ideas underlying the formulation of the model.
4. PCA is merely a transformation of the data. No assumptions are made about the form of the covariance matrix from which the data comes. On the other hand, factor analysis supposes that the data come from a well-defined model, where the underlying factors satisfy the assumptions. If these assumptions are not met, then factor analysis may give spurious results.

Factor analysis is useful in exploratory data analysis. It has the following general objectives: (1) to study the intercorrelations of a large number of variables by clustering them into common factors such that variables within each factor are highly correlated, (2) to interpret each factor according to the variables with high loadings belonging to it, and (3) to summarize many variables by new factor or summary scores. Using a correlation or covariance matrix, factor analysis extracts the dimensions that account for the variability in a space defined by the variables,

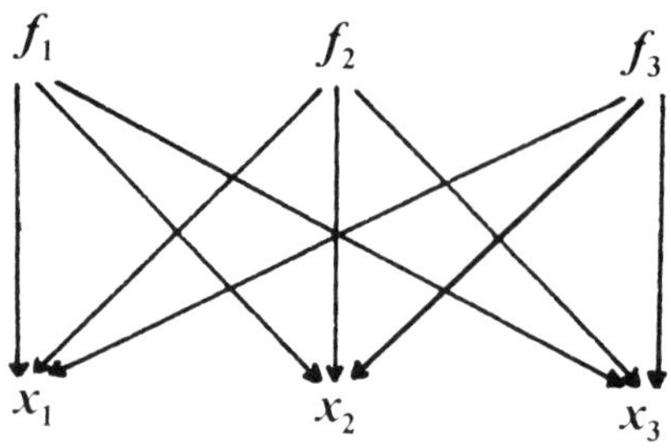

Fig. 1. Factor analysis model

rotates the factors to obtain a simple interpretation, and provides factor scores for each case. The assumption is that the number of factors will be appreciably less than the original number of variables. A successful factor analysis explains a large proportion of variance with very few factors (Dixon et al. 1990).

The basic premise of factor analysis is that there is an underlying order which, when found, will permit a simpler explanation of phenomena that may seem to be the result of a very large number of variables (Takeuchi et al. 1982).

A factor analysis model assumes that a set of observed variables x can be interpreted as dependent on a set of unobserved variables f. Figure 1 illustrates the factor analysis model. An arrow starting at f and pointing to x represents that variable f is a determiner of variable x.

In factor analysis is also assumed that each variable can be decomposed into a sum of a small number of common factors (Takeuchi et al. 1982).

A very clear example could be as follows:

$$x_1 = 0.9f_1$$
$$x_2 = 0.1f_1 + 0.7f_2$$
$$x_3 = 0.9f_3,$$

where, x_1, x_2 and x_3 are the observed variables; f_1, f_2, and f_3 the factors, and the numbers in front of the factors express the degree of contribution of the factors to the variables, these contributions are called factor loadings. In this example, factor f_1 explains, in some way, what x_1 and x_2 have in common, the influence of f_1 on x_1 being bigger than its influence on x_2. The variability of x_2 depends both on f_1 and f_2 (but to a larger extent on f_2), while the variable x_3 is completely accounted for by f_3.

The overall effect of the analysis is a regrouping of the data into patterns which can sometimes be interpreted meaningfully (Press 1972).

5 Why Are Factors Rotated? Types of Rotation

Factor analysis permits the researcher to rotate the "common factor axes" to any more convenient position (Seal 1964). The object of rotation is to obtain a simpler interpretation of the factors making each factor reflect information in only a few of the variables. The rotation of the factor axes is done in such a way that the loadings for each factor is either large or small (loadings attached to factors that should not appear in the specified group of test are approximately zero), not intermediate (Seal 1964; Dixon et al. 1990).

A comparison of the rotated factor loadings with the unrotated ones shows that the larger loadings are larger than before and the smaller loadings are smaller (Dixon et al. 1990).

As indicated above, the objective of factor analysis is to reduce the dimensionality of the vector space. Van de Geer (1971) explained the rotation of factors very clearly in the following way: suppose now that, on the basis of a PCA, we have decided which dimensions can be neglected, and suppose that we retain some vectors for an acceptable description of the vector bundle. PCA then also indicates the positions of these vectors in this reduced space. However, there is no logical reason to adopt the eigenvectors as reference directions: any other (orthogonal) set of reference directions would describe the bundle as well. Eigenvectors are nice for a first identification of the reduced vector space, but once it has been found, we need not prefer the eigenvectors as a reference system to any other set of coordinates.

For instance, in Figs. 2 and 3 the coordinates represent the first and the second eigenvectors. The bundle of test vectors in Fig. 2 could be identified as well if we rotate the coordinates throught about 45°, which results in Fig. 3. The two pictures are completely equivalent, of course, but we see that the interpretation of the factors changes. The latter set of factors is much more meaningful than the first and therefore should be preferred. In others words, the factors can be rotated to find the most meaningful set of factors.

To develop this idea further, according to van de Geer (1971), criteria for rotation should be specified. In fact, there are two different principles that apply here:

1. A factor is meaningful if it identifies a cluster of vectors: a cluster of vectors indicates that the variables in the cluster have things in common and are

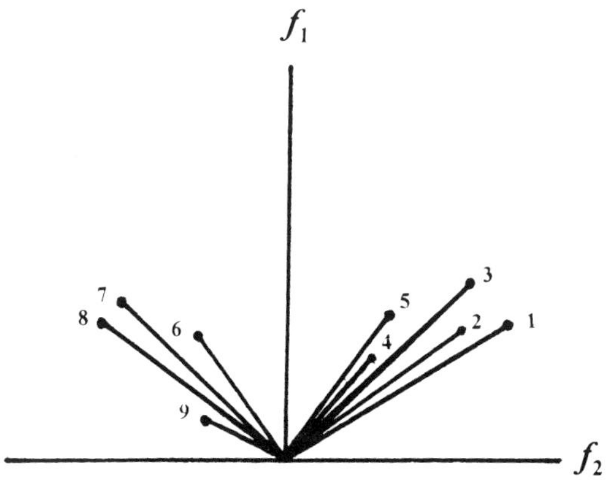

Fig. 2. Graph of factor loadings first two factors, nine variables, in two clusters

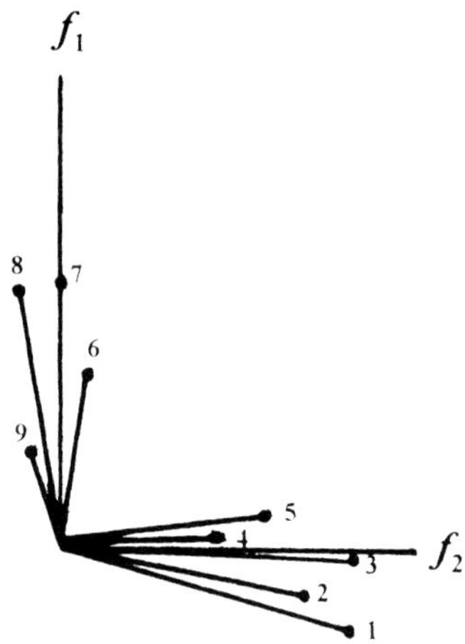

Fig. 3. Rotated version of Fig. 2

distinguished from variables outside the cluster – the clustered variables are "indicators" or a common underlying factor. This principle implies that the factor can be easily interpreted, since it will have high loadings for all variables in the cluster.

2. The other principle starts at the other end, by trying to find factors that are unrelated to as many observed variables as possible. The interpretation problem is then simplified, because we have to take into consideration only those remaining variables that have loadings in the factor.

We want the rotation to give the vectors' maximum projection on one factor and minimum projection on the other, for all pairs of factors.

In *orthogonal rotations*, the factor scores for the various factors are mutually uncorrelated. When the rotation is orthogonal, the eigenvalue is the variance explained by the factor, and the rotated loadings are the correlations of the variables with the factors (Takeuchi et al. 1982; Dixon et al. 1990). *Varimax rotation* is only a variety of factor analysis (van de Geer 1971). The *varimax method* of orthogonal rotation proposed by Kaiser (1958) is a convenient analytical choice of rotation. Its rationale is to provide axes with a few large loadings and as many near-zero loadings as possible (Mardia et al. 1979). In 1953, Carroll proposed the *Quartimax method*, in which the maximizing criterion is given as the sum of the fourth power of the factor loadings (Takeuchi et al. 1982).

In contrast to orthogonal rotations, in *oblique rotations* the factors are allowed to be correlated, which may result in "simpler" loadings than those for an orthogonal rotation. With oblique rotation there is a greater tendency for each variable to be associated with a single factor. When the rotation is not orthogonal, the eigenvalue is not the variance explained by the factor; however, large values of eigenvalues still indicate important factors and small values indicate less important ones (Dixon et al. 1990).

Comparing the loadings from oblique rotation with those from orthogonal rotation, we see in which the goal of having each variable associated with a single factor is better met (Dixon et al. 1990). It is necessary, however, to take into consideration that, as indicated above, when an oblique method is used to do a rotation, the estimation of the correlations among factors is obtained. Then, the interpretation of factors is more difficult because the correlations have to agree, in some way, with the estimated correlations.

6 Interpretation of Factors

An extremely important, but often troublesome, aspect of factor analysis is the interpretation of results. The only way open is to make a guess, on the basis of the observed variables, which factor is correlated with it (it also helps to see which variables the factor is not correlated with). Apparently the factor stands for what they all have in common. Interpretation of the factor can then be based on inspection of what these variables have in common (van de Geer 1971).

Problems of interpretation are compounded when nonlinear models are used (Press 1972).

7 The Case of *Thymus* Essential Oil

7.1 Introduction

Many papers have already been published concerning the analysis of essential oil composition of the family Lamiaceae, but as far as I know, only in two of them (Arrebola et al. 1994a, b) was a principal components analysis applied.

In this chapter, the results obtained in the case of *Thymus serpylloides* Bory subsp. *serpylloides* (Arrebola et al. 1994b) will be used. *Th. serpylloides* subsp. *serpylloides* is a protected (by Spanish and International authorities) endemic subspecies to the highland areas in the north side of Sierra Nevada (Granada, Spain). It is a summer-flowering plant, which grows abundantly on acid soils, which are very rocky and poor in nutrients.

7.2 Studied Material

One population was selected for *Th. serpylloides* subsp. *serpylloides* based on ecological conditions, aromatic and melliferous yield of the community where it is developed, and the degree of erosion of the soil on which this plant grows. The population observed in this area was pure and did not overlap with other species of thyme. The collection site was Granada: Sierra Nevada, near crossroads Estación of Borreguiles, 30SVG6604, 2750 m, 20.VII. 1991, O. Socorro and M.L. Arrebola, GDA 25 539.

The aerial parts of some specimens of this subspecies, with its sexual characteristics, were collected throughout the complete vegetative cycle over 3 consecutive years. This material was dried for 48 h in the shade, after which the stems were separated from the flowers, bracts, and leaves.

7.3 Extraction, and Qualitative and Quantitative Analyses of the Essential Oil

Plant material was distilled in a Clevenger-type apparatus for 4 or 5 h. The isolated oil was dried over Na_2SO_4 and stored at 4–5 °C.

Qualitative data were determined by GC and GC-MS (for conditions of analysis see Arrebola et al. 1994b). Components were identified by the peak enrichment technique and by comparing their mass spectra with those of authentic samples and/or data reported in the literature.

The GC quantitative analysis was performed by the relative proportion method under the same conditions as in the qualitative analysis. The comparative study was done by quantitative analysis. We compared essential oils corresponding to different phenological stages, in the same population and during 3 consecutive years. The statistical analysis was done with BMDP Statistical package version PC-90 from BMDP Statistical Software (Cork, Ireland; Dixon et al. 1990). For the analysis of the essential oils of this subspecies during 3 consecutive years, the Tukey test at significance levels 5, 1, and 0, 1% was carried out after a one-way variance analysis. For the global study of the samples, an orthogonal factor analysis model was estimated to the normalized (or typified) data. We used the normalized data instead of the original data because there is a large difference among the variances of these original variables. The normalized variables ($z_1, \ldots, z_{35}$) were obtained from the original variables ($x_1, \ldots, x_{35}$) in the following way:

$z_1 = [x_1 - me(x_1)]/sd(x_1)$, where $me(x_1)$ is the mean of x_1 and $sd(x_1)$ the standard deviation of x_1. With the typification of the variables, the range of variation of the original variables is equated. If we do not typify the variables, results obtained by PCA could be spurious.

Estimation of the model that uses orthogonal factor analysis was performed by PCA (the estimations of factors are then called principal components). The number of principal components was determined after applying asymptotic tests on the hypothesis related to the characteristics roots (eigenvalues) of the correlation matrix, in such a way that, at a confidence level of 95%, these components account for a rate of variability higher than 92%.

This model reduces the dimension, and after realizing the rotations in the space of the principal components selected, we can interpret the nonobservable variables. We have performed two types of orthogonal rotations obtained by the varimax and quartimax methods, in the same way as an oblique rotation obtained by the direct oblimin method. In our case, principal components analysis gave a better interpretation than those given by the other methods.

7.4 Results

The essential oil content of stems was always lower than 0.5% v/w, i.e., traces. Independently of its sexual characteristics, the highest yield in essential oil was always obtained from plants collected during the full flowering period. In hermaphrodite individuals the oil content was higher than in female.

The data obtained in the qualitative and quantitative analyses of the oil samples show a high degree of compositional homogeneity. Minor differences were found only for the minor constituents. The main groups detected were: monoterpenes, oxygenated monoterpenes, sesquiterpenes, and aromatic alcohols.

The essential oils obtained had a predominance of carvacrol, which in all cases reached a content greater than 44% (for further details see Arrebola et al. 1994b). Apart from this phenol, both thymol (from traces to 5%) and their biosynthetic precursors (γ-terpinene and *p*-cymene) appeared in the essential oil. This fact allowed us to affirm that this subspecies can be classified as a phenolic thyme, in particular, carvacrolic. In this group are included: *Th. capitatus* (L.) Hoffmanns. & Link (Papageorgiou and Vassilios 1980; Papageorgiou and Argyriadou 1981; Solinas et al. 1981; Lawrence 1982; Falchi-Delitala et al. 1983). *Th. fominii* Klok. et Shost (Kasumov 1981), *Th. karamarianicus* Klok. et Shost (Kasumov and Farkhadova 1986; Novruzova and Kasumov 1987), and others.

Carvacrol content variated in the different phenological stages during 3 consecutive years, but always kept the character of main component. There were no significant differences in its amount between hermaphrodite and female individuals if we consider the mean of the 3 years (hermaphrodite: 53.56 ± 4.52, and female: 54.98 ± 5.00). The annual variation of carvacrol could be due to variations in the climate experienced during the 3 years of harvesting (Arrebola 1992).

Besides this notable carvacrolic character, there were other components, whose quantity represents or exceeds 1%: α-pinene, α-terpinene, γ-terpinene, *p*-cymene, linalool, terpinen-4-ol, γ-caryophyllene, borneol, unidentified sesquiterpene, and thymol.

For the global study of the samples by principal components analysis, 12 factors that accounted for 96.2% of the variability of the components detected in the samples of *Th. serpylloides* subsp. *serpylloides* were considered (Table 1). The largest factor (f_1) mainly affected (with values higher than 0.5) components or variables belonging to different chemical groups, such as: monoterpene hydrocarbon (MTHC) derivatives of the menthane (α-terpinene, γ-terpinene, and *p*-cymenene); oxygenated MTHC (terpinen-4-ol and β-terpineol); phenolic aromatic derivatives (2,3,5,6-tetramethylphenol, thymol, and methyleugenol); oxygenated

Table 1. Factor loadings obtained by principal components analysis. (Arrebola et al. 1994b)

Component	Factor											
	f_1	f_2	f_3	f_4	f_5	f_6	f_7	f_8	f_9	f_{10}	f_{11}	f_{12}
Unidentified sesquiterpene	0.969											
Geranial	0.940											
Methylionone	0.881						−0.276					
Borneol	0.879	−0.302										
Linalool	0.876		0.298									
Methyleugenol	0.872											
α-Terpinene	−0.821	−0.265			0.351							
2,3,5,6-Tetramethylphenol	0.789	0.322	−0.375					0.256				
γ-Terpinene	0.746		0.304								0.325	
Terpinen-4-ol	0.741	−0.355				0.478						
Thymol	0.734	0.286	0.305			−0.369						
Fenchone/2-methylpentan-3-one	0.723	−0.504		0.362								
Myrcene	−0.721		0.344		0.341					0.254		
β-Terpineol	0.680	−0.508			0.370			−0.272				
Geranyl acetate	0.663		0.311	−0.345			0.266	−0.302				
ρ-Cymenene	0.579	−0.303	0.415	0.255	−0.377			0.347				

ρ-Cymen-8-ol		0.847					−0.293				
Lavandulol		0.749		−0.374	−0.323						
α-Thujene		0.710		−0.274	−0.340	0.291					0.299
δ-3-Carene		0.634		0.491	−0.363						
Thymil acetate		0.584	0.575			−0.357				0.259	
β-Pinene		0.578	0.578				−0.282	−0.370			
Geraniol			0.790	−0.454	0.288						
trans-Sabinene hydrate	−0.255		0.681	0.473			−0.328				
1,8-Cineol			0.652		−0.428	−0.312		0.368			
Carvacrol			−0.647	−0.380	−0.390				−0.282		
Linalyl acetate	0.298	−0.441	0.641	0.363	−0.263						
Terpenyl acetate			0.622	−0.573		−0.389					
γ-Caryophyllene		0.463		0.646	0.251	0.399	−0.264				
Camphor	0.312	0.353		0.572	0.511						
ρ-Cymene		0.536			0.631						
Limonene	−0.260		0.475	0.530	−0.552						
α-Pinene	−0.486		0.351			0.636	0.329		0.259		
Camphene		0.542	0.426	−0.289		0.567					
α-Terpineol	−0.435			0.396	−0.372				0.461	0.365	

Loadings less than 0.25 are not expressed.

MTHC derived from bicycles (borneol and fenchone); acyclic MTHC (myrcene); acyclic oxygenated MTHC (geranial, linalool, and geranyl acetate); and other compounds such as methylionone and an unidentified sesquiterpene. The factor f_2 accounted for the variability of: *p*-cymen-8-ol; thymil acetate (another aromatic derivative); three MTHC derivatives of distinct bicycles (α-thujene, δ-3-carene, and β-pinene) and lavandulol, an acyclic oxygenated MTHC. In the same degree as f_2, f_3 affected thymil acetate and β-pinene, and had a notorious incidence on geraniol (acyclic MTHC), *trans*-sabinene hydrate (menthane derivative), 1,8-cineol (epoxi derivative of mentane), carvacrol (aromatic alcohol), linalyl acetate, and terpenyl acetate. The factor f_4 could be considered as a relevant factor, but in a small degree, of the 16 identified components in the samples of *Th. serpylloides* subsp. *serpylloides*, reaching values higher than 0.5 in three cases. Like f_4, f_5 had a certain incidence in most cases on 16 of the detected components in the essential oil. The factor f_6 affected two bicyclic MTHC (α-pinene and camphene) with values higher than 0.5, and partially determined the variability of the other six components. Factors f_7–f_{10} are related to a scarce number of variables which belong to distinct chemical groups.

Aromatic derivatives and their precursors (γ-terpinene and *p*-cymene) deserve special consideration, because their concentration determines the classification of the different species of the genus *Thymus* in phenolics and nonphenolics.

The variability of γ-terpinene largely depended both on f_1 and f_3 (but to a larger extent on f_1), while *p*-cymene depended on f_2 and f_5; the variations of *p*-cymenene (intimately related to *p*-cymene) depended on both the factors affecting γ-terpinene and *p*-cymene, and others not affecting the aforementioned compounds, such as f_4 and f_8. Thymol, like γ-terpinene, depended on f_1 and f_3, and to a lesser extent on f_2 and f_6, whereas carvacrol variability was defined by f_3, and to a lesser extent by f_4 and f_5, and even less by f_9. *p*-cymen-8-ol, a nonphenolic derivative of *p*-cymene, largely depended on f_2 and less on f_9. Thymil acetate, a thymol derivative, was affected by the same factors as thymol, with the exception of f_1, while 2,3,5,6,-tetramethylphenol depended on the same factors as thymol, with the exception of f_6.

Generally speaking, it might be said that none of the considered factors affected the variability of the whole of the phenolic derivatives and their precursors, f_2 and f_3 being the factors that affected a lager number of compounds in these groups. This fact is very interesting for the study of the essential oils of the species of the *Thymus* genus, since these factors affected, to a larger or smaller extent, the variations of seven of the eight compounds in both groups.

8 Conclusion

Principal components analysis is a suitable statistical method for the analysis of essential oil composition data. Difficulty lies in the interpretation of factors, because the only way open is to make a guess, and sometimes this is difficult.

Acknowledgments. The author is indepted to Francisco Antonio Ocaña Lara (Departamento de Estadística e Investigación Operativa, Facultad de Farmacia, Universidad de Granada, Spain) for his advice and corrections in the statistical part of this chapter, as well as to Arie C. Hoek (Division of Pharmacognosy, Leiden/Amsterdam Center for Drug Research, Leiden University, The Netherlands) for checking the English manuscript.

References

Arrebola ML (1992) PhD Thesis. University of Granada, Granada

Arrebola ML, Navarro MC, Jiménez J, Ocaña FA (1994a) Variations in yield and composition of the essential oil of *Satureja obovata*. Phytochemistry 35:83–93

Arrebola ML, Navarro MC, Jiménez J, Ocaña FA (1994b) Yield and composition of the essential oil of *Thymus serpylloides* subsp. *serpylloides*. Phytochemistry 36:67–72

Arrebola ML, Navarro MC, Jiménez J (1995) Variations in yield and composition of the essential oil of *Thymus serpylloides* Bory subsp. *gadorensis* (Pau) Jalas. J Essent Oil Res 7:369–374

Benigni R, Giuliani A (1991) Multivariate Analyses in Genetic Toxicology. In: Devillers J, Karcher W (eds) Applied multivariate analysis in SAR and environmental studies. Kluwer, Dordrecht

Bruneton J (1987) Éléments de phytochimie et de pharmacognosie. Technique et Documentation (Lavoisier), Paris

Burt C, Banks C (1947) A factor analysis of body measurements for British adult males. Ann Eugen 13:238–256

Dixon WJ, Brown MB, Engelman L, Jennrich RI (1990) BMDP statistical software manual. University of California Press, Berkeley

Falchi-Delitala L, Solinas V, Gessa C (1983) Variazioni stagionali quantitative e qualitative di olio essenziale e dei suoi fenoli in *Thymus capitatus* Hofmgg. et LK ed *Thymus herba-baroni* Loisel. Fitoterapia 54:87–96

Holzinger KJ, Harman HH (1941) Factor analysis. University of Chicago Press, Chicago

Hotelling H (1933) Analysis of a compolex of statistical variables into principal components. J Educ Psychol 24:417–441

Jalas J (1971) Notes on *Thymus* L. (Labiatae) in Europe. I. Supraspecific classification and nomenclature. Bot J Linn Soc 64(2):199–215

Jalas J, Kaleva K (1970) Supraspezifische Gliederung und Verbreitungstypen in der Gattung *Thymus* L. (Labiatae). Feddes Repert 81:93–106

Janssen AM, Scheffer JJC, Baerheim Svendsen AB (1987) Antimicrobial activity of essential oils: a 1976–1986 literature review. Aspect of test methods. Planta Med Rev: 395–398

Kaiser HF (1958) The varimax criterion for analytic rotation in factor analysis. Psychometrika 23:187–200

Kasumov F Yu (1981) Composition of essential oils of *Thymus fominii* and *Thymus transcaucasicus*. Khim Prir Soedin 5:665–666

Kasumov F Yu, Farkhadova MT (1986) The composition of *Thymus karamarianicus* essential oil. Khim Prir Soedin 5:642–643

Kendall MG (1957) A Course in Multivariate Analysis. Charles Griffin, London

Lawrence BM (1982) Progress in essential oils. Perfumer Flavorist 7:35–40

Mardia KV, Kent JT, Bibby JM (1979) Multivariate analysis. In: Birnbaum ZW, Lukacs E (eds) Probability and mathematical statistics. Academic Press, London

Morales R (1986) Taxonomía de los géneros *Thymus* (excluída la sección *Serpyllum*) y *Thymbra* en la Península Ibérica. Ruizia 3:7–11

Muñoz F (1987) Plantas medicinales y aromáticas. Estudio, cultivo y procesado. Mundi-Prensa, Madrid

Novruzova ZA, Kasumov F Yu (1987) Anatomy of Caucasian representatives of the genus *Thymus* L. (Lamiaceae) in relation to the essential oil composition. Izv Akad Nauk Az SSR, Ser Biol Nauk 6:18–24

Papageorgiou VP, Argyriadou N (1981) Trace constituents in the essential oil of *Thymus capitatus*. Phytochemistry 20:2295–2297

Papageorgiou VP, Vassilios P (1980) GLC-MS computer analysis of the essential oil of *Thymus capitatus*. Planta Med 40:29–33

Pearson K (1901) On lines and planes of closet fit to systems of points in space. Philos Mag 2:559–572

Press SJ (1972) Applied multivariate analysis. Holt, Rinehart and Winston, New York

Seal HL (1964) Multivariate statistical analysis for biologists. Methuen, London

Solinas V, Gessa C, Falchi-Delitala L (1981) High-performance liquid chromatographic analysis of carvacrol and thymol in the essential oil of *Thymus capitatus*. J Chromatogr 219:332–337

Stahl E (1982) The essential oil from *Thymus praecox* ssp. *articus*. In: Margaris N, Koedam A, Vokou D (eds) World crops: production, utilization, and description, vol 7. Aromatic plants: basic and applied aspects. Martinus Nijhoff, The Hague, pp 203–206

Takeuchi K, Yanai H, Murherjee BN (1982) The foundations of multivariate analysis. A unified approach by means of projection onto linear subspaces. Wiley, New Delhi

Trease GE, Evans WC (1986) Tratado de farmacognosia 12 edn. Interamericana, Madrid

van de Geer JP (1971) Introduction to multivariate analysis for the social sciences. WH Freeman, San Francisco

Velenovsky J (1906) Vorstudien zu einer Monographie der Gattung *Thymus* L. Bot Zentralbl Beih 19(B 2):217–287

Adsorption-Thermal Desorption-Gas Chromatography Applied to the Determination of Wine Aromas

M.R. SALINAS FERNANDEZ and G.L. ALONSO DIAZ-MARTA

1 Introduction

Headspace techniques have been frequently used to study volatile components in foods. These techniques are nondestructive and, since headspace chemical composition is quite similar to that of food aromas detected by the human nose, a correlation can be established between chemical and sensorial analyses.

In headspace sampling, two major techniques arise: static and dynamic. In the former, the vapor phase is found to be in equilibrium with the solute; in the latter this equilibrium is continually being altered. Dynamic techniques can be further divided into sweep techniques, in which carrier gas circulates on the surface of the sample, and purge techniques, where the gas circulates throughout the sample.

With dynamic techniques, the isolated volatiles are first retained in a trap and then concentrated. The retention phase can take place in cryogenic traps and the concentration phase in different varieties of traps, with capillary traps subject to low temperatures often being used (Badings et al. 1985a,b; García-Martín et al. 1993; García-Jares et al. 1995). Other methods of retaining volatiles also exist, especially the frequently used adsorbent trap. Here, the concentration process can be carried out by using a cryogenic trap previous to chromatographic injection (Arora et al. 1995) or by using the chromatographic column itself as a cryogenic trap (on column method). When this second method is used, the chromatographic column acts as a concentration trap and this same column can function as a retention-concentration trap. In any of the above techniques, the column allows the cryofocusing of solutes (Jennings and Rapp 1983; Burger and Munro 1986; Werkhoff and Bretschneider 1987a,b; Reglero et al. 1990; Coleman 1992; Manura and Hartman 1992; Leino et al. 1993; Vallejo-Cordoba and Nakai 1993; Linforth et al. 1994; Yang and Min 1994). Another way of concentrating volatiles retained in adsorbent traps is through the introduction of another adsorbent trap at low temperature (Salinas et al. 1994).

Volatiles concentrated in cryogenic or adsorbent traps require posterior recuperation, usually achieved through thermal desorption.

In wine sampling, static headspace techniques have been used in various research studies (Bertuccioli and Montedoro 1975; Noble et al. 1980; Russo and Goretti 1987; Shimoda et al. 1993), although these techniques have not been widely accepted. This is partly because these static techniques are appropriate only for

Modern Methods of Plant Analysis, Vol. 19
Plant Volatile Analysis
Edited by H.-F. Linskens and J.F. Jackson

highly volatile compounds found in headspace in detectable concentrations. Nevertheless, small concentrations of low-volatility compounds in wines are present in headspace, but these compounds cannot be detected with these techniques, even though they do contribute decisively to the olfactory perception of wine aromas. Furthermore, water and ethanol present in important proportions during the vapor phase negatively affect the chromatographic process.

In contrast, different varieties of dynamic headspace techniques have been commonly used in wine sampling or in similarly composed synthetic matrices. One reason is because the introduction of traps allows us to choose the group of compounds to be determined and also permits the elimination of compounds which interfere with chromatographic analysis, e.g. ethanol and water. Important studies in this area have been carried out by Rapp and Knipser (1980); Ducruet (1984); Burger and Munro (1986); Kallio (1991); Leino et al. (1993); Darriet (1993); Lubbers et al. (1993, 1994); Fondville-Bagnol et al. (1995) and García-Jares et al. (1995). Studies on the advantages and disadvantages of using headspace techniques for the isolation of volatiles and their comparison with other techniques have also been carried out by Etievant et al. (1986) and by Blanch et al. (1991), on wine and grape juice, respectively.

Techniques which utilize the isolation of volatiles by static or dynamic headspace (sweeping or purging), concentration in cryogenic or adsorbent traps and thermal desorption with direct transfer to the chromatography column are generally termed purge and trap techniques. They imply the availability of functional automated systems, a wide range of which are presently found on the market. Although traditionally termed as headspace techniques, a recent trend refers to these techniques independently (Núñez and Maarse 1986; Coleman 1992; Abeel et al. 1994; Cadwallader and Xu 1994; Voice and Kolb 1994; Arora et al. 1995).

The technique dealt with in this chapter is called adsorption-thermal desorption-gas chromatography. It includes the processes of isolation and concentration of volatiles performed in the three classic stages of purging, concentration, and injection.

In the first stage, the volatiles are separated by bubbling an inert gas (purging) through the sample. They are then adsorbed in an initial trap containing an adsorbent.

In the second stage, the temperature increase in this initial trap (first thermal desorption) releases the volatiles that are then concentrated in the next trap subjected to low temperatures.

In the third stage, the temperature in this second trap is raised (second thermal desorption) to desorb the volatiles that are immediately injected into the gas chromatograph.

The term adsorption is used because it describes the net physical process that takes place in the purge and concentration part of the analysis.

2 Adsorption

2.1 Adsorbent

Many adsorbents which meet the necessary requirements for correct thermal desorption are available on today's market. These requirements include: an affinity for a wide range of compounds, a high capacity for adsorption at room temperature, reverse adsorption upon temperature increase, high thermal stability, low affinity for humidity and carbon dioxide, chemical inertia, uniform particle size, and no aptitude for dust formation.

The adsorbents most used for thermal desorption techniques are those belonging to the Tenax series (polymer based on 2,6-diphenyl-p-phenylene oxide). Unlike other polymers, e.g., the Chromosorbs and Porapaks, these adsorbents remain stable at temperatures of up to 380–400 °C (Sakodynskii et al. 1974; Bertsch et al. 1975) and are therefore particularly suited for heat desorption. They are also appropriate for samplings which analyze a broad group of organic compounds, as in the case of wine. Preparation for use is also quite simple, since the only requirement necessary is an elevated temperature with a current of inert gas (McLeod and Ames 1986). Tenax is commercialized as Tenax GC, Tenax TA, and Tenax GR (this last composed of 70% Tenax and 30% graphitized carbon approximately). The main difference between Tenax CG and Tenax TA is the lower production of artifacts causing background noise when Tenax Ta is used (Hutte et al. 1984; MacLeod and Ames 1986). Sunesson et al. (1995) point out that Tenax TA and Tenax GR have equivalent adsorption properties, though they consider Tenax TA as having the best qualities for complex mixtures of volatiles.

Adsorption in Tenax is reduced as analite polarity increases. Thus, alcohols, especially those with low molecular mass, are little retained; and water, besides having a minimum effect on other compounds, is practically not adsorbed (Janák et al. 1974; Brown and Purnell 1979; Sunesson et al. 1995). Tenax has a very small specific surface, between 20 and 35 mg^{-1}, so adsorption of the most volatile solutes is also small. All these properties make the polymers in the Tenax series quite appropriate for hydroalcoholic dissolutions such as wine or analogous drinks. Proof of this is the large number of published articles which have used Tenax (Noble et al. 1980; Rapp and Knipser 1980; Núñez and Maarse 1986; Loyola et al. 1989; Blanch et al. 1991; Tabera et al. 1991; Salinas et al. 1994).

2.2 Theoretical Aspects

If we assume that the adsorption tube functions as a chromatographic column, the volume of carrier gas needed for the complete desorption of a determined compound is related to the retention volume (Rv) of this compound. In turn, the retention volume is related to the temperature as follows:

$$\log Rv = A/T + B,$$

where Rv is the retention volume expressed in l/g at temperature T; T is the desorption temperature in K; and A and B are constants (Glover 1991). The measurement of the Rv of a compound is derived from the area of the highest peak obtained in the chromatographic analysis (Vidal Madjar 1978).

Assuming that the adsorption tube behaves as a chromatographic column, the passage of an adsorbed compound through this tube results in the complete loss of this compound. The volume of gas passed through the tube when the adsorbed compound begins to emerge is known as breakthrough volume (BTV). BTV varies with such parameters as adsorption temperature, vapor concentration, and flow speed. The BTV of each compound can be measured directly by using model dissolutions (Brown and Purnell 1979; Namiesnik et al. 1981; Przyjazny et al. 1982; Witkiewicz et al. 1991). However, this method is quite laborious and time-consuming, and is therefore only used to confirm the results of methods which use indirect measurements. BTV and Rv are related, and BTV can be measured indirectly from Rv when the number of theoretical plates in the tube is known (Brown and Purnell 1979).

Many authors have used the terms BTV and Rv indistinctly, even though BTV may be less than Rv. While BTV indicates the volume of gas necessary for desorption of the retained compound to begin, the Rv indicates the point where desorption is at a maximum.

When performing a quantitative analysis, carrier gas volume should be lower than Rv. If the contrary is true, we could obtain peaks with the same areas in the chromatographic analysis of samples that contain different concentrations of a certain compound. Thus, we could erroneously conclude that the concentration of this compound is the same in both analyses. This is why, when doing quantitative analyses, we should work with a volume of carrier gas that avoids this saturation. Brown and Purnell (1979) labeled this volume the safe sampling volume, defining it as 50% of Rv. These authors also demonstrated a relation between the boiling temperature of a compound and its Rv in the equation:

$$\mathrm{BP} = \mathrm{A}' \log \mathrm{Rv} - \mathrm{B}',$$

where A′ and B′ are constants and BP is the boiling temperature in °C. This study was done using Tenax GC and, by extrapolation, the authors were able to obtain the Rv of a large number of compounds. The authors also deduced that retention volumes are doubled each time temperature falls by 10 °C. Table 1 shows results different researchers have obtained for Rvs in Tenax from a group of compounds normally present in wines.

Flow speed of carrier gas affects the efficiency of the adsorption process. Brown and Purnell (1979) found the best flow rate to be 50 ml min^{-1} for adsorption tubes of 4.5 mm i.d. containing 0.13 g of Tenax GC. They advise the use of flow speeds between 10 and 100 ml min^{-1} and also consider values of 5 to 600 ml min^{-1} as acceptable if the safe sampling volume is used. In addition, they observed that there is not an important Rv variation of any compound in this wide range upon changing the flow speed.

Bertoni et al. (1981) found one relation between concentration in the vapor phase and the BTV of a compound as such:

Table 1. Retention volumes in Tenax for a group of compounds normally present in wines

Compound	Retention volume ($l g^{-1}$)	Temperature (°C)	Reference
Acetaldehyde	1.23	20	Brown and Purnell (1979)
	2	21	Krost et al. (1982)
Acetone	5.4	20	Brown and Purnell (1979)
	12	21	Krost et al. (1982)
	2.9	25	Janák et al. (1974)
Methanol	0.148	25	Janák et al. (1974)
	0.362	20	Manura (1995)[a]
Ethanol	1.8	20	Brown and Purnell (1979)
	0.83	25	Janák et al. (1974)
1-Propanol	8.4	20	Brown and Purnell (1979)
	5.7	25	Janák et al. (1974)
	11.0	20	Manura (1995)[a]
1-Pentanol	2.10	20	Manura (1995)[a]
Isobutanol	28	20	Brown and Purnell (1979)
1-Hexanol	1.80	20	Manura (1995)[a]
Octanol	13.8	20	Brown and Purnell (1979)
	31.60	20	Manura (1995)[a]
Methyl acetate	5.4	20	Brown and Purnell (1979)
Ethyl acetate	35.4	20	Brown and Purnell (1979)
	27.6	20	Janák et al. (1974)
	30.1	25	Holzer et al. (1977)
Propyl acetate	192	20	Brown and Purnell (1979)
Butyl acetate	846	20	Brown and Purnell (1979)
Ethyl propionate	117	25	Holzer et al. (1977)
d-Limonene	10.05	25	Salinas et al. (1994)
Benzaldehyde	3507	21	Krost et al. (1982)

[a] Breakthrough volume or retention volume according to the author.

$$BTV = -K \log C + D,$$

where K and D are constant and C is the concentration of a compound in the vapor phase. These authors found important changes in BTV when concentration is very high (more than 100 ppm for the compounds studied), or when large quantities of other compounds, capable of being strongly adsorbed, are found together with low analite concentrations.

3 Thermal Desorption

The reversibility of adsorption due to temperature increase is a condition that adsorbents must meet when working with thermal desorption. Simple thermal desorption, i.e., heating of the adsorption tube through which carrier gas passes in the direction opposite to adsorption, normally leads to the presence of a volume of analite too great to be injected directly into the gas chromatograph column. To increase the efficiency of the capillary column, it is of particular concern that the sample volume be as small as possible. This can be achieved by the introduction of a second adsorbent trap where analite can be concentrated. Using the theoretical considerations mentioned in the previous section, it can be seen that when the first trap is kept at a high enough temperature, all its analites are desorbed. Likewise, if the second trap is kept at a sufficiently low temperature, a small amount of adsorbent could retain all the analites desorbed in the first trap. Hence, the necessary reduction in sample volume will be achieved with this second trap.

Werkhoff and Bretschneider (1987a,b) using 100 ng of various aromatic compounds, including d-limonene, injected into 0.6 g of Tenax TA, obtained recuperations of almost 100% for the majority of compounds when desorption was carried out at 250 °C. Cao and Hewith (1993), when working with 40–50 ng of C5–C10 hydrocarbons injected into 0.16 g of Tenax TA and into 0.2 g of Tenax GR, achieved complete desorption at temperatures of 200 and 240 °C, respectively.

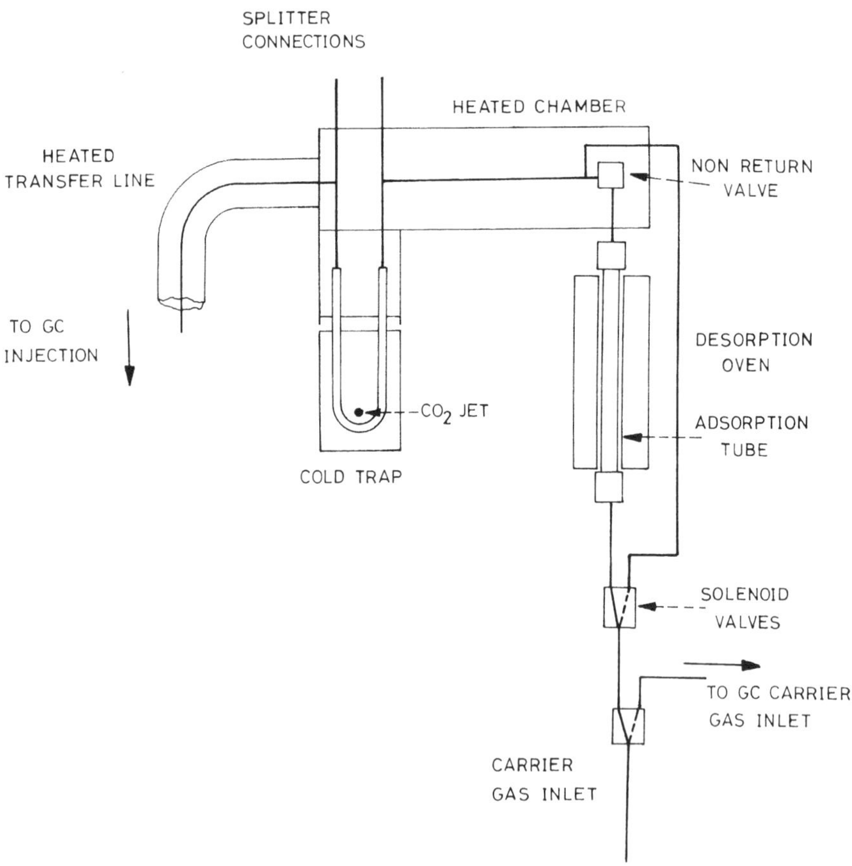

Fig. 1. Flow diagram of TD-4 thermal desorption unit. (After Glover 1991)

Instrumental devices are available on today's marker for thermal desorption of analites previously retained in an adsorption tube, their concentration in an adsorbent trap and a second thermal desorption with direct injection into the chromatographic column. A Spantech TD 4-Thermal Desorber (Perkin-Elmer) was used for our analyses. Figure 1 shows the gas flow pattern of the thermal desorber.

4 Adsorption-Thermal Desorption-Gas Chromatography Applied to the Determination of Wine Aromas

The many volatile compounds encountered in wines pertain to an ample group of chemical families and are found in very different concentrations. When using gas chromatography to analyze these volatiles, the sensitivity of the detector will determine whether the isolation and concentration of these compounds is indeed necessary. One direct injection of wine into the appropriate chromatographic column is enough to determine a large number of volatiles (González Raurich 1985; Reglero et al. 1986; Salinas et al. 1990a,b). These directly analyzed compounds are those which are usually found in high concentrations, superior to the sensitivity of the detector. Ethanol and water belong to this group of majority components, in addition to the superior alcohols and other compounds such as ethyl acetate, acetaldehyde, and dietilacetal. These compounds, which have boiling points of up to 145 °C, have low retention volumes in Tenax. In wines, however, a significant part of the volatile compounds is found in concentrations lower than the sensitivity of the detector. These, called minority volatiles, pertain to the medium volatility group of wines (with boiling points between 145 and 275 °C). The retention volumes of these minority volatiles are superior to those of majority volatiles, which is why saturation of the latter can be reached if they are analyzed together. In order to avoid this saturation, a large quantity of Tenax can be utilized, something which instrumental devices do not permit. Moreover, a problem would arise, since the complete liberation of volatiles in the first thermal desorption could not take place.

With this in mind, we decided to analyze only one group of these minority volatiles which play such an important role in the evaluation of aromatic quality in wines. The study we carried out to perfect the technique of analyzing minority volatiles is explained as follows.

When isolating volatiles in a sample, one of the main parameters influencing results is the quantity of the sample. This quantity will depend on the concentration of the scarcest compounds. After various tests with different volumes, we utilized 50 ml of wine deposited in a 150-ml purging vessel. This volume was considered as most appropriate for the quantity of adsorbent and the dimensions of the tube. The gas used was N_2, inert and very pure in order to avoid contamination or reactions with the sample.

The higher the temperature, the higher the proportion of volatiles in the vapor phase. Nevertheless, in wine samples, we find that temperatures higher than 30 °C can alter original characteristics (Di Stefano, 1992). In our analyses, temperature for the entire purging vessel was kept at 30 °C.

The flow of carrier gas and purging time are determined by adsorption, but very low flows lengthen the process unnecessarily, while excessively high ones hamper the retention of volatiles in the adsorbent. Purging time depends on gas flow and, for the same flow, a loss of volatiles may take place when retention volume is surpassed due to excessively long purging time. For our analyses, we selected a flow of carrier gas of 67 ml min^{-1}. We also obtained an appropriate purging time of 20 min for this flow, as described below.

The metal adsorption tube measured 88.9 mm × 6.35 mm o.d. × 4.50 mm i.d. and contained 0.2 g Tenax GC (35–60 mesh; Alltech, Illinois, USA). The adsorbent was conditioned by passing N_2 through the tube at 350 °C with a flow rate of 30 ml min^{-1} for 6 h. For isolation of the volatile, the tube was connected to the instrumental device outlined in Fig. 2. The temperature of the tube was 25 °C. After the isolation period, dry N_2 was passed through the tube to eliminate any remains of ethanol and water. The flow rate of the gas was always 67 ml min^{-1}, which is within the range recommended for the tube dimensions used and for Tenax (Brown and Purnell 1979).

The packed tube was introduced into a thermal desorber coupled to a PE 8310 gas chromatograph with a PTV injector and flame ionization detector. The chromatograph was equipped with an SGE 50 m × 0.22 mm i.d. fused silica capillary column coated with a 0.25-μm layer of cross-linked BP-21. Table 2 shows programmed parameters for thermal desorption and gas chromatography. Methyl octanoate (Merck, Darmstadt, Germany) was used as the internal standard in a proportion of 3 μl of a 5% solution in ethanol (Merck, Darmstadt, Germany) to 50 ml of sample. Identification was made by comparison of the retention times with those of the reference compounds (Integration system PE Nelson-2600) and Q-Mass-910 (IE 70 eV) (PE) by comparison of spectra with those of the NIST/EPA/MSDC mass spectral database version 3.02.

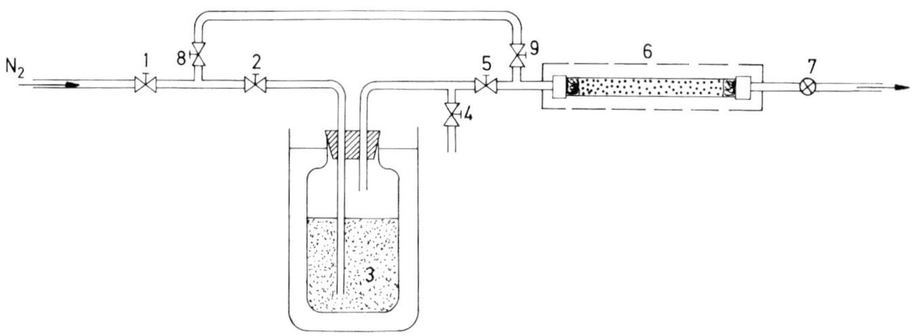

Fig. 2. Adsorption scheme: *1,2,4,5,8,* and *9* stop valve; *3* 50 ml of sample in a 30 °C bath; *6* metal tube containing 0.2 g of Tenax GC at 25 °C; *7* flow control. *Line A* Volatiles adsorption, 1 → 3 → 5 → 6 → 7; *line B* ethanol and water elimination, 1 → 8 → 9 → 6 → 7

Table 2. Program parameters in thermal desorption and gas chromatography

Thermal desorption
Oven temperature: 300 °C
Desorption time: 5 min
Cold trap low temperature: –30 °C
Cold trap high temperature: 350 °C
Injection time: 40 s
Transfer line temperatue: 200 °C
Trap adsorbent: Tenax GC (0.02 g)
Flow of carrier gas: 31 ml min^{-1}
Gas chromatography
Temperature programme: 50 °C, 2.5 °C min^{-1}, 180 °C, 8 min
Injector program: –0.02 relay 1 on, –0.01 PTV 200 °C
Gas carrier: He at 30 psig
Split ratio: 1:70
Detector: 230 °C

The adsorption tube can be reused after each analysis. We have observed that adsorption tubes used over 200 times show no significant differences to tubes used only once (Huerta 1996). To assure the absence of possible residues, we use a simple instrument presently in the process of being patented (Salinas and Alonso 1995). This instrument enables rapid cleaning and activation of the adsorbent.

All the automatic systems involved in analysis (concentration trap, transference line, injector, column, etc.) should be meticulously cleaned to avoid the presence of residues and artifacts.

In order to optimize analytic conditions, we used a dissolution of d-limonene (Chem Service, West Chester, California, USA) in ethanol 12% (v/v) since it has one of the lowest boiling points (178 °C) of the compounds present in the fraction of minority volatiles of wines. First of all it is necessary to know the bubbling time of gas in the sample (flow rate of 67 ml min^{-1}) which allows us to determine this compound without adsorbent saturation. Figure 3 shows the area of peaks obtained with solutions with a minimum of 3 ppm against the nitrogen of adsorption time. Retention volume was calculated to be 2010 ml for 0.2 g of Tenax at 25 °C.

These results were applied to dissolutions containing d-limonene, benzaldehyde, 2-phenethyl acetate, and octanoic acid (Chem. Service, West Chester, California, USA) in ethanol 12% v/v. These compounds are located all along the chromatogram obtained with the same column as described in this chapter. The d-limonene is found at the beginning, benzaldehyde and 2-phenethyl acetate in the center, and octanoic acid at the end. To understand whether different concentrations of the compounds remain within the linearity zone or in the saturation zone, four solutions of 12, 6, 4, and 3 ppm were adsorbed using adsorption of 10, 20, and

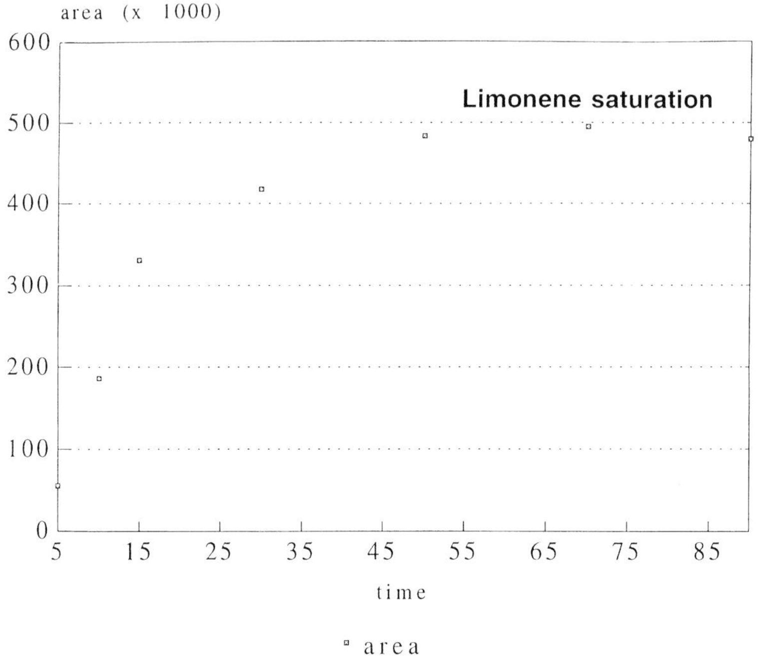

Fig. 3. Evolution of peak areas for d-limonene with increasing adsorption times

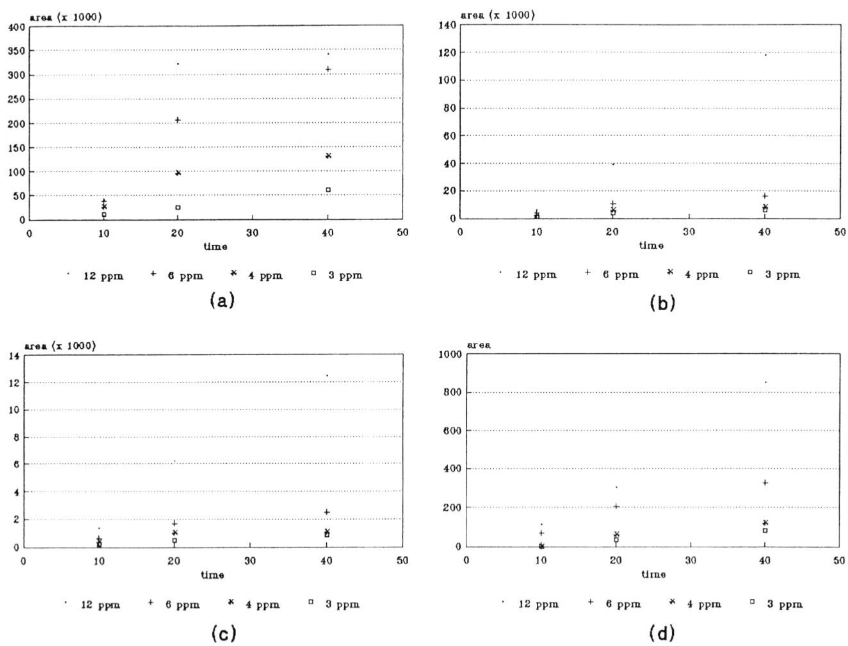

Fig. 4a–d. Evolution of peak areas for **a** d-limonene; **b** benzaldehyde; **c** 2-phenethylacetate; and **d** octanoic acid with increasing adsorption times and different concentrations

40 min. In all cases, the variation coefficients were in the order of 3% (n = 3), although less for the octanoic acid (12%). Figure 4 shows graphs for the above compounds. Octanoic acid was not detectable after 10 min of adsorption in the solutions of 3 and 4 ppm. The rest of the compounds showed linear increases of peak areas when the time was increased, except d-limonene in the 12-ppm solution after 20 min of adsorption. The graphs of compound area/internal standard against concentration for each of the adsorption times gave calibrations with correlation coefficients greater than 0.99 and significance levels lower than 0.1% for all compounds, except d-limonene at 40 min and octanoic acid at 10 min.

Table 3. Concentrations in $mg\,l^{-1}$ (c) and variation coefficients in % (cv) (n = 3) for A, B, and C dissolutions (see page 187)

Compound	A		B		C	
	c	cv	c	cv	c	cv
C_6 Compounds						
1-Hexanol	0.97	2.30	0.48	1.07	0.24	0.21
Z 3-hexen-1-ol	0.99	3.87	0.50	3.79	0.25	5.50
E 2-hexen-1-ol	0.98	2.67	0.49	2.06	0.24	3.06
Alcohols						
1-Pentanol	0.96	5.29	0.48	1.61	0.24	9.65
1-Octanol	0.98	4.40	0.49	3.44	0.25	10.13
2-Phenyl ethanol	2.40	10.73	1.20	11.73	0.60	10.74
Esters						
Isoamyl acetate	1.04	4.63	0.52	4.63	0.26	3.14
Ethyl hexanoate	1.04	1.93	0.52	2.60	0.26	3.46
Hexyl acetate	1.04	1.77	0.52	2.36	0.26	3.71
Ethyl pyruvate	1.25	6.47	0.62	6.80	0.31	5.50
Ethyl heptanoate	1.03	2.30	0.52	1.07	0.26	0.21
Ethyl lactate	1.34	4.01	0.62	4.32	0.31	6.63
Ethyl octanoate	1.03	2.92	0.51	2.00	0.26	3.95
Isoamyl lactate	1.13	0.89	0.56	3.61	0.28	0.89
Diethyl succinate	2.47	8.87	1.23	2.88	0.62	9.01
Ethyl decanoate	2.05	4.18	1.03	4.37	0.51	5.07
2-Phenylethylacetate	2.53	9.63	1.26	4.04	0.63	4.69
Ethyl dodecanoate	2.05	7.93	1.02	3.69	0.52	2.23
Terpenes						
d-Limonene	0.98	0.23	0.49	1.09	0.24	6.50
Citronellol	0.95	12.96	0.47	12.66	0.24	12.65
Linalool	1.01	0.49	0.51	4.17	0.25	4.08
α-Terpineol	2.22	4.07	1.11	4.51	0.55	8.57
Geraniol	2.09	2.91	1.04	7.27	0.52	3.44
Acids						
Butyric acid	1.14	3.70	0.57	8.81	0.28	6.86
Hexanoic acid	2.20	3.93	1.10	4.34	0.55	4.42
Heptanoic acid	2.18	10.89	1.10	10.47	0.54	10.44
Octanoic acid	2.14	9.93	1.07	10.20	0.53	9.44
Others						
Benzaldehyde	1.22	2.57	0.61	2.59	0.31	0.74

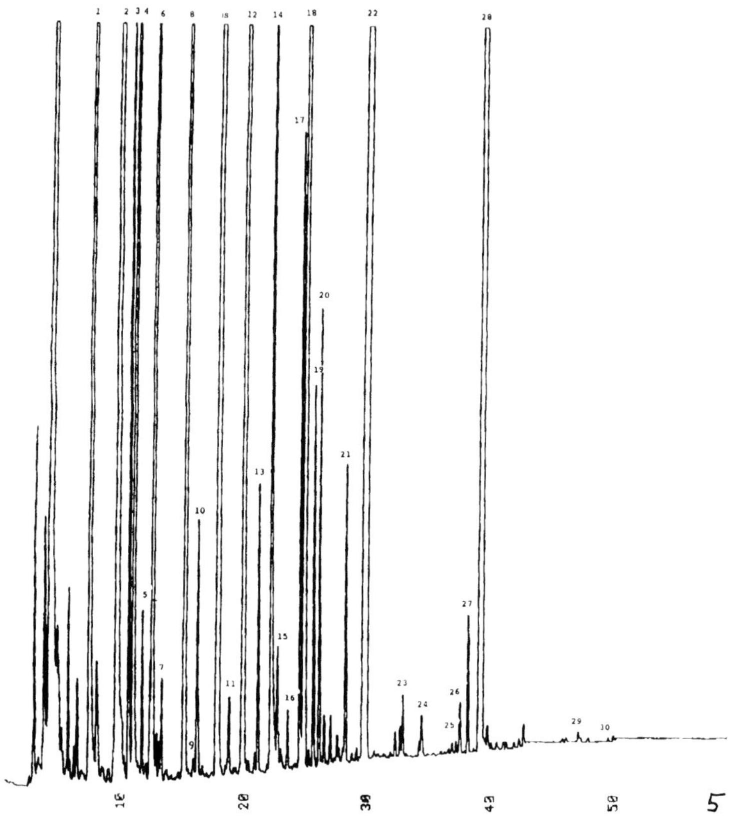

Fig. 5. Chromatogram for a standard solution composed of the following: *1* isoamyl acetate; *2* d-limonene; *3* *trans*-2-hexenal; *4* ethyl hexanoate; *5* 1-pentanol; *6* hexyl acetate; *7* ethyl piruvate; *8* ethyl heptanoate; *9* 1-hexanol; *10* ethyl lactate; *11* Z 3-hexen-1-ol; *12* E 2-hexen-1-ol; *13* ethyl octanoate; *14* 1-heptanol; *15* citronellol; *16* benzaldehyde; *17* linalool; *18* isoamyl lactate; *19* 1-octanol; *20* hexanoic acid; *21* butyric acid; *22* ethyl decanoate; *23* diethyl succinate; *24* α-terpineol; *25* 2-phenethylacetate; *26* ethyl dodecanoate; *27* geraniol; *28* 2-phenyl ethanol; *29* heptanoic acid; *30* octanoic acid

On the basis of the above results, a time of 20 min was chosen as the most suitable for purposes of this study, although this is slightly superior to the safe sampling volume defined, since the fraction compounds of wine aromas seldom exceed 6 ppm and there should be no risk of saturating the Tenax. By using the 20-min time, smaller peak areas were more easily detectable.

The parameters previously selected were applied to a dissolution containing the following compounds: 1-hexanol, 1-pentanol, 1-octanol, 2-phenylethanol, citronellol, α-terpineol, ethyl hexanoate, diethyl succinate, 2-phenylethylacetate, d-limonene, benzaldehyde, butyric acid, hexanoic acid, heptanoic acid, octanoic acid (Chem. Service, West Chester, California, USA), Z 3-hexen-1-ol, E 2-hexen-1-ol, linalool, geraniol, isoamyl acetate, hexyl acetate, ethyl pyruvate, ethyl heptanoate, ethyl lactate, ethyl octanoate, isoamyl lactate, ethyl dodecanoate (Aldrich, Madrid, Spain) and ethyl decanoate (Merck, Darmstadt, Germany).

We used three dissolutions containing concentrations between 0.95 and 2.47 ppm (dissolution A), between 0.47 and 1.26 ppm (dissolution B), and between 0.24 and 0.63 ppm (dissolution C). Figure 5 shows one of the chromatograms obtained with dissolution B. In Table 3 we can observe the concentration of each compound in each one of the dissolutions and the coefficient of variation (n = 3) obtained from the relation between the chromatographic peak areas and the peak area of the internal standard. Variation coefficients are seen to be close to 5% in the majority of cases, but they increase by over 10% in compounds such as citronellol, 2-feniletanol, and heptanoic and octanoic acids. Greatest variations are found in

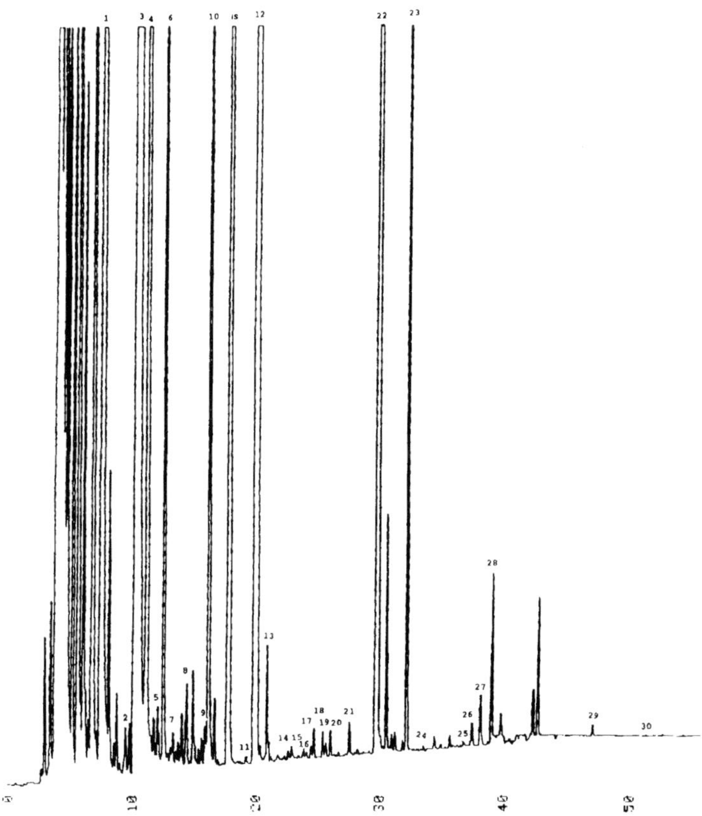

Fig. 6. Chromatogram for a traditional red wine composed of the following: *1* isoamyl acetate; *2* d-limonene; *3* *trans*-2-hexenal+ amylic alcohols; *4* ethyl hexanoate; *5* 1-pentanol; *6* hexyl acetate; *7* ethyl piruvate; *8* ethyl heptanoate; *9* 1-hexanol; *10* ethyl lactate; *11* Z-3-hexen-1-ol; *12* E 2-hexen-1-ol; *13* ethyl octanoate; *14* 1-heptanol; *15* citronellol; *16* benzaldehyde; *17* linalool; *18* isoamyl lactate; *19* 1-octanol; *20* hexanoic acid; *21* butyric acid; *22* ethyl decanoate; *23* diethyl succinate; *24* α-terpineol; *25* 2-phenethylacetate; *26* ethyl dodecanoate; *27* geraniol; *28* 2-phenyl ethanol; *29* heptanoic acid; *30* octanoic acid

the most diluted dissolutions since the chromatographic peaks are quite small and therefore their resolution is more difficult. In a later study, dissolutions containing concentrations between 0.25 and 0.10 ppm were used, resulting in variations similar to those obtained with previous dissolutions. The graphic representation of compound area/internal standard versus concentration of each compound in the three dissolutions indicated gives correlation coefficients higher than 0.99 in all cases and significance levels lower than 0.1%.

Table 4. Mean composition of volatiles (n = 3) in mgl^{-1} (x) and variation coefficients in % (cv) of a traditional red wine (1) and another wine undergoing carbonic maceration (2)

Compound	1		2	
	x	cv	x	cv
C_6 Compounds				
1-Hexanol	2.40	2.37	3.66	4.53
Z 3-hexen-1-ol	0.02	5.43	0.44	2.52
E 2-hexen-1-ol	0.87	4.89	0.02	8.71
Alcohols				
1-Pentanol	0.15	11.59	nd	
1-Octanol	0.02	8.32	0.49	3.44
2-Phenyl ethanol	0.08	10.02	0.10	8.67
Esters				
Isoamyl acetate	0.63	1.61	0.45	3.23
Ethyl hexanoate	0.86	0.93	1.13	2.04
Hexyl acetate	0.16	2.70	0.13	2.86
Ethyl pyruvate	0.05	8.64	1.80	4.68
Ethyl heptanoate	0.06	4.16	0.73	0.64
Ethyl lactate	8.83	4.80	9.10	7.63
Ethyl octanoate	0.61	3.92	0.05	6.14
Isoamyl lactate	0.06	1.25	0.02	3.38
Diethyl succinate	0.43	9.21	0.27	5.62
Ethyl decanoate	0.30	6.17	0.01	8.58
2-Phenylethylacetate	0.12	9.58	0.17	4.61
Ethyl dodecanoate	0.16	1.95	0.05	2.60
Terpenes				
d-Limonene	0.02	4.81	0.13	1.56
Citronellol	0.05	11.04	0.03	14.84
Linalool	0.02	4.46	nd	
α-Terpineol	0.16	9.07	0.06	6.51
Geraniol	0.29	5.41	0.20	4.35
Acids				
Butyric acid	0.07	10.80	0.12	2.68
Hexanoic acid	0.54	2.95	0.24	5.30
Heptanoic acid	1.93	11.13	nd	
Octanoic acid	1.30	7.57	1.12	9.46
Others				
Benzaldehyde	0.12	1.54	0.48	1.53

Our studies in the field of enology focus on red wines produced by carbonic maceration (Flanzy et al. 1990). The evolution of aromas throughout the fermentation process (Salinas et al. 1990a,b, 1992) and also throughout the aging process in wooden barrels (Salinas et al. 1996) are two of the many aspects included in our research. Table 4 shows results of aroma analyses for wines elaborated with traditional vinification and carbonic maceration. Figure 6 gives one of the chromatograms for wine of traditional vinification.

The attempt to combine organoleptic analysis with aromatic composition, coupled with the possibility of carrying out aroma analysis at the winery itself, has led us to the use of a nondestructive sampling technique, easy to apply in little time. The technique developed in this chapter requires a continual effort on our part for improvement. Current goals seek to introduce a new adsorbent which would increase the number of volatile compounds determined, optimize analytic conditions for the compounds present in low concentrations, and combine chromatography with olfactometric analysis.

Acknowledgments. The authors would like to thank the Comisión Interministerial de Ciencia y Tecnología of Spain for their help in financing.

References

Abeel SM, Vickers AK, Decker D (1994) Trends in purge and trap. J Chromatogr Sci 32:328–338
Arora G, Cormier F, Lee B (1995) Analysis of odor-active volatiles in cheddar cheese headspace by multidimensional GC/sniffing. J Agric Food Chem 43:748–752
Badings HT, de Jong C, Dooper RPM, de Nijs RCM (1985a) Rapid analysis of volatile compounds in food products by purge and cold trapping capillary gas chromatography. In: Adda J (ed) Progress in flavor research. Elsevier. Amsterdam
Badings HT, de Jong C, Dooper RPM (1985b) Automatic system for rapid analysis of volatile compounds by purge and cold trapping-capillary gas chromatography. J High Resolut Chromatogr Commun 8:755–763
Bertoni G, Bruner F, Liberti A, Perrino C (1981) Some critical parameters in collection, recovery and gas chromatographic analysis of organic pollutants in ambient air using light adsorbents. J Chromatogr 203:263–270
Bertsch W, Anderson E, Holzer G (1975) Determination of organic volatiles in air pollution studies for characterization of profiles. J Chromatogr 112:701–712
Bertuccioli M, Montedoro G (1975) Impiego di polimeri per la concentrazione di costituenti volatili dello "spazio di testa" dei vini e loro diretta analisi gas-cromatografica. Riv Sci Tec Aliment Nutr Um 5:39–43
Blanch GP, Reglero G, Herraiz M, Tabera J (1991) A comparison of different extraction methods for the volatile components of grape juice. J Chromatogr Sci 29:11–15
Brown RH, Purnell CJ (1979) Collection and analysis of trace organic vapour pollutants in ambient atmospheres. J Chromatogr 178:79–90
Burger BV, Munro Z (1986) Quantitative trapping and thermal desorption of volatiles using fused-silica open tubular capillary traps. J Chromatogr 370:449–464
Cadwallader KR, Xu Y (1994) Analysis of volatile components in fresh grapefruit juice by purge and trap/gas chromatography. J Agric Food Chem 42:782–784
Cao XL, Hewitt N (1993) Thermal desorption efficiencies for different adsorbate/adsorbent systems typically used in air monitoring programmes. Chemosphere 27:695–705

Coleman WM (1992) Automated purge-and-trap-gas chromatography analysis of headspace volatiles from natural products. J Chromatogr Sci 30:159–163

Darriet P (1993) Recherches sur l'arôme et les précurseurs d'arôme du sauvignon. Applications technologiques. Thèse, Univ Bordeaux

Di Stefano R (1992) Problemas relacionados con la estabilidad del aroma de los vinos y de las bebidas aromáticas derivadas del vino. En: Segundo Congr Int de Tecnología y Desarrollo Alimentarios, Murcia, España

Ducruet V (1984) Comparison of the headspace volatiles of carbonic maceration and traditional wine. Lebensm Wiss Technol 17:217–221

Etievant P, Maarse H, Van den Berg F (1986) Wine analysis: study and comparison of techniques developed for the study of volatile constituents. Chromatographia 21:379–386

Flanzy C, Flanzy M, Benard P (1990) La vinificación por maceración carbónica. Madrid Vicente, Madrid

Fondville-Bagnol A, Sarris J, Bayonove C, Flanzy C (1995) Méthode d'analyse de composés volatils de baies de raisin en hypoxie. Sci Aliment 15:359–366

García-Jares C, García-Martín S, Cela-Torrijos R (1995) Analysis of some highly volatile compounds of wine by means of purge and cold trapping injector capillary gas chromatography. Application to the differentiation of Riax Baixas Spanish white wines. J Agric Food Chem 43:764–768

Carcía-Martín MS, García-Jares CM, Cela-Torrijos R (1993) Optimization of purge and cold trap injection capillary gas chromatography techniques in the analysis of wine aromas. In: 15th Symp on Capillary chromatography, vol 1. Hüthig, Heidelberg, pp 421–426

Glover JH (1991) Thermal desorption in industrial hygiene and environmental analysis. Spantech Scientific Series 1. Spantech, New Delhi

González Raurich M, Reglero G, Herraiz M, Cabezudo MD (1985) El análisis íntegro de los vinos. I. Componentes volátiles mayoritarios. Aliment Equip y Tecnol, Madrid marzo-abril: 131–135

Holzer G, Shanfield H, Zlatkis A, Lubich HM (1977) Collection and analysis of trace organic emissions from natural sources. J Chromatogr 142:755–764

Huerta MD (1996) Evaluación de parámetros para la diferenciación de vinos de la Comunidad Autónoma de Madrid. Tesis doctoral, Universidad de Alcalá de Henares, España

Hutte RS, Williams EJ, Staehelin J, Hawthorne SB, Barkeley RM, Sievers RE (1984) Chromatographic analysis of organic compounds in the atmosphere. J Chromatogr 302:173–179

Janák J, Ruzickova J, Novák J (1974) Effect of water vapour in the quantitation of trace components concentrated by frontal gas chromatography on Tenax-GC. J Chromatogr 99:689–696

Jennings WG, Filsoof M (1977) Comparison of sample preparation techniques for gas chromatographic analysis. J Agric Food Chem 25:440–445

Jennings WG, Rapp A (1983) Sample preparation for gas chromatographic analysis. Hüthig, Heidelberg

Kallio H (1991) Method of sensitive analysis of wine headspace volatiles based on selective capillary column trapping. J Chromatogr Sci 29:438–443

Krost KJ, Pellizzari ED, Walburn SG, Hubbard SA (1982) Collection and analysis of hazardous organic emissions. Anal Chem 54:810–817

Leino M, Francis IL, Kallio H, Williams PJ (1993) Gas chromatographic headspace analysis of chardonnay and semillon wines after thermal processing. Z Lebensm Unters Forsch 197:29–33

Linforth RS, Savary I, Pattenden B, Taylor A (1994) Volatile compounds found in expired air during eating of fresh tomatoes and in the headspace above tomatoes. J Sci Food Agric 65:241–247

Loyola E, Martín-Alvarez PJ, Herraiz T, Reglero G, Herraiz M (1989) A contribution to the study of the volatile fraction in distillates of wines made from muscat grapes (pisco). Z Lebensm Unters Forsch 190:501–505

Lubbers S, Charpentier C, Feuillant M, Voilley A (1993) Interactions arôme-autres constituants du vin. Mise au point d'un système de dialuse à l'équilibre. In: Actes du Symp Int sur Connaissance aromatiques des cépages et qualité des vins, 3–14 février, Montpellier, pp 312–317

Lubbers S, Charpentier C, Feuillant M, Voilley A (1994) Influence of yeast walls on the behavior of aroma compounds in a model wine. Am J Enol Vitic 45:29–33

MacLeod G, Ames JM (1986) Comparative assessment of the artefact background on thermal desorption of Tenax GC and Tenax TA. J Chromatogr 355:393–398

Manura J (1995) Selection and use of adsorbent resins for purge and trap thermal desorption applications. Short path thermal desorption. Application Note SIS:32

Manura J, Hartman J (1992) Applications of a short path thermal desorption GC accessory. Am Lab, May: 46–52

Namiesnik J, Torres L, Kozlowski E, Mathieu J (1981) Evaluation of the suitability of selected porous polymers for preconcentration of volatile organic compounds. J Chromatogr 208:239–252

Noble AC, Flath RA, Forrey RR (1980) Wine headspace analysis. Reproducibility and application to varietal classification. J Agric Food Chem 28:346–353

Núñez AJ, Maarse H (1986) Headspace methods for volatile components of grapefruit juice. Chromatographia 21:44–48

Przyjazny A, Janicki W, Dhrzanowski W, Staszewski R (1982) Simple method for determination of breakthrough volumes for trace concentration of volatile organic compounds in liquids. J Chromatogr 245:256–260

Rapp A, Knipser W (1980) Eine neue Methode zur Anreicherung von Dampfkomponenten. Dargestellt am Beispiel des Weines. Chromatographia 13:698–702

Reglero G, Herraiz T, Herraiz M, Cabezudo MD (1986) Mixed micropacked columns designed for selective separation of fermentation products. Chromatographia 22:358–362

Reglero G, Herraiz T, Herraiz M (1990) Direct headspace sampling with on-column thermal focusing in capillary gas chromatography. J Chromatogr Sci 28:221–224

Russo MV, Goretti G (1987) Headspace gas chromatography analysis of table wine in glass and tetrabrik container. Ann Chim 77:755–761

Sakodynskii K. Panina L, Klinskaya N (1974) A study of some properties of Tenax, a porous polymer sorbent. Chromatographia 7:339–344

Salinas MR, Alonso GL (1995) Patente de invención Instrumento de limpieza y activación de adsorbentes de volátiles orgánicos N° solicitud: P 9500236. Universidad de Castilla-La Mancha

Salinas MR, Alonso GL, Navarro G, Pardo F (1990a) Aportación al estudio de la evolución de componentes aromáticos mayoritarios en vinificación por maceración carbónica de uvas monstrell. I. Alcoholes. Anal Bromatol 42:209–217

Salinas MR, Alonso GL, Navarro G, Pardo F (1990b) Aportación al estudio de la evolución de componentes aromáticos mayoritarios en vinificación por maceración carbónica de uvas monastrell. II. Esteres. Anal Bromatol 42:219–226

Salinas MR, Navarro G, Alonso GL, Esteban-Infantes FJ, Pardo F (1992) Constituyentes volátiles de un vino de maceración carbónica y de vinificación tradicional. Vitic Enol Prof 22:46–51

Salinas MR, Alonso GL, Esteban-Infantes FJ (1994) Adsorption-thermal desorption-gas chromatography applied to the determination of wine aromas. J Agric Food Chem 1328–1331

Salinas MR, Alonso GL, Navarro G, Pardo F, Jimeno J, Huerta MD (1996) Evolution of the aromatic composition of wines undergoing carbonic maceration in different ageing conditions. Am J Enol Vitic 47(2):134–144

Shimoda M, Shibamoto T, Noble AC (1993) Evaluation of headspace volatiles of cabernet sauvignon wines sampled by an on-column method. J Agric Food Chem 41:1664–1668

Simpson RF (1980) Volatile aroma components of Australian port wines. J Sci Food Agric 31:214–222

Sunesson AL, Nilsson CA, Andersson B (1995) Evaluation of adsorbents for sampling and quantitative analysis of microbial volatiles using thermal desorption-gas chromatography. J Chromatogr 699:203–214

Tabera J, Reglero G, Herraiz M, Blanch GP (1991) Optimization of the dynamic headspace sampling for the analysis of trace volatiles components of grape juice: use of PTV injector for intermediate trapping. J High Resol Chromatogr 14:392–396

Vallejo-Cordoba B, Nakai S (1993) Using a simultaneous factor optimization approach for the detection of volatiles in milk by dynamic headspace gas chromatography. J Agric Food Chem 41:2378–2384

Vidal-Madjar C, Gonnard MF, Beuchah F, Guiochon G (1978) Performance of various adsorbents for the trapping and analysis of organohalogenated air pollutants by gas chromatography. J Chromatogr Sci 16:190–196

Voice TC, Kolb B (1994) Comparison of European and American technic compounds in environmental matrices. J Chromatogr Sci 32:306–311

Werkhoff P, Bretschneider W (1987a) Dynamic headspace gas chromatography: concentration of volatile components after thermal desorption by intermediate cryofocusing in a cold trap. I. Principle and applications. J Chromatogr 405:87–98

Werkhoff P, Bretschneider W (1987b) Dynamic headspace gas chromatography: concentration of volatile components after thermal desorption by intermediate cryofocusing in a cold trap. II. Effect of sampling and desorption parameters on recovery. J Chromatogr 405:99–106

Witkiewicz A, Gajek H, Choma J (1991) Chromatographic determination of the physico-chemical parameters of adsorption on active carbons. J Chromatogr 556:441–456

Yang WT, Min DB (1994) Dynamic headspace analyses of volatile compounds of cheddar and Swiss cheese during ripening. J Food Sci 59:1309–1312

Stable Isotope Dilution Gas Chromatography-Mass Spectrometry for Determination of Methoxypyrazines ("Green" Aroma) in Wine

M.S. Allen and M.J. Lacey

1 Introduction

Methoxypyrazines are strikingly potent odorants. 2-Methoxy-3-(2-methylpropyl) pyrazine (**1**) (isobutylmethoxypyrazine) has an odour threshold of 1–2 $ng\,l^{-1}$ in water (Buttery et al. 1969a,b; Seifert et al. 1970, 1972) and has been widely used in studies of odorant binding to olfactory mucosa (Pelosi et al. 1982), including isolation of odorant-binding proteins (Bignetti et al. 1985; Pes and Pelosi 1995), and in studies of the response of olfactory neuronal cultures to odorant binding (Ronnett et al. 1991). The aroma of this methoxypyrazine is described as bell pepper-like (Buttery et al. 1969a,b; Seifert et al. 1970, 1972; Parliment and Epstein 1973; Murray and Whitfield 1975; Takken et al. 1975), green (Takken et al. 1975), and slightly musty or leafy (Maga 1989). The methoxypyrazine occurs in the berries of some grape varieties, particularly those of Sauvignon blanc and Cabernet Sauvignon vines, and it is incorporated into the wines of these grape varieties in a concentration that can provide distinctive aroma. Odours described as green are also contributed by other odorants, such as the C_6 alcohols generated by oxidative breakdown of fatty acids during fruit processing. However, isobutylmethoxypyrazine provides an earthy, musty, bell pepper or gooseberry-like aroma that is quite distinct. In some winemaking regions of the world, this distinctive aroma is considered important to the regional style of Sauvignon blanc wines; in others, it is disliked. It can be overpowering and unpleasant if at high concentration, but at lower concentration it has an important role in distinguishing wines of the grape varieties Sauvignon blanc and Cabernet Sauvignon from those of other grape varieties (Allen et al. 1991).

Isobutylmethoxypyrazine (**1**), its isomer 2-methoxy-3-(1-methylpropyl)-pyrazine (**2**) (*sec*-butylmethoxypyrazine), and the homologue 2-methoxy-3-(1-methylethyl)pyrazine (**3**) (isopropylmethoxypyrazine) (Fig. 1) are a group of methoxypyrazines that have been found in varying proportions and at very low concentrations in some plants and insects. They occur in a broad selection of raw vegetables (Murray and Whitfield 1975); they also occur widely in brightly colour-patterned insects and the host plants of these insects, and some insects sequester methoxypyrazines from their host plants. The wide distribution of these pyrazines, their importance in some insects, and their occurrence in some plants with high toxin levels have led to the suggestion that such methoxypyrazines are widely used in nature as alerting signals (Moore et al. 1990). An attracting role in

Modern Methods of Plant Analysis, Vol. 19
Plant Volatile Analysis
Edited by H.-F. Linskens and J.F. Jackson

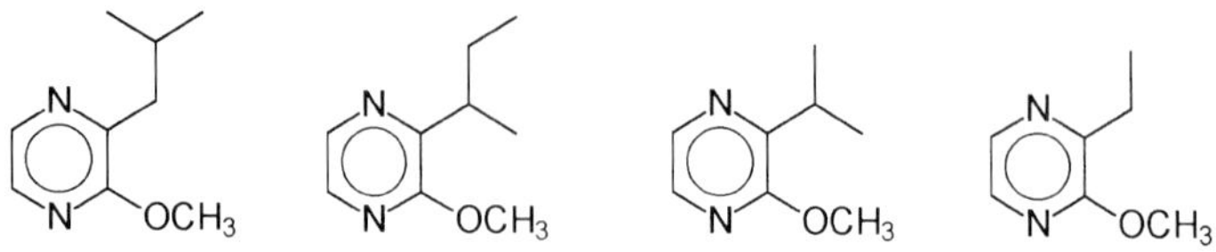

(1) (2) (3) (4)

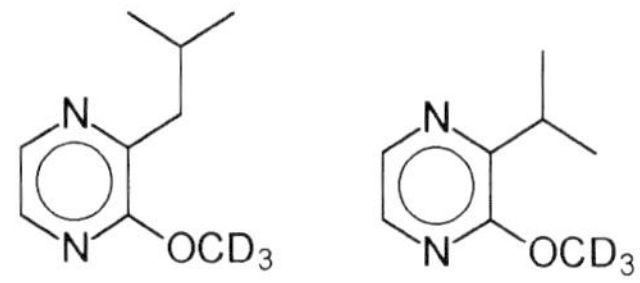

(5) (6)

Fig. 1. Natually occurring methoxypyrazines in wine **1–4** and the trideuterated internal standards **5** and **6** that are used to quantify them (D = ^{2}H)

migrating eels has also been identified (Sola 1995). Methoxypyrazines are also components produced by some microorganisms, particularly species of *Pseudomonas*, which have sometimes produced comparatively high concentrations that have allowed biosynthetic studies (Gallois et al. 1988; Leete et al. 1992).

In origin, the methoxypyrazines or methoxyalkylpyrazines are quite different from the alkylpyrazines, such as tetramethylpyrazine (Maga and Sizer 1973). Non-enzymatic processes, such as the Maillard reaction during roasting and cooking, appear to be the source of alkylpyrazines, and they are characteristic of products such as roasted coffee and cooked meat, to which they are frequently important odorants. This contrasts with the biosynthetic origin of the methoxypyrazines and the occurrence of the latter group of components in raw plant material.

2 Characterisation

Identification of methoxypyrazines in plant material has been hampered by their extremely low ($ng\,l^{-1}$) trace or even ultratrace level. Identification proceeded rapidly in the late 1960s and early 1970s with the development of gas chromatography-mass spectrometry (GC-MS). Studies focused on bell peppers (Buttery et al. 1969a,b; Seifert et al. 1970; Murray et al. 1970), potatoes (Buttery et al. 1970; Buttery and Ling 1973; Meigh et al. 1973; Nursten and Sheen 1974) and other vegetables (Murray and Whitfield 1975). Structures of the methoxypyrazines **1**, **2** and **3** were identified, their extremely low sensory detection thresholds were established, and their structures were confirmed by synthesis. These developments have been reviewed (Maga and Sizer 1973; Maga 1982).

As might be expected for such potent odorants, and particularly for studies with wine, the use of odour in identification has been important. The first confirmation of the occurrence of methoxypyrazine **1** in Cabernet Sauvignon grapes (Bayonove et al. 1975) relied heavily on agreement of the aroma retention time with that of synthetic material on five columns of differing polarity, with further confirmation by identification of characteristic fragment peaks in an electron ionisation (EI) mass spectrum. In that study, the concentration of **1** was speculated to be $5 \mu g l^{-1}$, a concentration that we now know to be at least a hundredfold higher than the concentration present in such grapes. The low concentration of methoxypyrazines in grapes and wine has frustrated some attempts at identification (Slingsby et al. 1980) and has required use of techniques suited to trace analysis. Analysis of these trace components in complex sample matrices has almost invariably required GC-MS instrumentation. This combines the high resolving power of GC with the excellent selectivity and sensitivity of MS. Additional sensitivity is usually gained through selected ion monitoring (SIM) mass spectrometry, a technique that was used in positive ion EI mode to analyse a basic fraction from freon extracts of Sauvignon blanc grapes and identify the existence of isobutylmethoxypyrazine (**1**), 2-ethyl-3-methoxypyrazine (**4**) and, tentatively, isopropylmethoxypyrazine (**3**) in Sauvignon blanc grapes (Augustyn et al. 1982).

However, more was required for wine flavour study than just verification of the occurrence of methoxypyrazines in wine. It was necessary to determine their concentration in wines, and to compare it with sensory data, preferably data determined with real wines. To those in the wine industry, it was clear also that the earthy, bell pepper-like aroma was characteristic not just of some grape varieties but also of the vine-growing and berry maturation conditions; it was more developed in some regions than others, and it showed a dependence on soil, climate and vine canopy. If methoxypyrazines were the source of the aroma, the dependence of methoxypyrazines on such growing conditions needed to be confirmed and the impact of these growing conditions evaluated. If this were to be achieved, quantitative data were required.

3 Trace Quantitative Analysis

3.1 Advantages of GC-MS

Analysis of the headspace volatiles above raw vegetables by GC-MS allowed Murray and Whitfield (1975) to identify methoxypyrazines **1**, **2** and **3** and to estimate the amounts of these components in the headspace. Good quantitative estimation was not claimed, but a carefully standardised procedure was followed, so the relative amounts found in the headspace samples were expected to represent the comparative methoxypyrazine levels in the different vegetables. Importantly, their findings did indicate low ng levels in their headspace samples; levels in the realm of ultratrace analysis. These levels confirmed the need for a quantitative analysis technique that can determine methoxypyrazines with good precision and accuracy at levels in the low $ng\,l^{-1}$ range and in the presence of other matrix components.

For routine quantitative analysis of trace materials, immunoassay methods are rapid and useful (Robins 1986). However, development of the methodology is not always straightforward and the precision, accuracy and specificity of the technique are not intrinsically good. With such techniques as GC, AA and HPLC, even at the part per million level, the coefficient of variation established in detailed interlaboratory validation is typically about 16%. At lower levels, the problems of achieving consistency become more complex, and the sensitivity of these techniques is often inadequate. For trace levels, MS techniques are usually required, particularly GC-MS for volatile components. However, MS itself is a technique that does not readily lend itself to quantitative precision, and at the level appropriate to methoxypyrazines in wine, the task is formidable. The chosen analytical strategy should be able to detect below the sensory detection threshold of $1\text{–}2\,ng\,l^{-1}$, preferably by one order of magnitude. On this basis, with use of a moderate sample size and an allowance for losses in isolation, it is evident that a detection limit of a few picograms is likely to be required. Furthermore, this needs to be combined with methodology that provides good quantitative precision and accuracy as this level. Many factors interfere with the precision and accuracy of quantitative GC-MS analysis, and at trace levels these factors become more important (Millard 1980); they need to be understood and controlled if quantitative accuracy is to be achieved.

3.2 Stable Isotope-Labelled Internal Standards

As it is virtually impossible to introduce an exact amount of a sample into the mass spectrometer ion source, quantitative analysis is always performed with an internal standard that is added before the analysis. Critical to accuracy and precision is the choice of the internal standard (Giovannini et al. 1991; Boyd 1993). It must behave, towards all factors that influence the analyte's recovery, as nearly identically as possible to the analyte. Factors that influence recovery include losses in sampling, extraction, evaporation and derivatisation; adsorption onto glass; decomposition during chromatography; adsorption to active sites on the column; and variation of mass spectrometer operating conditions. To control these factors, the internal standard must be at least of the same class of compound as the analyte; it is also better if it is a close homologue of the analyte. Widely recognised to be preferable to even a close homologue is a stable isotope-labelled analogue of the analyte. Mass spectrometry is unique in that it can distinguish mass differences arising from stable isotope labelling, allowing use of the technique of isotope dilution-mass spectrometry (ID-MS), the technique that provides the closest approach yet to a definitive method (Giovannini et al. 1991).

A benefit of the use of a stable isotope-labelled analogue as internal standard is that there is no possibility of the labelled analogue occurring naturally in any of the samples, provided that the nature of the isotope and the number of isotopically substituted atoms are appropriate. A further benefit of ID-MS is that quantitative recovery of the analyte during isolation is not required, and recovery studies are not necessary; it is the mole ratio of analyte to internal standard that is important, not the absolute amount of analyte. Reduced recovery increases the difficulty of gaining adequate signal-to-noise ratio in the MS analysis, but it does not directly alter the determined analyte concentration. Furthermore, the retention time of a stable isotope-labelled analogue is very close to that of the analyte, reducing the opportunity for variation of instrumental parameters such as ionisation yield. However, a drawback to the use of a stable isotope-labelled analogue is the cost, both financially and in time, of its synthesis. In addition, the stability of the isotopic label in the molecule and the extent of contamination by unlabelled material must also be considered. Interferences from natural isotopic abundances of ^{2}H, ^{13}C and ^{18}O are best minimised by use of a standard with a mass three or more units greater than the analyte; for this reason, trideuterated analogues are widely used if stability of the isotopic label is assured.

Quantitative analysis requires comparison of measured analyte and internal standard peak area ratios in the sample with ratios determined over a relevant concentration range with calibration standards (Giovannini et al. 1991; Boyd 1993). As the analyte concentration in the sample decreases, the problem of intrusion of other components into the signal of interest increases. This places increased demands upon selectivity of component isolation, chromatographic resolution, and verification of quantitation under different chromatographic and/or ionisation conditions. Analytical sensitivity can be improved and selectivity gained by using SIM to monitor only one or a few ions of the analyte and internal standard. Chemical ionisation (CI) frequently increases selectivity by concentrating the ion mass profile into molecular parent ions such as $(M+H)^{+}$ ions, reducing the likelihood of intrusion by ions of other sample components. Careful choice of reagent gas proton affinity can also provide selectivity of ionisation (Milne and Lacey 1974). In trace analysis, effective transfer of sample to the chromatography system becomes crucial, to avoid molecular discrimination and to ensure that the analyte is not lost during injection. No less important is that the results can have an accuracy no better than the accuracy of preparation of the internal standard solution and the integrity with which its concentration is maintained. Changes of concentration due to solvent evaporation or internal standard decomposition need to be controlled and monitored. Finally, the results can have an accuracy and precision no better than that of the addition of the internal standard to the sample. For trace component analysis, careful use of techniques that are suited to the accurate transfer of small volumes of internal standard solution is required.

3.3 Methoxypyrazine Quantitative Analysis

A Sauvignon blanc wine was the subject of the first methoxypyrazine determination by ID-MS, reported in 1987 (Harris et al.). This used 2-($^{2}H_3$)methoxy-3-(2-methylpropyl)pyrazine (**5**) as internal standard to quantify the occurrence of

35 (± 2) $ng\,l^{-1}$ of isobutylmethoxypyrazine (**1**) in the wine. In addition, 6 $ng\,l^{-1}$ of isopropylmethoxypyrazine (**3**) and less than 0.5 $ng\,l^{-1}$ of the *sec*-butylmethoxypyrazine (**2**) were also identified. The retention time difference of several minutes between **3** and the internal standard **5** is a potential source of error. In this time, mass spectrometer operating conditions may intrinsically vary. Furthermore, the changing nature of the other components that are eluted may influence the chemical ionisation process. These variables can lead to changes of response between **3** and **5** that reduce quantitative precision and accuracy. For this reason, quantitation of **3** was improved by the combined use of two trideuterated internal standards, **5** and **6**. Application of this technique to a wide range of Sauvignon blanc and Cabernet Sauvignon grapes and wines clearly established **1** as the dominant methoxypyrazine, with a typical concentration of up to 56 $ng\,l^{-1}$ (Lacey et al. 1991; Allen et al. 1994). Some aspects of ID-MS analysis of methoxypyrazines have been reviewed (Williams and Allen 1996).

The danger of not using ID-MS techniques is highlighted by the reported occurrence of **1** at 500 (± 70) $ng\,l^{-1}$ in a Bordeaux wine (Boison and Tomlinson 1990) in work that lacked the rigour of an internal standard throughout the isolation procedure. Identification of a wide range of components was sought so that, where an internal standard was used, it was structurally unrelated to methoxypyrazines. Finally, the method was not verified with methoxypyrazines and verification was performed only on components at comparatively much higher concentration. By comparison, examination of a range of Australian, New Zealand and French red wines by ID-MS, including Bordeaux wines (Allen et al. 1994), has established that the concentration of **1** is typically 50-fold lower than 500 $ng\,l^{-1}$.

4 Method of ID-MS Analysis of Methoxypyrazines

4.1 Method Outline

The key steps of the experimental approach are indicated in Fig. 2. Before commencing the synthesis of stable isotope-labelled methoxypyrazines, it is wise to identify problems, such as poor instrument sensitivity, excessive column contamination, or column bleed, that may interfere with the capability of the mass spectrometer to detect trace levels of methoxypyrazines. This can be achieved by adding a trace level of a commercially available unlabelled methoxypyrazine to a wine that is not expected to contain methoxypyrazines, such as a Pinot noir or Chardonnay wine, then isolating and analysing the recovered component. Use of the *sec*-butylmethoxypyrazine (**2**) is useful, as this has so far always been a comparatively minor methoxypyrazine in wine.

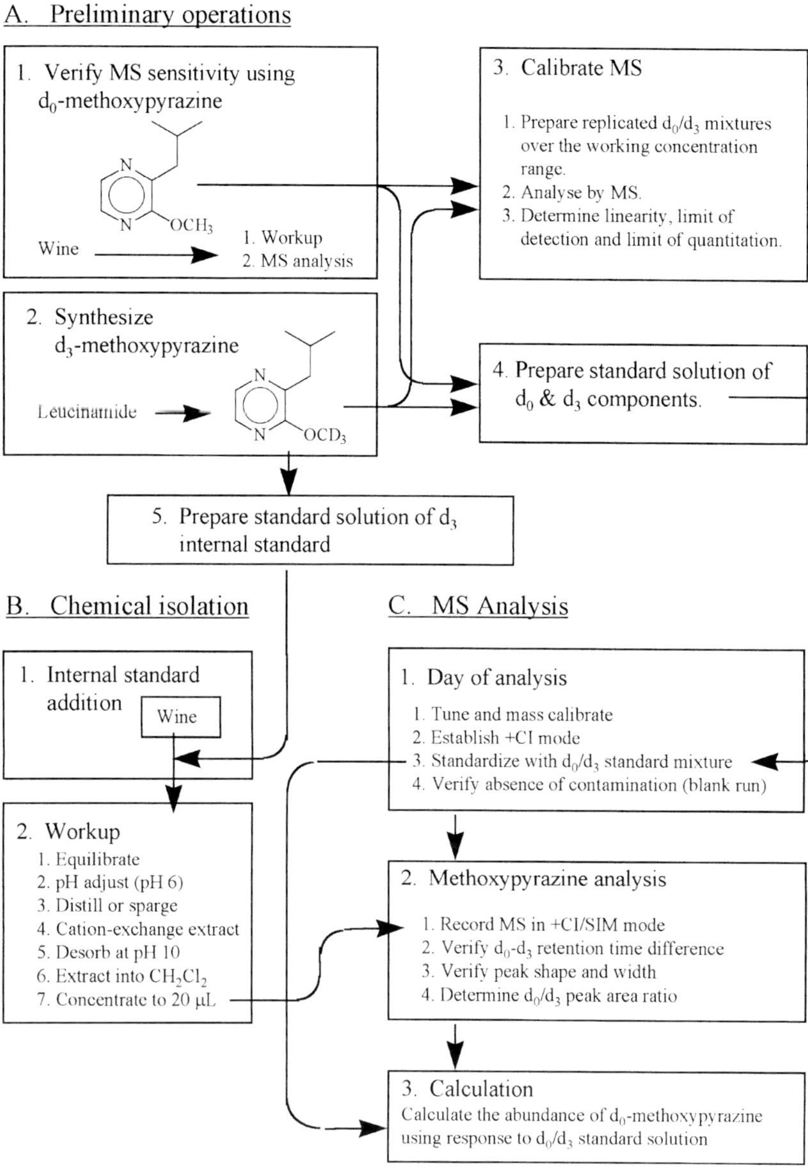

Fig. 2A–C. Key steps of the determination of methoxypyrazines by ID-MS ($d_0 = {}^2H_0$, $d_3 = {}^2H_3$)

4.2 Precautions

In ID-MS, erroneous results will occur if a sample contaminant provides ions in a monitored mass channel at essentially the same retention time as a methoxypyrazine, even if the contaminant is present in trace quantity. Signals from these ions contribute to the signal intensity recorded, leading to higher than true measurements of peak areas for the natural or trideuterated methoxypyrazines, to the detriment of quantitative accuracy. To minimise contamination, contact with plastic is minimised at all times, except where the material is a fluoropolymer. Solvents are thoroughly purified (CH_2Cl_2 is shaken repeatedly with conc. H_2SO_4, washed with H_2O, $NaHCO_3$ aq, then H_2O, dried with Na_2SO_4 then distilled in glass from P_2O_5). Critical glassware in the later stages of methoxypyrazine isolation is washed with chromic acid (24 h, ca. 25 °C; caution: corrosive and oxidising agent) then 3 M HCl (3 h, 80 °C), 5% $NaHCO_3$ (24 h, 25 °C) then high purity (ca. 18 MΩ conductivity) water (Allen et al. 1994).

4.3 Instrumentation

Analysis has been performed with both a double-focussing, magnetic sector instrument (VG Micromass 70–70 F; trap current of 100 μA, EI; emission current of 500 μA, CI; interfaced to a Hewlett Packard 5790A gas chromatograph and a VG 11–250 data system) and with a quadrupole instrument (VG Trio 2; emission current of 200 μA, EI; or 700–800 μA, CI; incorporating a Hewlett Packard 5890 Series II gas chromatograph). Mass spectra and selected ion monitoring are recorded in positive-ion mode at an electron energy of either 70 eV (EI) or 50 eV (CI) at a source temperature of 150–180 °C. Ammonia (anhydrous) was used as reagent gas (source pressure of 60 Pa, 70–70 F; or source envelope pressure of 3×10^{-3} Pa, Trio 2). For isobutylmethoxypyrazine (**1**), the monitored ions were *m/z* 167.1 and 170.1 (CI), or 124.1 and 127.1 (EI) with a dwell time of 0.05–0.10 s. Scan spectra were recorded over the range 60–200 u, with scan time of 0.5 s and interscan delay of 0.1 s. Relative ion abundances were assessed from peak areas.

Gas chromatography was performed with fused silica capillary columns of dimethylsilicone (J&W, DB-1), 5% phenyl and 95% dimethylsilicone (SGE, BP5), 6% phenyl and 6% cyanopropyl dimethylsilicone (J&W, DB-1701), or Carbowax (J&W, DB-Wax; SGE, BP20) of 25–60 m length, 0.32 mm i.d., and either 0.5- or 1.0-μm film thickness. Columns were inserted directly into the ion source. Injections were made at 30 °C into either a cool on-column injector (HP 19290B) or a temperature-programmable cool on-column capillary inlet with electronic pressure control (HP 19245C), using a deactivated fused silica retention gap (J&W or SGE, 1–3 m) to reconcentrate components at the beginning of the stationary phase. Large injected volumes of 6–10 μl (ca. 30–50% of sample) enabled the full dynamic range of sensitivity to be explored. They were achieved without loss of chromatographic resolution, and without exceeding the capacity of the mass spectrometer vacuum system to accept the volume of injected solvent, by accumulating the solutes from successive injections of small aliquots (1–3 μl), within a few minutes, onto the retention gap (Lacey and Sanders 1992).

The oven was programmed to provide a 2-min delay at 30 °C followed by a temperature rise of either (1) 25 °C min^{-1} to 90 °C then 5 °C min^{-1} to 200 °C (DB-1), or (2) 25 °C min^{-1} to 110 °C, 5 °C min^{-1} to 150 °C, 25 °C min^{-1} to 210 °C, then held at

210 °C for 5 min (DB-1701 and DB-Wax). Helium was used as carrier gas at either a flow rate of 1 ml min^{-1} (70–70 F), or a constant (69 kPa, 30 m column) or programmed (69 kPa for 0.5 min, then 138 kPa min^{-1} to 138 kPa, 60 m column) inlet pressure (Trio 2).

4.4 Internal Standard Synthesis

Unlabelled 2-methoxy-3-alkylpyrazines are commercially available (Pyrazine Specialities, Inc., Atlanta, Ga, USA). The trideuterated methoxypyrazine internal standards were synthesised essentially by the methods of Seifert et al. (1970) and Murray and Whitfield (1975). The retention time difference between trideuterated isobutylmethoxypyrazine (**5**) and isopropylmethoxypyrazine (**3**) requires use of both the trideuterated isobutyl- and trideuterated isopropylmethoxypyrazines, **5** and **6**, respectively, for accuracy of quantitative analysis. Commercially available starting materials are the hydrochlorides of the α-amino acid amides, leucinamide (**7a**) and valinamide (**7b**) (Fig. 3). Condensation with glyoxal produces the corresponding 2(1*H*)-pyrazinone (**8a** and **8b**), chlorination with phosphorus oxychloride generates the chloro compounds (**9a** and **9b**), and these are converted to the trideuterated methoxypyrazines (**5** and **6**), respectively, by the action of sodium ($^2H_3$)methoxide (Harris et al. 1987; Lacey et al. 1991). Yields are moderate to good, except for the condensation step, which has a low and variable yield and appears to require tight control of the low reaction temperature for success.

As methoxypyrazines are potent aroma components, methoxylation of the chloropyrazine is performed in a good fume hood using techniques that minimise release of the methoxypyrazine into the laboratory. To avoid contamination, synthesis is completed in a different laboratory and with different equipment to that used for methoxypyrazine isolation. Furthermore, as methoxypyrazines undergo photodegradation (Heymann et al. 1986), the final synthetic step is performed in the dark or in subdued light where possible. The following procedure, for synthesis of the trideuterated isobutylmethoxypyrazine **5**, is readily adapted to the trideuterated isopropyl compound **6**.

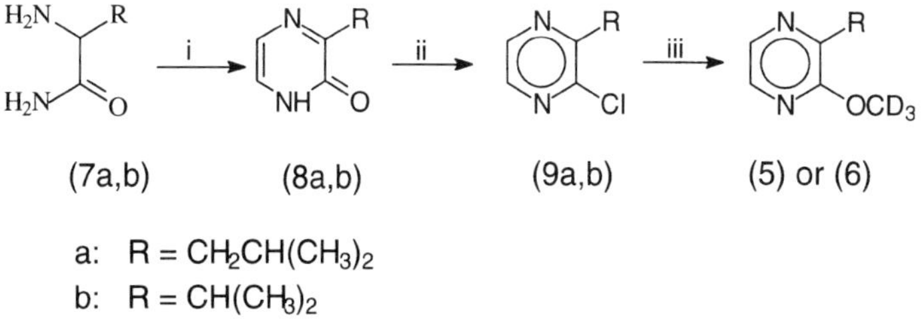

Fig. 3. Synthesis of trideuterated methoxypyrazine internal standards **5** and **6** (i = glyoxal bis(bisulphite)/NaOH; ii = $POCl_3$; iii = NaH/CD_3OD)

Leucinamide.HCl (5.0 g) in H_2O (30 ml) and CH_3OH (30 ml) was stirred at – 25 °C while adding glyoxal bis(bisulphite) adduct (8.0 g) then a solution of NaOH (4.20 g) in H_2O (10 ml) at a rate such that the temperature did not exceed –20 °C. The mixture was then stirred at –20 °C for 1 h. Water (120 ml) was added over 15 min, with stirring, maintaining the temperature below –10 °C. The mixture was stirred for 3 h at 10 °C then adjusted to pH 6 with conc. HCl while maintaining this temperature. The volume was reduced by half (rotary evaporator) and the residue was extracted with boiling $CHCl_3$ (4 × 150 ml). Recrystallisation of the evaporated organic extracts from hexane provided 1.37 g (32%) of 2(1*H*)-pyrazinone **8a**.

2(1*H*)-Pyrazinone **8a** (2.3 g), phosphorus oxychloride (9.6 g) and one drop conc. H_2SO_4 were refluxed for 2 h, then cooled and added to ice (75 g) and diethyl ether (40 ml) chilled in ice water. The mixture was adjusted to pH 7 (20% aq NH_3) then to pH 12 (NaOH aq) at a temperature below 10 °C, then the ether phase was separated and the aqueous phase washed twice with ether (50 ml). The combined ether phases were dried (Na_2SO_4), filtered through silica gel (Kieselgel 60, 230–400 mesh), concentrated (distillation, vigreux column) and the chloropyrazine **9a** (1.1 g) isolated by microdistillation in vacuo.

THF (anhydrous, 20 ml) and CD_3OD (1.0 ml) were added to NaH (1.7 g) in glassware that had been oven-dried and was protected from moisture. After stirring for 15 min, the chloropyrazine **9a** (1.1 g) was added and the mixture was refluxed until conversion was complete (ca. 10 h) as assessed by TLC (silica gel G, CH_2Cl_2), H_2O (10 ml) was then added slowly to decompose excess hydride (care: H_2 evolution). The mixture was extracted (pentane, 3 × 10 ml), the organic phase washed (H_2O, 10 ml), dried (Na_2SO_4), concentrated (distillation with a vigreux column, then microdistillation at atmospheric pressure), then purified by chromatography (Kieselgel 60, 230–400 mesh, 25 g, eluted pentane then 1 : 1 CH_2Cl_2 : ether). Fractions were monitored by TLC (silica gel G, CH_2Cl_2). Appropriate fractions were concentrated (distillation with a vigreux column) and the methoxypyrazine **1** distilled in vacuo.

4.5 Instrument Calibration

Calibration is required to determine the response of the mass spectrometer to ($^2H_0$) and ($^2H_3$)methoxypyrazines and to measure and compensate for the trace of ($^2H_0$)methoxypyrazine that exists in the internal standard (Fig. 2). It requires GC-MS analysis of mixtures of ($^2H_0$) and ($^2H_3$)methoxypyrazines to measure the $^2H_0/^2H_3$ peak area ratio at defined $^2H_0/^2H_3$ mole ratios over the full mole ratio working range that is expected. This allows calculation of the linearity of the response and, from the standard deviation of the intercept, the limit of detection and the limit of quantitation. Calibration is necessary for each of the principal analytes and its internal standard, and needs to be performed under the operating conditions to be used for methoxypyrazine quantitative analysis. Where internal standards **5** and **6** are both available, it requires GC-MS analysis of mixtures of **1** and **5**, and mixtures of **3** and **6**. In calibration mixtures, the concentration of the internal standard should be close to that expected in the samples for methoxypyrazine determination. The concentration of the unlabelled methoxypyrazine should range from zero to higher than the maximum concentration expected in the samples for methoxypyrazine determination. It is desirable to have a minimum of six mole ratios covering the useful concentration range, with each calibration standard prepared and analysed in duplicate. The instrumental response (i.e. weighted linear regression of methoxypyrazine $^2H_0/^2H_3$ peak area ratio on methoxypyrazine $^2H_0/^2H_3$ mole ratio) is calculated using the reciprocal of the variance as weighting (Miller 1991) to give adequate emphasis to the samples at low methoxypyrazine $^2H_0/^2H_3$ mole ratios which critically examine the limit of detection. The change of the variance with mole ratio can be estimated from at least six replicate analyses of mixtures at each of high and low levels of undeuterated material. The limit of detection is determined as intercept plus three standard deviations; the limit of quantitation as intercept plus ten standard deviations. The $^2H_0/^2H_3$ peak area ratio at zero $^2H_0/^2H_3$ mole ratio indicates the extent of contamination of the synthetic internal standard by ($^2H_0$)methoxypyrazine. Compensation for this is performed in calculating the amount of natural methoxypyrazine in the wine samples. If a linear relationship between peak area ratio and mole ratio is not obtained, a number of approaches may be used to circumvent the problem (Giovannini et al. 1991).

4.6 Determination of Standard Response

It is not practical to perform a full instrumental calibration on each day of GC-MS analysis. However, if very good instrument linearity is found in calibration, day-to-day variation of instrumental performance can be accommodated by determining the methoxypyrazine $^2H_0/^2H_3$ peak area ratio on that day with a standard of known methoxypyrazine $^2H_0/^2H_3$ mole ratio in which both the mole ratio and the methoxypyrazine concentrations reflect the typical concentrations encountered. The response with the standard is then used to calculate the sample methoxypyrazine concentrations (Fig. 2). The standard solution contains each of the analytes to be determined and internal standards to be used, allowing the instrumental response to each analyte/internal standard pair to be considered. A dichloromethane solution of **1**, **3**, **5** and **6** at 200 $\mu g\,l^{-1}$, and **2** at 20 $\mu g\,l^{-1}$ is suitable. On the day of analysis, 1 µl is injected after mass calibration and mass spectrometer tuning. Quantitative analysis is attempted only if the response to the standard is within 10% of the mean of previous satisfactory determinations.

4.7 Preparation and Handling of Internal Standard Solutions

The accuracy of ID-MS work is limited ultimately by the accuracy with which the purity of the internal standard, the concentration of its solution, and the volume of internal standard added to the sample are known. The structural purity of synthesised trideuterated methoxypyrazines needs to be ascertained by NMR and GC, and the isotopic purity by GC-MS. As methoxypyrazines are light-sensitive (Heymann et al. 1986), they are stored in the dark and are best handled in amber-coloured containers. Determination of the UV absorbance of solutions prepared at an appropriate concentration is a simple means of monitoring for any change in methoxypyrazine concentration. Evaporation of internal standard solutions is a possible source of error, and one that is complicated by the volatility of the methoxypyrazines themselves. It is a problem that can be minimised by use of vials with closures that provide a positive seal with a valve yet, in the open valve position, allow sample removal by syringe through a rubber septum (e.g. Pierce Mininert valve). Measurement of weight immediately after use and immediately prior to the next use allows any inadvertent evaporation during storage to be detected. Dilutions are performed serially with calibrated microlitre syringes, and calibrated microlitre syringes are used for internal standard addition.

4.8 Isolation of Methoxypyrazines

The trideuterated methoxypyrazines **5** and **6** are added as internal standards to the wine (typically 12.0 μl of 2.0 ng μl^{-1} solution in purified dichloromethane added to 240 ml wine) with agitation and allowed to stand for several hours to ensure equilibration. After adjustment to pH 5–6 (10% NaOH), the sample is distilled at atmospheric pressure into an ice-cooled receiver. Methoxypyrazines in the distillate (50 ml) are trapped by standing (1 h) with occasional stirring with a strongly acidic ion-exchange resin (AG50WX4, 200 mg). After filtration and washing of the resin (H_2O, 2 × 1 ml), methoxypyrazines are removed by dilute alkali (1 ml, pH 10), which in turn is extracted with dichloromethane (2 × 0.5 ml). The dichloromethane extracts are slowly concentrated over 6–7 h to ca. 20 μl in tapered microvials, under nitrogen in the dark at room temperature, then transferred to glass capillaries that are sealed then stored in the dark at –20 °C until analysis by mass spectrometry.

An alternative to distillation is to sparge the wine (330 ml) with high-purity nitrogen (300 ml min^{-1}; 5 h) with passage of the headspace gas through the ion exchange resin. Where comparison of sparging and distillation was made, the determined concentrations were the same by both techniques (see Table 1; Harris et al. 1987).

The ability of the resin to adsorb methoxypyrazines from the gaseous phase highlights the need to scrupulously minimise contact of the resin with the laboratory atmosphere, particularly where methoxypyrazine synthesis has taken place or methoxypyrazine standards have been used. Analyses without wine are performed occasionally to identify any methoxypyrazine contamination of the resin.

Although a volume of 240 ml of wine is usual for this analysis, lesser quantities may be used if the instrument sensitivity allows. Particular care must be taken in the step of evaporation to ca. 20 μl; if allowed to proceed to dryness, methoxypyrazines are not recovered by washing the walls of the vessel with solvent.

4.9 GC-MS Analysis of Methoxypyrazines

Quantitative analysis of trace components demands that attention be given to optimising instrument performance; this process will vary with the type of GC-MS equipment used. With the VG Trio 2 system, the mass spectrometer is tuned and mass calibrated each day of analysis. Manual tuning is preferred to the auto-tune program to allow the tuning to be optimised for sensitivity. Selectivity is gained by analysis in positive ion CI mode. In this mode, fragmentation is minimised, the mass channel for the $(M+H)^+$ ion is used and, consequently, a higher *m/z* is analysed; this reduces the likelihood of sporadic intrusion by ions of contaminants. Furthermore, the higher proton affinity of ammonia, by comparison with, e.g. isobutene or methane, provides selectivity of ionisation (Milne and Lacey 1974). Further tuning in +CI mode can further optimise the mass spectrometer operation. To gain further selectivity and sensitivity, the analysis is performed in SIM mode, monitoring the mass channel of the $(M+H)^+$ ion of each component. Before analysis itself, the suitability of the system is verified by an injection of solvent alone, to test the absence of system contamination; analysis of the methoxypyrazine standard mixture, to measure the response factor for calculating the sample methoxypyrazine concentrations, and to verify that the methoxypyrazine $^2H_0/^2H_3$ peak area ratio lies within satisfactory limits; and a further injection of solvent alone, to ensure the absence of carryover of methoxypyrazine standards. To avoid contamination, a different injection syringe is always used for methoxypyrazine standards.

4.10 Data Interpretation

Despite the selectivity achieved in isolation, chromatography, ionisation and selected monitoring of mass channels, further scrutiny of the data is required before the existence of a methoxypyrazine-derived signal is accepted and a concentration is determined.

Isotopic substitution slightly reduces the retention time of the $(^2H_3)$methoxypyrazine internal standard by comparison with its $(^2H_0)$methoxypyrazine analogue. The retention time difference between the two components (typically 3 s with a 60-m column) is highly consistent for a given GC column and operating condition. As the internal standard is readily identifiable, comparison of the retention time difference between a peak tentatively assigned to a $(^2H_0)$methoxypyrazine and that of its $(^2H_3)$-labelled analogue is a more precise test than reliance upon the absolute analyte retention time. A peak that is outside a narrow window (typically $\pm$ 0.5 s) must be treated with caution.

Although peak tailing frequently occurs, the peak shape of a pure component will be essentially Gaussian; shoulders or other peak shape irregularities indicate contamination of the mass channel by intrusive components. Additionally, the peak width at half height should be identical to that of the internal standard. Where contamination is suspected, analysis with a column of differing polarity is required. Analysis in +EI mode is also useful as it allows monitoring of different mass channels.

If peak homogeneity is found, calculation of the natural methoxypyrazine content in the wine sample can be determined as:

$$\mathrm{Conc}^{\mathrm{Sample}} = \left[\mathrm{Conc}^{\mathrm{IntStd}} \times \mathrm{Mwt\ ratio}\left({}^{2}\mathrm{H}_{0}/{}^{2}\mathrm{H}_{3}\right) \times \mathrm{Area\ ratio}\left({}^{2}\mathrm{H}_{0}/{}^{2}\mathrm{H}_{3}\right)\right] / \mathrm{Slope}^{\mathrm{Stdn}},$$

where

$\mathrm{Conc}^{\mathrm{Sample}}$	=	Wine methoxypyrazine concentration ($\mathrm{ng\,l^{-1}}$)
$\mathrm{Conc}^{\mathrm{IntStd}}$	=	Internal standard concentration in sample ($\mathrm{ng\,l^{-1}}$)
Mwt ratio($^2H_0/^2H_3$)	=	Ratio of molecular weights of unlabelled and trideuterated methoxypyrazines
Area ratio($^2H_0/^2H_3$)	=	Ratio of measured peak areas for unlabelled and trideuterated methoxypyrazines in analysis

and

$$\mathrm{Slope}^{\mathrm{Stdn}} = \mathrm{Area\ ratio}\left({}^{2}\mathrm{H}_{0}/{}^{2}\mathrm{H}_{3}\right)^{\mathrm{Stdn}} / \left[\mathrm{Mole\ ratio}\left({}^{2}\mathrm{H}_{0}/{}^{2}\mathrm{H}_{3}\right)^{\mathrm{Stdn}} + \mathrm{Mole\ ratio}\left({}^{2}\mathrm{H}_{0}/{}^{2}\mathrm{H}_{3}\right)^{\mathrm{Resid}}\right]$$

and

$$\mathrm{Mole\ ratio}\left({}^{2}\mathrm{H}_{0}/{}^{2}\mathrm{H}_{3}\right)^{\mathrm{Resid}} = \mathrm{Int}^{\mathrm{Calib}} / \mathrm{Slope}^{\mathrm{Calib}},$$

where

Area ratio($^2H_0/^2H_3$)^{Stdn}	=	Ratio of measured peak areas of unlabelled and trideuterated methoxypyrazines in determination of the response to the standard mixture
Mole ratio($^2H_0/^2H_3$)^{Stdn}	=	Known mole ratio of unlabelled to trideuterated methoxypyrazine in the standard mixture
$\mathrm{Int}^{\mathrm{Calib}}$	=	Intercept of regression line in calibration
$\mathrm{Slope}^{\mathrm{Calib}}$	=	Slope of regression line in calibration

5 Application to Methoxypyrazines of Wine

Correctly performed, ID-MS provides quantitative data of high precision, as evident in the findings of the first ID-MS analysis of a Sauvignon blanc wine in Table

1 (Harris et al. 1987). This precision allows concentration differences of even a few parts per trillion to be meaningfully interpreted.

Amongst plants, insects and microorganisms, there are many qualitative or semi-quantitative data to indicate that a wide variation exists in the relative proportions of methoxypyrazines **1**, **2** and **3**. With ID-MS, data of high precision on the relative proportions of these three methoxypyrazines is available, even when the methoxypyrazine is below its sensory detection threshold of ca. $2\,\mathrm{ng\,l^{-1}}$. These quantitative data indicate that, for those grape varieties that produce methoxypyrazines at a level above their sensory detection threshold, there is a high degree of similarity in the relative proportions of methoxypyrazines **1**, **2** and **3** even though there can be wide variation in the total methoxypyrazine concentration. This can be observed in Table 2, which compares representative wines selected to be each largely from a single grape variety. It must be noted, however, that there can be exceptions such as isopropylmethoxypyrazine (**3**) being as abundant as isobutylmethoxypyrazine (**1**) in one wine (Allen et al. 1995), a situation suggesting that the former component may arise occasionally from sources other than the grape berry and normal winemaking processes.

The sensory role of isobutylmethoxypyrazine (**1**) has been confirmed through the use of ID-MS (Allen et al. 1991), but the precision of ID-MS also allows the sensory role of the less abundant components to be considered.

Table 1. Concentration ($\mathrm{ng\,l^{-1}}$) of isobutylmethoxypyrazine (**1**) in replicate determinations of a Sauvignon blanc wine

Isolation	GC column	
	5% Phenyl methylsilicone	Carbowax
Distillation	34.4	33.9, 34.3, 38.0
Sparging		33.0, 34.6

Table 2. Percentage contribution of methoxypyrazines **1**, **2** and **3** to the total methoxypyrazine content

Origin	Grape variety	Total ($\mathrm{ng\,l^{-1}}$)	Percentage of total		
			1	2	3
France, Pouilly-Fumé	Sauvignon blanc	23.8	86	2.5	11.3
France, Sancerre	Sauvignon blanc	42.8	89	3.0	7.9
New Zealand	Sauvignon blanc	21.9	86	1.9	12.5
California, USA	Cabernet Sauvignon	39.7	94	1.7	4.2
California, USA	Cabernet Sauvignon	9.7	90	1.3	8.5
Australia	Cabernet Sauvignon	11.8	72	3.0	25.2
Australia	Cabernet Sauvignon	5.0	90	3.2	6.4
	Mean value:		87%	2.4%	10.9%

Isopropylmethoxypyrazine (**3**) (range 0.32–3.4 $ng\,l^{-1}$, mean 2.1 $ng\,l^{-1}$, in Table 2) often occurs in wine at a level comparable to the 1–2 $ng\,l^{-1}$ sensory detection threshold of the component in water (Murray et al. 1970; Seifert et al. 1970) or wine (Allen et al. 1991; Maga 1989). Consequently, it may well have the sensory effect of modifying the principal aroma perception created by isobutylmethoxypyrazine (**1**).

The precision of ID-MS has been used to identify the effect of viticultural conditions on methoxypyrazine level. Part of the variation of total methoxypyrazine level can be attributed to climate, with cool grape ripening conditions increasing the concentration of methoxypyrazine components (Lacey et al. 1991; Allen et al. 1994). Vine management can also affect grape methoxypyrazine levels; a contributing factor is the ease of light penetration through the leaf canopy which leads to deeply shaded fruit having threefold higher levels than exposed fruit on the same vines (Allen 1993).

Although not strictly ID-MS, the principles of ID-MS and the use of a stable isotope-labelled internal standard can be applied to quantitative analysis of an isomer or a homologue of the standard. This is the case with determination of *sec*-butylmethoxypyrazine (**2**), using trideuterated isobutylmethoxypyrazine (**5**) as internal standard, and determination of ethylmethoxypyrazine (**4**), using trideuterated isopropylmethoxypyrazine (**6**) as internal standard. With ethylmethoxypyrazine (**4**), the quantitative precision of this approach has demonstrated convincingly that this component has a different biosynthetic origin to that of isobutylmethoxypyrazine (**1**) in grapes and wine (Allen et al. 1996).

6 Conclusion

The quantitative precision of ID-MS has enabled for the first time not only the identification of various methoxypyrazines in wine but also the precise and accurate measurement of such levels. This is achieved despite their ultratrace level in wine, a level that provides the mass spectrometer with only a few picograms of material for analysis. The technique allows a detailed insight into the production, behaviour and sensory impact of methoxypyrazines in wine, providing the analytical criteria essential to evaluate meaningfully the approximately tenfold range of concentration from the imperceptible to the distinctly "green" and "herbaceous" in wine.

Acknowledgments. The authors would like to thank F. Brochet, Dr. R. Harris and Dr. W. Vance Brown for assistance and research input into various aspects of this work. The Institute of Oenology, University of Bordeaux, France, provided some financial support. The major part of this work was made possible by either the Grape and Wine Research and Development Corporation, Australia or its predecessor the Grape and Wine Research Council, Australia.

References

Allen MS (1993) Viticultural effects on methoxypyrazine grape flavour: current research directions. Aust Grapegrower and Winemaker, No 354, Ryan Publications, Adelaide, pp 10–12

Allen MS, Lacey MJ, Harris RLN, Brown WV (1991) Contribution of methoxypyrazines to Sauvignon blanc wine aroma. Am J Enol Vitic 42:109–112

Allen MS, Lacey MJ, Boyd S (1994) Determination of methoxypyrazines in red wines by stable isotope dilution gas chromatography-mass spectrometry. J Agric Food Chem 42:1734–1738

Allen MS, Lacey MJ, Boyd SJ (1995) Methoxypyrazines in red wines: occurrence of 2-methoxy-3-(1-methylethyl)pyrazine. J Agric Food Chem 43:769–772

Allen MS, Lacey MJ, Boyd SJ (1996) Existence of different origins for methoxypyrazines of grapes and wines. In: Takeoka G (ed) Biotechnology for improved foods and flavors. ACS Symp Ser. 637, ACS, Washington, pp 220–227

Augustyn OPH, Rapp A, van Wyk CJ (1982) Some volatile aroma components of *Vitis vinifera* L. cv. Sauvignon blanc. S Afr J Enol Vitic 3:53–60

Bayonove C, Cordonnier R, Dubois P (1975) Etude d'une fraction caractéristique de l'arôme du raisin de la variété Cabernet-Sauvignon; mise en evidence de la 2-méthoxy-3-isobutylpyrazine. C R Acad Sci Paris Ser D 281:75–78

Bignetti E, Cavaggioni A, Pelosi P, Persaud KC, Sorbi RT, Tirindelli R (1985) Purification and characterization of an odorant-binding protein from cow nasal tissue. Eur J Biochem 149:227–231

Boison JOK, Tomlinson RH (1990) New sensitive method for the examination of the volatile flavor fraction of Cabernet Sauvignon wines. J Chromatogr 522:315–327

Boyd RK (1993) Quantitative trace analysis by combined chromatography and mass spectrometry using external and internal standards. Rapid Commun Mass Spectrom 7:257–271

Buttery RG, Ling LC (1973) Earthy aroma of potatoes. J Agric Food Chem 21:745–746

Buttery RG, Seifert RM, Guadagni DG, Ling LC (1969a) Characterization of some volatile constituents of bell peppers. J Agric Food Chem 17:1322–1327

Buttery RG, Seifert RM, Lundin RE, Guadagni DG, Ling LC (1969b) Characterisation of an important aroma component of bell peppers. Chem Ind (Lond):490–491

Buttery RG, Seifert RM, Ling LC (1970) Characterization of some volatile potato components. J Agric Food Chem 18:538–539

Gallois A, Kergomard A, Adda J (1988) Study of the biosynthesis of 3-isopropyl-2-methoxypyrazine produced by *Pseudomonas taetrolens*. Food Chem 28:299–309

Giovannini MG, Pieraccini G, Moneti G (1991) Isotope dilution mass spectrometry: definitive methods and reference materials in clinical chemistry. Ann Ist Super Sanità 27:401–410

Harris RLN, Lacey MJ, Brown WV, Allen MS (1987) Determination of 2-methoxy-3-alkylpyrazines in wine by gas chromatography/mass spectrometry. Vitis 26:201–207

Heymann H, Noble AC, Boulton RB (1986) Analysis of methoxypyrazines in wines. 1. Development of a quantitative procedure. J Agric Food Chem 34:268–271

Lacey MJ, Sanders CJ (1992) Chemical composition of sex pheromone of oriental fruit moth and rates of release by individual female moths. J Chem Ecol 18:1421–1435

Lacey MJ, Allen MS, Harris RLN, Brown WV (1991) Methoxypyrazines in Sauvignon blanc grapes and wines. Am J Enol Vitic 42:103–108

Leete E, Bjorklund JA, Reineccius CA, Cheng T-B (1992) Biosynthesis of 3-isopropyl-2-methoxypyrazine and other alkylpyrazines: widely distributed flavour compounds. In: Patterson RLS, Charlwood BV, MacLeod G, Williams AA (eds) Bioformation of flavours. Royal Society of Chemistry, Cambridge, pp 75–95

Maga JA (1982) Pyrazines in foods: an update. CRC Crit Rev Food Sci Nutr 16:1–48

Maga JA (1989) Sensory and stability properties of added methoxypyrazines to model and authentic wines. In: Charalambous G (ed) Flavors and off-flavors. Proc of 6th Int Flavour Conference. Rethymnon, Crete, 5–7 July 1989. Elsevier, Amsterdam, pp 61–70

Maga JA, Sizer CE (1973) Pyrazines in foods: a review. J Agric Food Chem 21:22–30

Meigh DF, Filmer AAE, Self R (1973) Growth-inhibitory volatile aromatic compounds produced by *Solanum tuberosum* tubers. Phytochemistry 12:987–993

Millard BJ (1980) Sources of error in quantitative mass spectrometry. In: Facchetti S (ed) Applications of mass spectrometry to trace analysis. Lectures of a course held at the Joint Research Centre, 29 Sept–3 Oct 1980. Ispra, Italy. Elsevier, Amsterdam, 1982, pp 163–179

Miller JN (1991) Basic statistical methods for analytical chemistry. Part 2. Calibration and regression methods. Analyst 116:3–14

Milne GWA, Lacey MJ (1974) Modern ionization techniques in mass spectrometry. CRC Crit Rev Anal Chem, July:45–104

Moore BP, Brown WV, Rothschild M (1990) Methylalkylpyrazines in aposematic insects their hostplants and mimics. Chemoecology 1:43–51

Murray KE, Whitfield FB (1975) The occurrence of 3-alkyl-2-methoxypyrazines in raw vegetables. J Sci Food Agric 26:973–986

Murray KE, Shipton J, Whitfield FB (1970) 2-Methoxypyrazines and the flavour of green peas (*Pisum sativum*). Chem Ind (Lond):897–898

Nursten HE, Sheen MR (1974) Volatile flavour components of cooked potato. J Sci Food Agric 25:643–663

Parliment TH, Epstein ME (1973) Organoleptic properties of some alkylsubstituted alkoxy and alkylthiopyrazines. J Agric Food Chem 21:714–716

Pelosi P, Baldaccini NE, Pisanelli AM (1982) Identification of a specific olfactory receptor for 2-isobutyl-3-methoxypyrazine. Biochem J 201:245–248

Pes D, Pelosi P (1995) Odorant-binding proteins of the mouse. Comp Biochem Physiol B-Biochem Mol Biol 112:471–479

Robins RJ (1986) The measurement of low-molecular-weight, non-immunogenic compounds by immunoassay. In: Linskens HF, Jackson JF (eds) Immunology in plant sciences. Modern methods in plant analysis, New Ser, ver. 4. Springer, Berlin Heidelberg New York

Ronnett GV, Parfitt DJ, Hester LD, Snyder SH (1991) Odorant-sensitive adenylate cyclase: rapid, potent activation and desensitization in primary olfactory neuronal cultures. Proc Natl Acad Sci USA 88:2366–2369

Seifert RM, Buttery RG, Guadagni DG, Black DR, Harris JG (1970) Synthesis of some 2-methoxy-3-alkylpyrazines with strong bell pepper-like odors. J Agric Food Chem 18:246–249

Seifert RM, Buttery RG, Guadagni DG, Black DR, Harris JG (1972) Synthesis and odor properties of some additional compounds related to 2-isobutyl-3-methoxypyrazine. J Agric Food Chem 20:135–137

Slingsby RW, Kepner RE, Muller CJ, Webb AD (1980) Some volatile components of *Vitis vinifera* variety Cabernet Sauvignon wine. Am J Enol Vitic 31:360–363

Sola C (1995) Chemoattraction of upstream migrating glass eels *Anguila anguila* to earthy and green odorants. Environ Biol Fishes 43:179–185

Takken HJ, van der Linde LM, Boelens M, van Dort JM (1975) Olfactive properties of a number of polysubstituted pyrazines. J Agric Food Chem 23:638–642

Williams PJ, Allen MS (1996) The analysis of flavouring compounds in grapes. In: Linskens H-F, Jackson JF (eds) Fruit analysis. Modern methods of plant analysis, vol 18. Springer, Berlin Heidelberg New York, pp 37–57

Comparison of Extraction Techniques for Characterizing Tea Aroma and Analysis of Tea by GC-FTIR-MS

M. KAWAKAMI

1 Introduction

The custom of drinking tea has spread all over the world. *Camellia sinensis* tea is the most popular beverage in the world, and mate tea (*Ilex paraguaryensis*) and rooibos tea (*Aspalathus linearis*) are locally consumed in South America and South Africa (Dahlgren 1968), respectively. The aroma is the major factor contributing to tea quality and enjoyment. Research into characterizing tea aroma has been carried out for past 100 years, and more than 600 volatile compounds have been reported or isolated by using GC or GC-MS, in spite of the small quantity of essential oil (600 ppm) (Straten and Maarse 1989; Yamanishi 1995). Although more than 99.9% of tea essential oil has been determined, characterizing tea aroma has remained unidentified. The problem has been caused by the severe preparation process for aroma concentrates. Tea aroma varies with current extraction methods such as SDE (simultaneous distillation and distraction) or SDR (steam distillation under reducing pressure), and the aroma concentrate obtained through these preparation processes no longer has the original tea aroma.

Brewed tea extract is usually drunk after the leaves have soaked in hot water for 1–3 min, and the analysis of the brewed tea extract is considered to be more important than that of the essential oil to elucidate the real tea aroma. These requirements for characterizing tea aroma led to comparing extraction techniques between SDE, SDR, and the brewed extraction method, by using high resolution gas chromatography (GC) combined with Fourier transform infrared spectrometry (FTIR) and mass spectrometry (MS).

2 Preparation of Aroma Concentrates

2.1 Simultaneous Distillation and Extraction (SDE)

The method by the modified Likens-Nickerson apparatus of simultaneous distillation and extraction (Schultz et al. 1977) is currently used to obtain essential oils from plants. Generally, 100 g of powdered tea are placed in a sample flask with 500 ml of deionized boiling water, and 50 ml of dichloromethane or ethyl ether in a solvent flask. The sample flask and the solvent flask, attached to the SDE appara-

Modern Methods of Plant Analysis, Vol. 19
Plant Volatile Analysis
Edited by H.-F. Linskens and J.F. Jackson

tus, are heated at 100 and 50 °C for 30–60 min, respectively. After the extract has been dried over anhydrous sodium sulfate for 12 h, sodium sulfate and solvent are removed carefully, using a Kunderna-Danish evaporative concentrator.

This method of preparation has the advantage of performing two processes simultaneously and increasing the yield of aroma concentrate; but the drawback of this method is the alteration in the aroma components. As a tea sample is boiled in hot water at 100 °C for 30–60 min, hydrolysis of the aroma compounds or aroma precursors easily occurs.

2.2 Steam Distillation Under Reduced Pressure (SDR)

Steam distillation under normal pressure such as SDE is a simple isolation method to obtain a large yield of essential oil; but the method has the disadvantage of changing the aroma quality. Steam distillation under reduced pressure improves the quality a little because overheating of the sample can be avoided.

Steam distillation by using a rotary evaporator apparatus is effective in obtaining essential oil (Yamanishi et al. 1970). One hundred g of powdered tea mixed with 500 ml of deionized water (60 °C) in a flask is evaporated under reduced pressure (20–30 mmHg). During distillation, another 300 ml of deionized water (60 °C) is added, and around 750 ml of distillate is trapped in a cooled flask. The distillate is saturated with sodium chloride and extracted with ethyl ether. After drying over anhydrous sodium sulfate for 12 h, the solvent is removed carefully, using a Kunderna-Danish evaporative concentrator.

This preparation method also alters the aroma composition and cannot avoid hydrolysis of aroma precursors. Figure 1 shows the comparison of the GC patterns between SDE and SDR sample preparations (Kawakami et al. 1987).

2.3 Brewed Extraction Method (BW)

Aroma studies of *Camellia sinensis* tea have been reported by various authors (Straten and Maarse 1989). However, GC aroma patterns of the vacuum steam volatile, steam distillation volatile, and headspace vapor have failed to reproduce the real tea aroma experienced by humans (Kobayashi and Kawakami 1991). The resolution of these questions depends on the analysis of the tea infusion that we drink.

The direct, brewed extraction method (Fig. 2) is more suitable for reproducing the real tea aroma as the result of model experiments (Fig. 3; Kawakami et al. 1993a,b). Under this preparation, alteration of aroma compounds can be minimized. The brewed extract accurately reproduced the GC pattern of the standard mixture, while the SDE extract lacked lactone compounds such as 4-butanolide and dihydroactinidiolide. The lactone ring is easily opened through hydrolysis by boiling in water.

One hundred g of powdered tea is brewed in 700 ml of deionized boiling water for 10 min. After filtration, the filtrate is saturated with sodium chloride and is

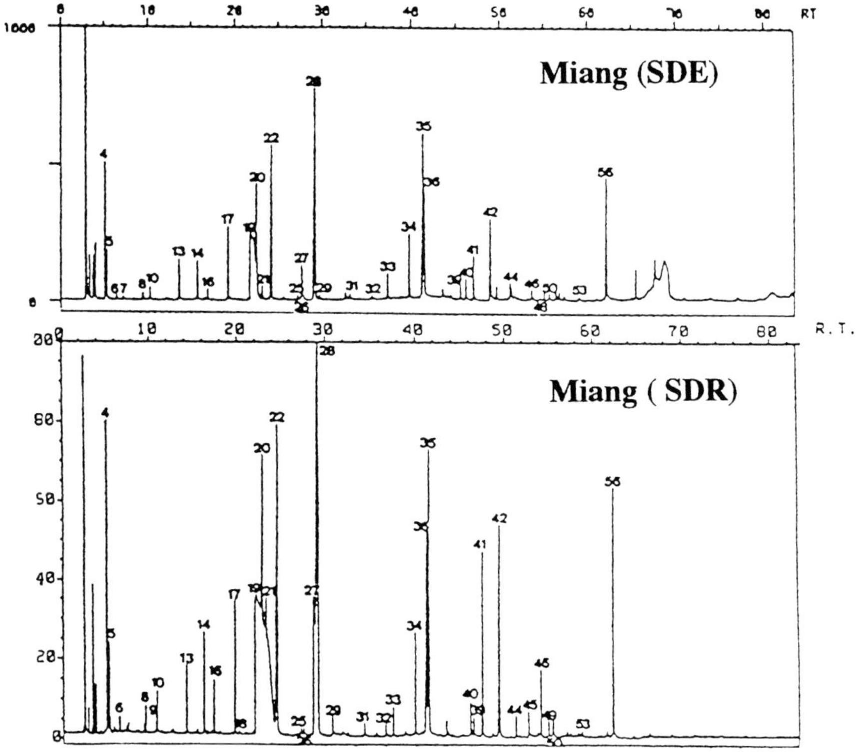

Fig. 1. Gas chromatograms of pickled tea Miang in Thailand: SDE extract and SDR extract. FFAP 0.25 mm i.d. × 50 m capillary column; 60 °C (4 min) → 2 °C/min → 180 °C. *4* 2-Butanol; *5* propanol; *10* 3-methylbutanol; *13* acetoin; *14* 2-heptanol; *17* (*Z*)-3-hexenol; *19* acetic acid; *20* linalool oxide I; *21* 1-octen-3-ol; *22* linalool oxide II; *27* propionic acid; *28* linalool; *29* isobutyric acid; *31* butyric acid; *32* isovaleric acid; *33* α-terpineol; *34* linalool oxide III; *35* linalool oxide IV; *36* methyl salicylate; *39* hexanoic acid; *40* geraniol; *41* benzyl alcohol; *42* 2-phenylethanol; *44* 4-methyguaiacol; *45* (*Z*)-3-hexenoic acid; *46* phenol; *56* 4-ethylphenol

Tea sample 100g

↓ powdered
soaked in hot water (700 ml), 10 min

Tea infusion

↓ filtration

Filtrated solution

↓ extraction with CH_2Cl_2

CH_2Cl_2 aroma solution

↓ centrifuging
concentration

Brewed extract (aroma concentrate)

Fig. 2. Preparation procedure by brewed extraction method

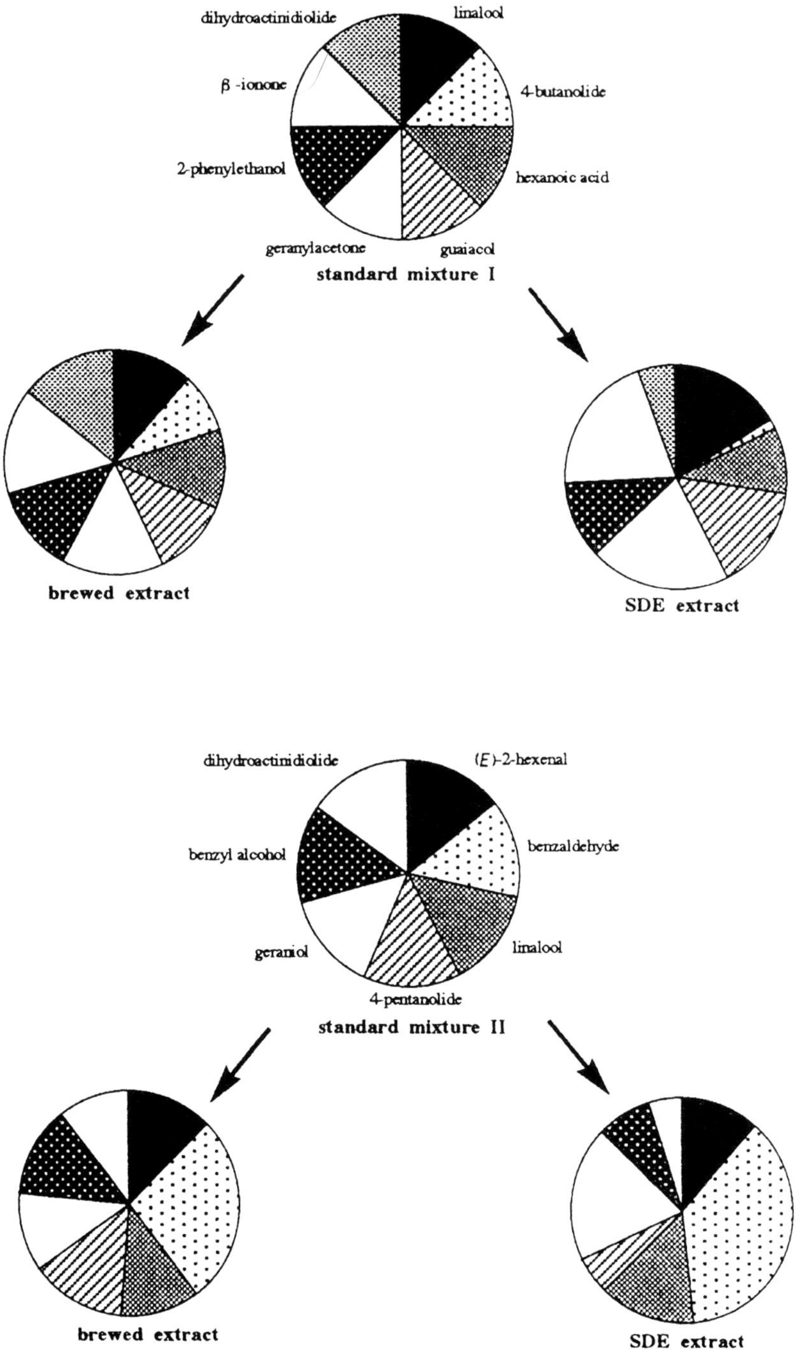

Fig. 3. Alteration of GC patterns by model experiment using standard mixture; brewed extraction method and SDE method

extracted by using 200 ml dichloromethane. If emulsion occurs, it is easily separated by using a centrifugal separator. The extract is dried over anhydrous sodium sulfate for 12 h. After the sodium sulfate is filtered out, the solvent is removed carefully, using a Kunderna-Danish evaporative concentrator. As the content of

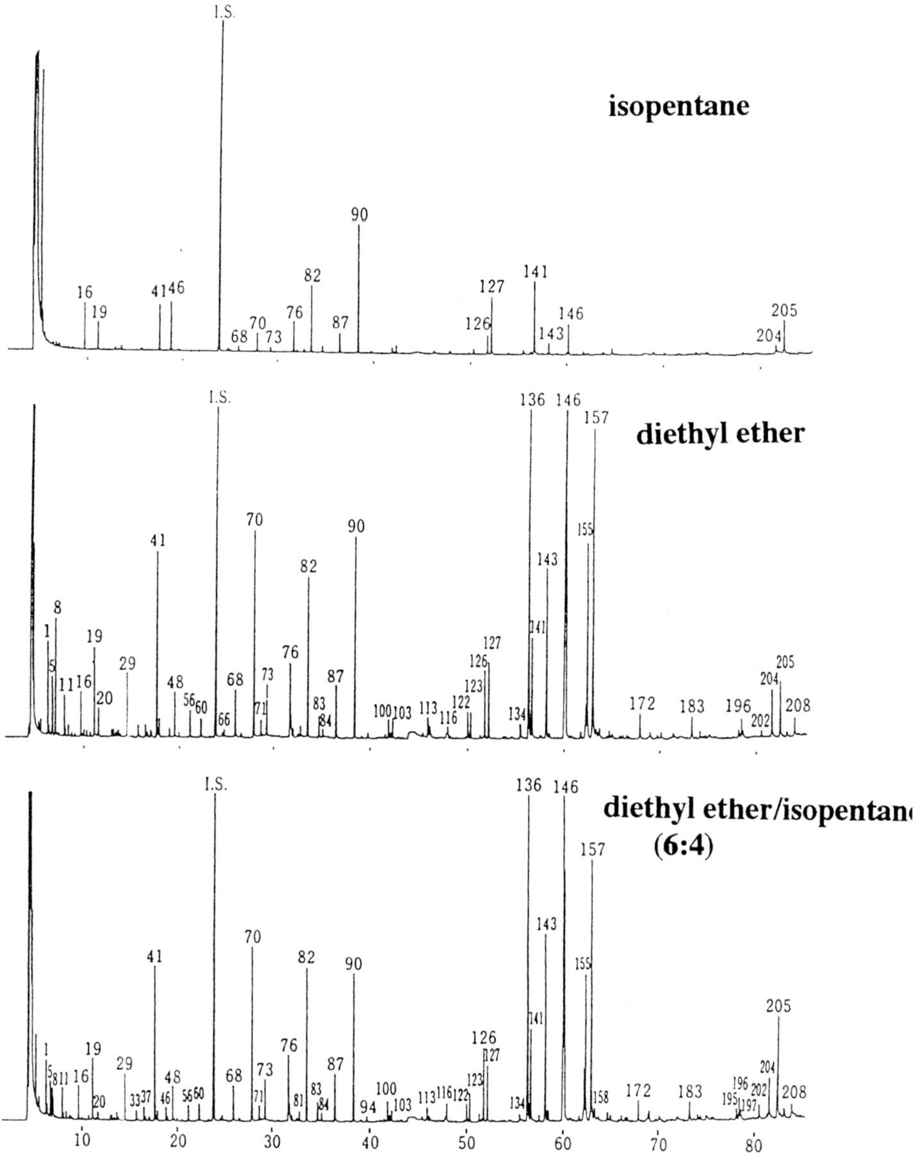

Fig. 4. Gas chromatograms of Darjeeling black tea from Porapak Q: eluted from isopentane, diethyl ether, and diethyl ether/isopentane (6:4) (Shigematsu et al. 1994). PEG 20 M 0.25 mm i.d. × 60 m capillary column; 60 °C (5 mm) → 2 °C/min → 230 °C; *I.S.* internal standard

essential oil in the brewed extract is not so high, the yield of aroma concentrate is less than one-tenth of the SDE method.

2.4 Porapack Q Column Adsorption Method

This method is one of modified brewed extractions. Under this method, the brewed extract is not directly extracted by solvent, but is passed through a Porapack Q column. Aroma compounds and catechins are adsorbed once in Porapack Q, then aroma compounds are released by solvent. The disadvantage of this method is the alteration of the aroma pattern by different extraction solvents (Shigematsu et al. 1994). The result of comparison of three kinds of solvent shows that diethyl ether is the most suitable for extraction, but catchins mix with the aroma concentration. A mixed solvent of isopentane and diethyl ether (4:6) also gives almost the same GC pattern of ethyl ether; but isopentane gives a quite different GC pattern, as shown in Fig. 4. The GC pattern of Porapack Q column adsorption diethyl ether extract agrees with that of the brewed extract (see Fig. 7).

Ten g of powdered tea are brewed in 400 ml of deionized boiling water for 3 min. After filtration, the filtrate is passed through 20 ml of Porapack Q column (50–80 mesh). After cleaning with 50 ml deionized water, aroma compounds are extracted by using ca. 50 ml of solvent. The extract is dried over anhydrous sodium sulfate. After sodium sulfate is filtered out, the extract is concentrated.

2.5 Headspace Gas Analysis

A Tenax GC trapping method for gas absorption (Tsugita et al. 1979), the purged and trap injection (PTI) method (Badings et al. 1985), and the column injection method are helpful for analysis of tea volatiles.

Vapor from tea infusion stimulates the human olfactory organ; but the GC pattern of headspace gas does not correspond to that of tea infusion. Because each compound has a characteristic volatility, highly volatile compounds occupy the headspace vapor. Therefore, analysis of headspace gas cannot clearly characterize the tea itself.

3 Instrumental Method: Separation and Identification of Components

3.1 Gas Chromatography

The flame ionization detector (FID) is most popular as GC detector and can give information about the quantity of constituents by peak area proportion. In order to know the GC peak's aroma, sniffing of the GC effluent is indispensable, using GC equipped with a thermal conductivity detector (TCD). A wide bore column (0.53 mm i.d.) is effective in sniffing.

3.2 Gas Chromatography-Fourier Transform Infrared Spectrometry-Mass Spectrometry

A combined system of GC-FTIR-MS is convenient for flavor analysis (Smith 1984). GC equipped with a capillary column provides clear separation of aroma compounds. Usually, 0.25 mm i.d. × 50–60-m-long chemically bonded, fused silica, and wall-coated open tube (FS-WCOT) is used for tea aroma analysis. The polyethylene glycol type of polar liquid phase (PEG 20 M) is mostly used because of its efficient separation.

Under GC-FTIR-MS, the GC effluent is divided into two detectors, FTIR and MS. Generally, the effluent FTIR-MS ratio is set up 10 : 1 on account of the difference in sensitivity. IR spectra provide much useful information on the functional group, the presence of a double bond, and the stereo-chemical formula. Mass spectra give information on the molecular ion and decomposition pattern of the compound by ion fragment analysis. A library search of mass spectrum or IR spectrum is enough to identify each compound. The comprehensive information from GC retention time (Kovats' index), IR spectrum, and MS spectrum has made it possible to identify a more complex structure. The 3D IR spectrum and MS spectrum are also available by sophisticated integrator or computer.

3.3 Isolation of Optical Isomers by Using a Capillary Column Coated with an Optically Active Liquid Phase, Permethylated β-Cyclodextrin

CP-cyclodextrin capillary column is effective in separating chiral compounds. The chirality of compounds determines human organoleptic evaluation, because each olfactory receptor can only accept the corresponding stereoisomer, and most of these stereoisomers differ in aroma quality and threshold value (Acree et al. 1985; Mosandl et al. 1986, 1990). Therefore it is very important for aroma research to clarify the absolute configuration of the stereoisomer.

For example, linalool oxides, the main components of tea aroma, have eight stereoisomers. As shown in Fig 5, each optical isomer from four kinds of diastereomer is separated clearly by using a CP-cyclodextrin capillary column (Wang et al. 1994). Those odors identified by sniffing at the GC exit port of the CP-cyclodextrin column are shown in Table 1 (Wang et al. 1994). (2*S*)-isomers have a sweet floral creamy note, while (2*R*)-isomers are earthy and leafy.

Figure 6 shows 3D IR and MS spectra of 2,6-dimethyl-3,7-octadiene-2,6-diol in Taiwanese Chan Pin oolong brewed extract, separated by using a CP-cyclodextrin (0.25 mm i.d. × 50 m) capillary column. This compound and the dehydration compound, 3,7-dimethyl-1,5,7-octatrien-3-ol, are key substances of special teas such as Taiwanese Chan Pin oolong, Taiwanese Pom Fon oolong, and Indian Darjeeling, made from tea leaves infested with green fly (*Emposca flavescens*). These two

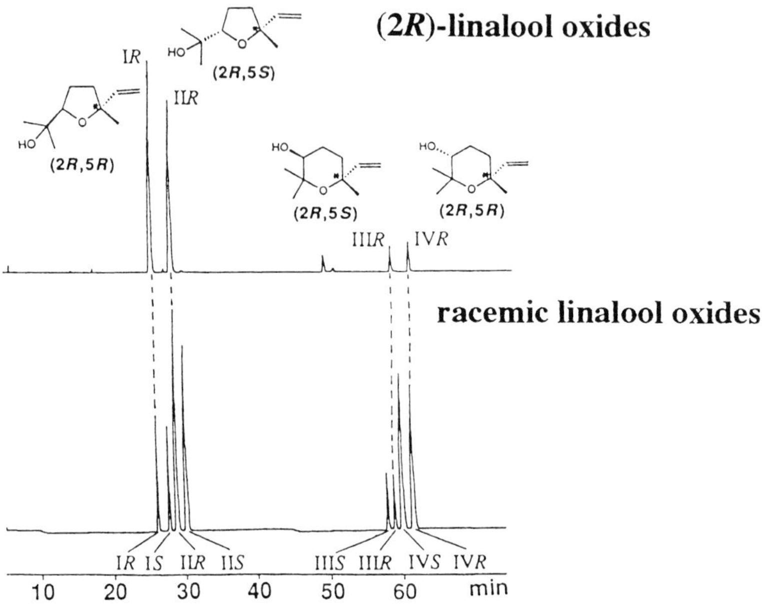

Fig. 5. Gas chromatograms of linalool oxide stereoisomers: (2*R*)-linalool oxides and racemic linalool oxides (Wang et al. 1994). CP-cyclo-dextrin-B-236-M-19 (0.25 mm i.d. × 50 m) capillary column; 100 °C (40 mm) → 1 °C/min → 130 °C

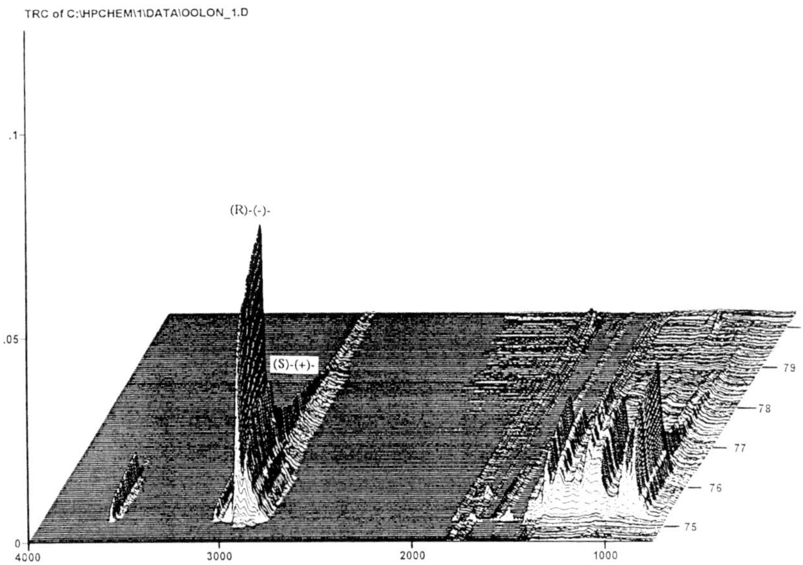

Fig. 6. 3D IR and MS spectra of (*R*)-(−)- and (*S*)-(+)-2,6-dimethyl-3,7-octadiene-2,6-diol in Chan-Pin oolong brewed extract. CP-cyclo-dextrin-B-236-M-19 (0.25 mm i.d. × 50 m) capillary column; 60 °C (8 min) → 1 °C/min → 200 °C

Table 1. Configuration and odor of linalool oxide enantiomers. (Wang et al. 1994)

Compound	Configuration	KI[a]	Odor
Linalool oxide I	2*R*, 5*R*	1191	Leafy, earthy
trans, furanoid	2*S*, 5*S*	1203	Sweet, floral, creamy
Linalool oxide II	2*R*, 5*S*	1210	Leafy, earthy
cis, furanoid	2*S*, 5*R*	1222	Sweet, floral, creamy
Linalool oxide III	2*S*, 5*R*	1357	Sweet, floral, creamy
trans, pyranoid	2*R*, 5*S*	1362	Earthy
Linalool oxide IV	2*S*, 5*S*	1367	Sweet, floral, creamy
cis, pyranoid	2*R*, 5*R*	1375	Earthy

[a] Kovats index on CP-cyclo-dextrin-B-236-M-19.

compounds appear to be related to the Darjeeling's characteristic muscat flavor. The ratio of (*R*) and (*S*)-configuration and (*S*)-configurational enantiomeric excess (*e.e.*%) of these compounds shows the relation between these two components (Table 2). (*S*)-configurational enantiomeric excess (*e.e.*%) is calculated from the value of GC peak areas according to the following equation.

$$e.e.\% = [(A_S - A_R)/(A_S + A_R)] \times 100$$

A

A_S: peak area of the (*S*)-isomer

A_R: peak area of the (*R*)-isomer

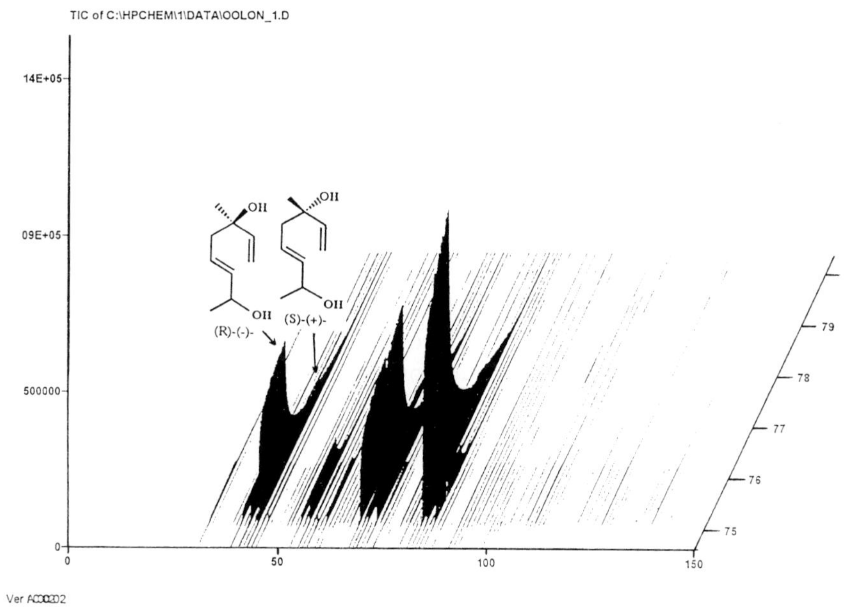

Fig. 6. Continued

Table 2. *e.e.*Percentage[a] of linalool, 3,7-dimethyl-1,5,7-octatrien-3-ol, and 2,6-dimethyl-3,7-octadiene-2,6-diol between Chan Pin oolong tea and Darjeeling black tea

	Chan Pin oolong	Darjeeling
Linalool	44.2	−4.8
3,7-Dimethyl-1,5,7-octatrien-3-ol	−26.0	100.0
2,6-Dimethyl-3,7-octadiene-2,6-diol	−83.6	85.2

[a] *e.e.*% = [(As − Ar)/(As + Ar)] × 100. **As** *S*-isomer peak area; **Ar** *R*-isomer peak area.

3,7-Dimethyl-1,5,7-cotatrien-3-ol has been reported to be produced from 2,6-dimethyl-3,7-octadiene-2,6-diol in vitro by dehydration at 120–130 °C (Hara and Hotta 1987) and has also been reported as the same thermal induction product in muscat grape juices (Williams et al. 1980). In black tea flavor, only (*S*)-(+)-2,6-dimethyl-3,7-octadiene-2,6-diol (Etho et al. 1980) and (*S*)-(+)-3,7,-dimethyl-1,5,7-octatrien-3-ol (Nakatani et al. 1969) have been previously isolated, and both compounds were estimated to be produced from (*S*)-(+)-linalool. In contrast, Chan Pin oolong tea mainly contains (*R*)-(−)-2,6-dimethyl-3,7-octadiene-2,6-diol and (*R*)-(−)-3,7,-dimethyl-1,5,7-octatrien-3-ol, regardless of the (*R*)/(*S*) ratio of linalool.

4 Comparison of Extraction Techniques: Brewed Extraction Method and SDE Method

4.1 *Camellia sinensis* Tea Aroma

Volatile components of oolong teas and black teas were prepared by using two different extraction methods, the brewed extraction method and the SDE method, and were compared by gas chromatography and gas chromatography/FTIR spectrometry/mass spectrometry (Kawakami et al. 1994, 1995a,b). The characteristic tea aroma of each was successfully clarified by the brewed extraction method. The aroma obtained from the brewed extract bore a strong resemblance in taste to each tea. The GC pattern of brewed tea extracted by dichloromethane (brewed extract) differs greatly from the GC pattern of SDE extract, as shown in Fig. 7.

→

Fig. 7. Gas chromatograms of Darjeeling black tea and clone DT-1 black tea: brewed extract and SDE extract PEG 20 M 0.25 mm i.d. × 50 m capillary column; 60 °C (4 min) → 2 °C/min → 180 °C. *1* (*Z*)-3-hexenol; *2* linalool oxide I; *3* linalool oxide II; *4* linalool; *5* 3,7-dimethyl-1,5,7-octatrien-3-ol; *6* linalool oxide III; *7* methyl salicylate; *8* linalool oxide IV; *9* hexanoic acid; *10* geraniol; *11* benzyl alcohol; *12* 2-phenylethanol; *13* 2,6-dimethyl-3,-7-ocatadiene-2,6-diol; *14* (*Z*)-3-hexenoic acid; *15* (*E*)-2-hexenoic acid; *16* *trans*-geranic acid

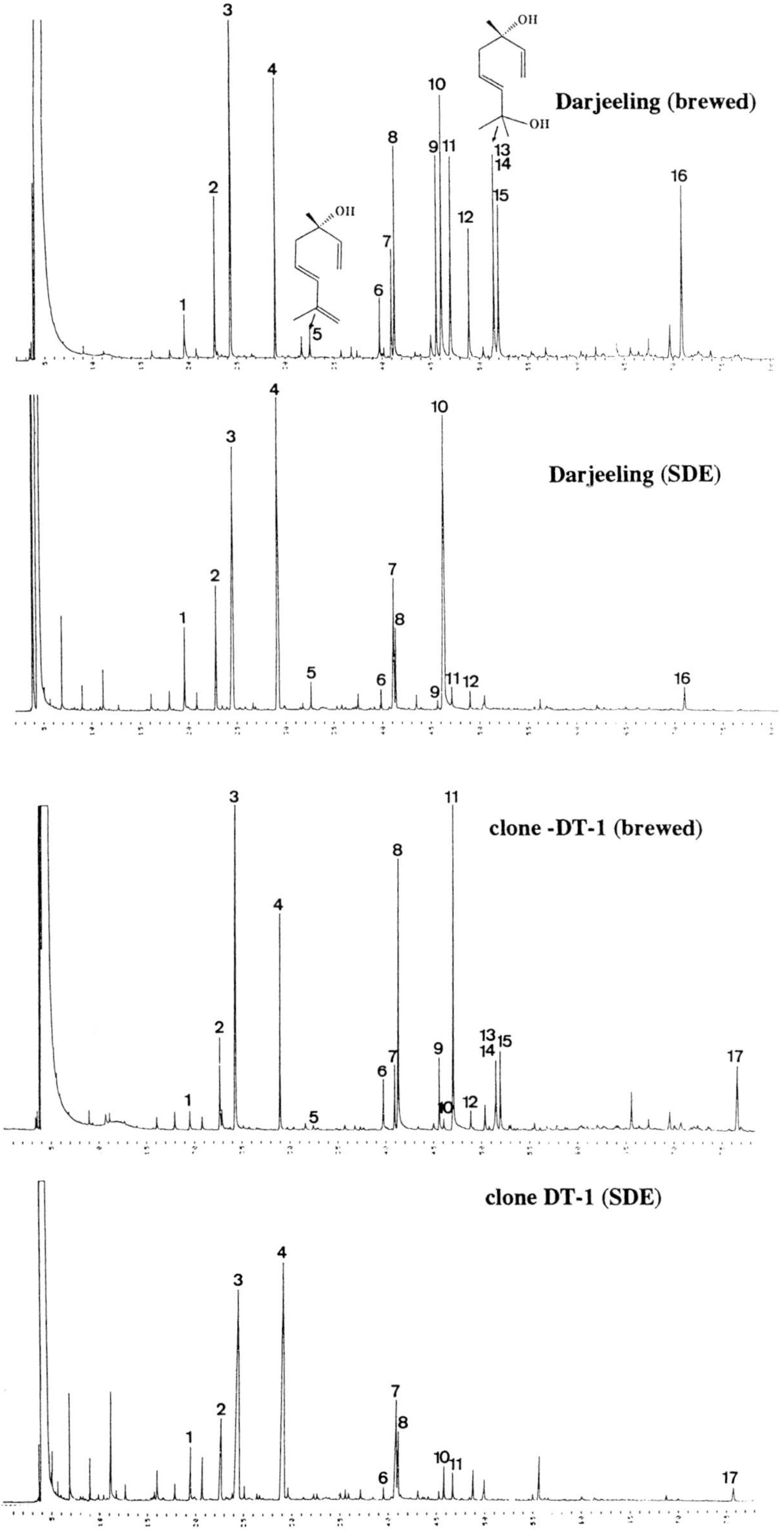

OH
OH
OH
Darjeeling (brewed)
1
2
3
4
5
6
7
8
9
10
11
12
13
14
15
16
Darjeeling (SDE)
1
2
3
4
5
6
7
8
9
10
11
12
16
clone -DT-1 (brewed)
1
2
3
4
5
6
7
8
9
10
11
12
13
14
15
17
clone DT-1 (SDE)
1
2
3
4
6
7
8
10
11
17

For example, 38 compounds were identified from the Darjeeling brewed extract, and 51 compounds from the Darjeeling SDE extract. The Darjeeling brewed extract, which has a more complicated aroma pattern, consists of four linalool oxides, linalool, geraniol, hexanoic acid, benzyl alcohol, 2-phenylethanol, *trans*-gernaic acid, (*E*)-2-hexenoic acid, (*Z*)-3-hexenoic acid, and 2,6-dimethyl-3,7-octadiene-2,6-diol, while the Darjeeling SDE extract consists of seven main components which include geraniol, linalool, four linalool oxides, and methyl salicylate. Terpene alcohols such as geraniol and linalool, which have been reported previously as the main Darjeeling tea flavor, were fewer in the brewed extract. These alcohols such as geraniol, linalool, benzyl alcohol, 2-phenylethanol, and linalool oxides are formed by the SDE process from the precursors, glycosidic derivatives, namely, β-primeverosides (6-*O*-β-D-xylopyranosyl-β-D-glucopyranosides) (Guo et al. 1994a,b). 2,6-Dimethyl-3,7-octadiene-2,6-diol, found in larger quantities only in brewed extract, seemed to become dehydrated during the SDE process. The relation between 2,6-dimethyl-3,7-octadiene-2,6-diol and the dehydration compound, 3,7-dimethyl-1,5,7-octatrien-3-ol, contributing to Darjeeling's characteristic muscat flavor, could not be clarified without the brewed extraction method.

The main components of the brewed Sri Lankan clone DT-1 extract, with a strong floral flavor, were linalool oxide II, linalool, linalool oxide IV, and benzyl alcohol, while the GC peak of benzyl alcohol become very small in the SDE extract. Generally, the relative quantity of benzyl alcohol and 2-phenylethanol decreased in the SDE tea extract on account of the increase in terpene alcohols.

Gas chromatograms of Huang Chin Kuei, which is a typical aromatic Chinese oolong tea produced in the Fujian Province of the People's Republic of China, are shown in Fig. 8. Fifty-two compounds were identified from the brewed extract, and 100 compounds from the SDE extract of Huang Chin Kuei. From the comparison of the brewed extract with the SDE extract of Huang Chin Kuei, the SDE extract shows thermal effects during distillation and extraction. Terpenes such as linalool and α-farnesene, especially nerolidol, reported previously as the main oolong tea flavor (Nobumoto et al. 1990), were found in only small amounts in brewed extract, similar to the result obtained for Darjeeling tea. Concerning nerolidol, the precursor has not been identified. The concentrate of brewed extract consists of high levels of jasmine lactone, 2-phenylethanol, and indole. The amount of jasmine lactone, making up 26.43% of the aroma composition, is worth especial note. Other important componets in brewed extract were benzyl alcohol, benzylcyanide, methyl jasmonete, 2-phenylethyl benzoate, hexanoic acid, and 2,6-dimethyl-3,7-octadiene-2,6-diol. The amounts of lactone compounds such as jasmine lactone, dihydroactinidiolide, 4-hexanolide, 5-decanolide, 4-butanolide, 4-nonanolide, and 2-hexen-4-olide were found to be decreased by the SDE process, as in rooibos tea analysis (Kawakami et al. 1993a,b), and these lactones appear to be easily hydrolyzed to the hydroxy-acid compound by SDE.

Important facts on tea aroma, especially infusion tea aroma, were clarified through this study by the use of the brewed extraction method.

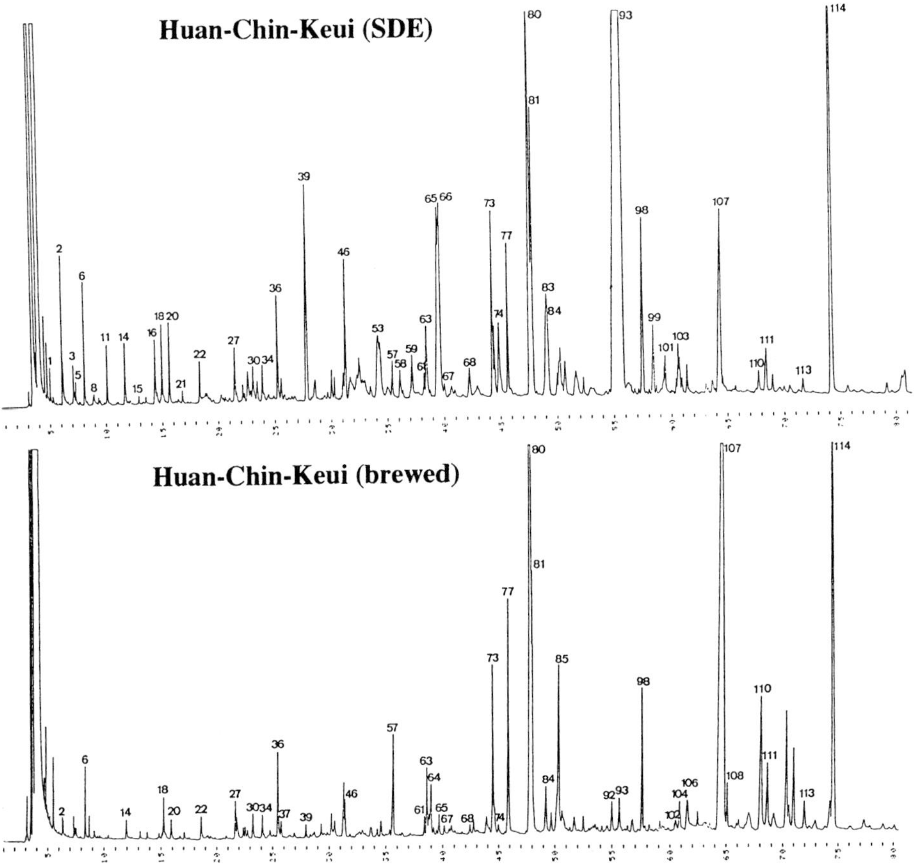

Fig. 8. Gas chromatograms of Huan Chin Kuei oolong tea: brewed extract and SDE extract. PEG 20 M 0.25 mm i.d. × 50 m capillary column; 60 °C (4 min) → 2 °C/min → 180 °C. *36* Benzaldehyde; *39* linalool; *46* 3,7-dimethyl-1,5,7-octatrien-3-ol; *57* 4-hexanolide; *65* methyl salicylate; *66* α-farnesene; *73* hexanoic acid; *77* benzyl alcohol; *80* 2-phenylethanol; *81* benzylcyanide; *85* 2,6-dimethyl-3,7-octadiene-2,6-diol *93* nerolidol; *98* 2-phenylethyl benzoate; *107* jasmine lactone; *110* dihydroactinidiolide; *111* methyl jasmonete; *114* indole

4.2 Comparison of the Various Black Tea Aromas by the Brewed Extraction Method

Table 3 shows aroma compositions of famous black teas prepared by the brewed extraction method (Kawakami et al. 1994, 1995a,b). The aroma of Sri Lankan clone 2025 consists mainly of linalool oxide II, linalool, methyl salicylate, hexanoic acid, and (*Z*)-3-hexenoic acid. The aroma of Chinese Keemun contains high amounts of benzyl alcohol, 2-phenylethanol, hexanoic acid, N-ethylsaccinimide, and dihydroactinidiolide. Hexanoic acid, (*E*)-2-hexenoic acid, and dihydroactinidiolide contribute to the aroma of Java Taloon. These aroma patterns differ from those of SDR extracts, as reported in a previous study (Aisaka et al. 1978).

Table 3. Composition of the brewed extracts of black tea, Darjeeling, clone DT-1, clone 2025, Java Taloon, and Keemun

KI[a]	Compound	Darjeeling	c-DT-1	c-2025	Java	Keemum
1192	(*E*)-2-Hexenal	0.57	0.35	0.69	2.05	1.00
1361	(*Z*)-3-Hexenol	3.00	0.90	0.91	3.35	0.82
1386	(*E*)-2-Hexenol	0.67	0.64	0.78	1.27	0.04
1415	Linalool oxide I	3.95	3.43	2.01	1.41	1.63
1440	Linalool oxide II	14.47	20.01	8.95	4.67	3.81
1522	Linalool	5.78	7.43	9.56	3.72	0.97
1573	3,7-Dimethyl-1,5,7-octatrien-3-ol	0.37	0.05	0.10	0.90	0.42
1588	4-Butanolide	0.74	0.16	0.38	2.14	2.78
1660	4-Hexanolide	0.52	0.23	0.99	1.40	0.63
1698	Linalool oxide III	1.65	2.44	0.82	0.61	1.79
1727	Methyl salicylate	3.16	2.40	7.72	1.75	0.39
1739	Linalool oxide IV	6.46	13.58	5.38	3.27	3.46
1807	Hexanoic acid	6.26	2.64	10.68	21.57	9.12
1812	Geraniol	9.34	0.58	0.64	0.07	3.10
1833	Benzyl alcohol	5.35	14.31	2.47	1.53	14.79
1836	N-Ethylsuccinimide	1.99	1.26	1.23	5.17	9.63
1863	2-Phenylethanol	4.74	0.75	1.80	2.93	15.99
1907	2,6-Dimethyl-3,7-octadiene-2,6-diol	3.29	1.08	1.04	0.35	1.82
1907	(*Z*)-3-Hexenoic acid	4.55	2.80	9.40	7.66	2.03
1922	(*E*)-2-Hexenoic acid	5.63	3.52	3.51	13.58	2.86
2157	Jasmine lactone	0.74	2.39	3.35	–	0.69
2251	Dihydroactinidiolide	1.35	1.40	3.76	4.46	4.97
2260	Methyl epijasmonate	–	0.61	1.18	–	–
2294	*trans*-Geranic acid	6.74	1.19	1.90	–	3.26
2365	Indole	–	4.49	4.52	–	0.47

[a] Kovats index on Carbowax 20M.

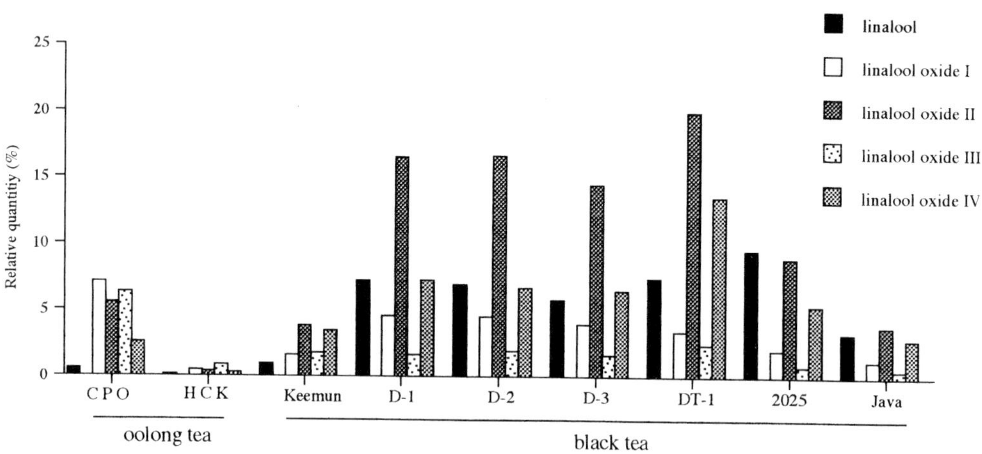

Fig. 9. Ratio of linalool and linalool oxides in brewed extract: oolong teas and black teas *CPO* Chan-Pin-oolong; *HCK* Huan-Chin-Keui-oolong; *Keemun* Keemun black tea; *D-1,D-2,D-3* Darjeeling black tea; *DT-1* clone DT-1 black tea; *2025* clone 2025 black tea; *Java* Java Taloon black tea

Figure 9 shows the ratio of the main tea aroma components, linalool and linalool oxide I, II, III, IV in oolong teas and black teas. It is worth noting that oolong tea contains higher amounts of the *trans* form of linalool oxides, such as I and III, than the *cis* form, while black tea contains higher amounts of the *cis* form, such as II and IV, than the *trans* form. This fact was not clarified in a previous study (Wang et al. 1994) using steam distillation under reduced pressure (SDR). The quantity of linalool and linalool oxides in brewed tea extract is also found not to be as large as SDR extract.

To investigate characteristics of tea aroma, usually SDE extracts have been used for principal components analysis or cluster analysis (Kobayashi et al. 1993). It is necessary to reexamine the analyses by using brewed extracts.

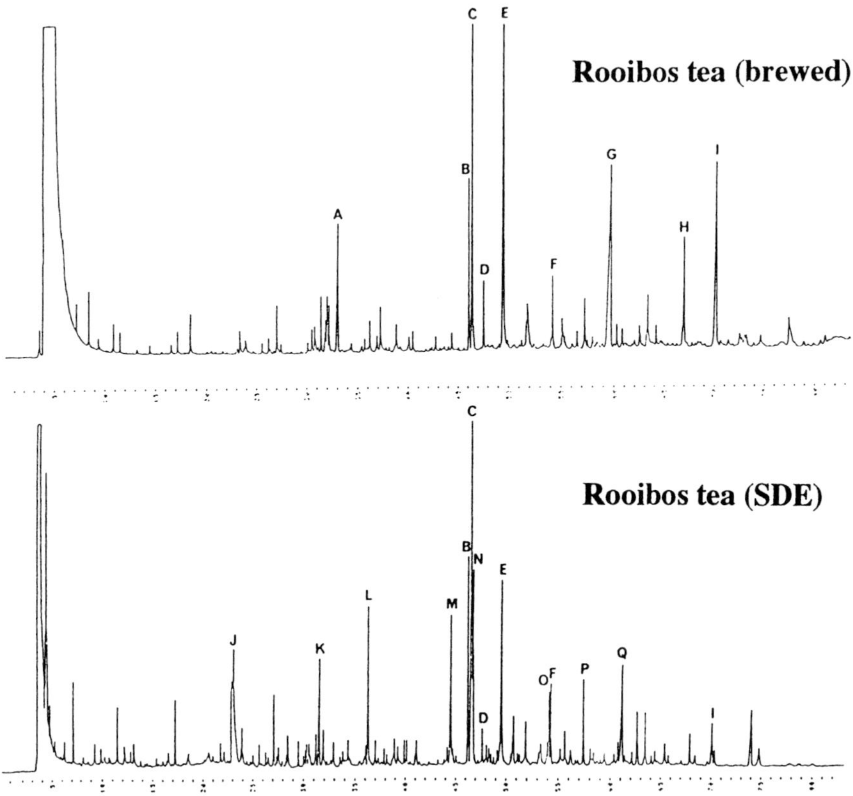

Fig. 10. Gas chromatograms of rooibos tea: brewed extract and SDE extract. PEG 20 M 0.25 mm i.d. × 50 m capillary column; 60 °C (4 min) → 2 °C/min → 180 °C. *A* 4-Butanolide; *B* hexanoic acid; *C* guaiacol; *D* benzyl alcohol; *E* 2-phenylethanol; *F* phenol; *G* 2-methoxy-2-buten-4-olide; *H* methylethylmaleimide; *I* dihydroactinidiolide; *J* acetic acid; *K* 6-methyl-(*E*)-3,5-heptadien-2-one; *L* 3-methylbutanoicacid; *M* β-damascenone; *N* geranylacetone; *O* epoxy-β-ionone; *P* octanoic acid; *Q* 6,10,14-trimethyl-2-pentadecanone

4.3 Rooibos Tea Aroma

As shown in Fig. 10, the concentrate of brewed extract of rooibos tea consists of high levels of 2-phenylethanol, guaiacol, methylethylmaleimide, and three kinds of lactones such as 2-methyoxy-2-buten-4-olide, dihydroactinidiolide, and 4-butanolide. Other important components in brewed extract are also lactones such as 4-pentanolide, 3-methyl-4-butanolide, 2-methyl-4-butanolide, 2-methyl-2-buten-4-olide, 4-hexanolide, 2-hexen-4-olide, 2-buten-4-olide, 5-pentanolide, 5-octanolide, and 4-nonanolide. These lactones are not identified in the SDE extract, except low levels of 4-nonanolide, dihydroactinidiolide, and 2-hexen-4-olide. While the main components in the SDE extract are guaiacol, 2-phenylethanol, ketones such as geranylacetone, β-damascenone, and 6,10-14-trimethyl-2-pentadecanone, and acids such as acetic acid, 3-methylbutanoic acid, hexanoic acid, and octanoic acid. Fused bicyclic hydrocarbons such as α-muurolene, δ-cadinene, calamene, calamenene, and cadalene, probably derived from terpenes, have also been identified (Kawakami et al. 1993a,b). The difference between these two compositions is primarily due to different preparation methods. The lactone skeleton is fragile in heating in the aqueous medium, and easily opened during the SDE process, as mentioned above. Ketones are also degradation products from carotene in heating in aqueous medium (Kawakami 1982).

Table 4. Comparison of main aroma components of roast mate between brewed extract and SDE extract

KI[a]	Compound	Brewed extract	SDE extract
1320	6-Methyl-5-hepten-2-one	2.0	1.3
1371	2-Butoxyethanol	0.7	1.2
1415	Linalool oxide I	5.0	2.2
1427	Furfural	4.8	3.2
1440	Linalool oxide II	5.3	1.3
1539	5-Methylfurlfural	5.2	4.3
1573	4-Butanolide	4.2	–
1807	Hexanoic acid	3.5	3.2
1833	Benzyl alcohol	2.6	0.9
1887	Heptanoic acid	3.2	1.2
1888	Maltol	3.0	–
1953	Phenol	2.8	0.9
1978	2-Formylpyrrole	2.8	0.9
2013	Octanoic acid	5.7	3.5
2096	Eugenol	2.2	2.5
2099	Nonanoic acid	2.8	2.0
2158	2,6-Dimethoxyphenol	3.7	–
2159	Methylethylmaleimide	6.3	–
2260	Dihydroactinidiolide	4.8	0.4
	5-Hydroxymethyl-2-furan-carboxyaldehyde	2.4	–
	Vanilline	5.7	–

[a] Kovats index on Carbowax 20M.

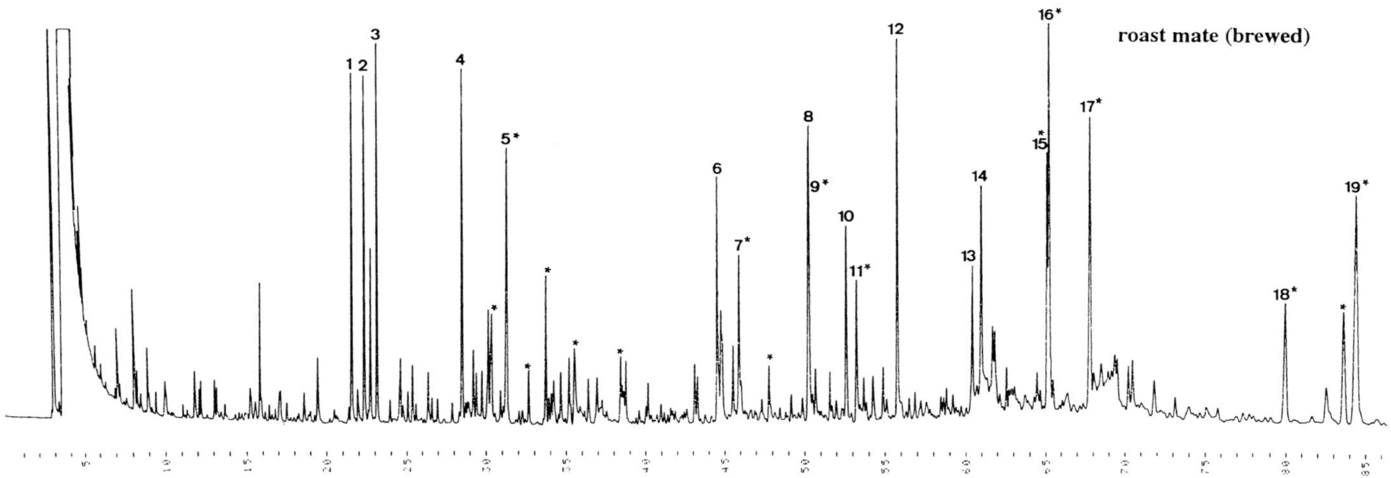

Fig. 11. Gas chromatogram of roast mate tea: brewed extract. PEG 20 M 0.25 mm i.d. × 50 m capillary column; 60 °C (4 min) → 2 °C/min → 180 °C. *1* Linalool oxide I; *2* furfural; *3* linalool oxide II; *4* 5-methylfurfural; *5* 4-butanolide; *6* hexanoic acid; *7* benzyl alcohol; *8* heptanoic acid; *9* maltol; *10* phenol; *11* 2-formylpyrrole; *12* octanoic acid; *13* eugenol; *14* nonanoic acid; *15* 2,6-dimethoxyphenol; *16* methylethylmaleimide; *17* dihydroactinidiolide; *18* 5-hydroxymethyl-2-furancarboxy aldehyde; *19* vanilline

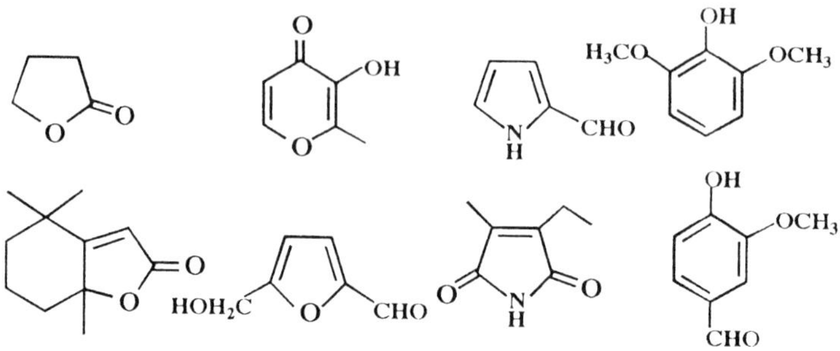

Fig. 12. Characteristic components of mate brewed extract

4.4 Mate Tea Aroma

Figure 11 shows a GC chromatogram of Brazilian roast mate tea extract, called Terere, which is brewed with cold water. A comparison of the relative quantities of main peaks between water extract and SDE extract (Kawakami and Kobayashi 1991) is listed in Table 4. Asterisked peaks are specific to water extract. Dihydroactinidiolide, 4-butanolide, vanilline and 2,6-dimethoxyphenol, 2-formylpyrrol, ethylmethylmaleimide, maltol, and 5-hydroxymethyl-2-furancarboxyaldehyde, whose structures are shown in Fig. 12, are only minimally extracted by the SDE method.

References

Acree T, Nishida R, Fukami H (1985) Odor thresholds of the stereoisomers of methyl jasmonate. J Agric Food Chem 33:425–427

Aisaka H, Kosuge M, Yamanishi T (1978) Comparison of the flavors of Chinese 'Keemun' black tea and Ceylon black tea. Agric Biol Chem 42:2157–2159

Badings HT, Dejong C, Dooper RPM, De Nigs RCM (1985) Rapid analysis of volatile compounds in food products by purge-and-cold-trapping/capillary gas chromatography. In: Adda J(ed) Progress in flavor research 1984. Elsevier, Amsterdam, p 523

Dahlgren R (1968) Revision of the genus Aspalathus. II. Bot. Notiser 121:165–208

Etoh, H, Ina K, Eguchi M (1980) 3*S*-(+)-3,7-Dimethyl-1,5-octadien-3,7-diol and ionone derivatives from tea. Agric Biol Chem 44:2999–3000

Guo W, Hosoi R, Sakata K, Watanabe N, Yagi A, Ina K, Luo S (1994a) (*S*)-Linalyl, 2-phenylethyl and benzyl disaccharide glycosides isolated as aroma precursors from oolong tea leaves. Biosci Biotechnol Biochem 58:1532–1534

Guo W, Hosoi R, Yamauchi K, Watanabe N, Usui T, Luo S, Sakata K (1994b) Premeverosidase as a main glycosidase concerned with the alcoholic aroma formation in tea leaves. Biosci Biotechnol Biochem 59:962–964

Hara T, Hotta H (1987) 3,7-Dimethyl-1,5,7-octatrien-3-ol formed and its precursor in the firing of green tea. Nippon Nogeikagaku Kaishi 61:353–356

Kawakami M (1982) Ionone series compounds from β-carotene by the theramal degradation in aqueous medium. Nippon Nogeikagaku Kaishi 56:917–921

Kawakami M, Kobayashi A (1991) Volatile components of green mate and roasted mate. J Agric Food Chem 39:1275–1279

Kawakami M, Chairote G, Kobayashi A (1987) Flavor consituents of pickled tea, miang, in Thailand. Agric Biol Chem 51:1683–1687

Kawakami M, Kobayashi A, Kator K (1993a) Volatile constituents of rooibos tea (*Aspalathus linearis*) as affected by extraction process. J Agric Food Chem 41:633–636

Kawakami M, Kobayshi A, Yamanishi T (1993b) The effect on tea aroma composition by the different extraction method. In: Proc Symp on the Agricultural Chemical Society of Japan, Sendai, p 351

Kawakami M, Kobayashi A, Yamanishi T, Ganguly SN, Banerjee J (1994) Tea flavor analysis by brewed extraction method and newly identified compound. In: Proc Symp, Agricultural Chemical Society of Japan, Tokyo, p 497

Kawakami M, Ganguly SN, Banerjee J, Kobayashi A (1995a) Aroma composition of oolong tea and black tea by brewed extraction method and characterizing compounds of Darjeeling tea aroma. J Agric Food Chem 43:200–207

Kawakami M, Yamanishi T, Kobayashi A (1995b) Aroma composition of original Chinese black tea, Zheng Shan Xiao Zhong and various other black teas. In: Proc '95 Int Tea-quality – human health Symp, Shanghai, pp 164–174

Kobayashi A, Kawakami M (1991) Analysis of essential oils of tea. In: Linskens HF, Jackson JF (eds) Essential oils and waxes. Modern methods of plant analysis, New Series, vol 12. Springer, Berlin Heidelberg New York, pp 21–40

Kobayashi A, Kubota K, Yano M (1993) Formation of some volatile components of tea. In: Teranishi R, Buttery RG, Sugisawa H (eds) Bioactive volatile compounds for plants. American Chemical Society, Washington, DC, pp 49–56

Mosandl A, Heusinger G, Gessner M (1986) Analytical and sensory differentiation of 1-octen-3-ol enantiomers. J Agric Food Chem 34:119–122

Mosandl A, Haner U, Hagenauer-Hener U, Kustermann A (1990) Stereometric flavor compounds 33. Multidimentional gas chromatography direct enantiomer separation of γ-lactones from fruits, foods, and beverages. J Agric Food Chem 38:767–771

Nakatani Y, Sato S, Yamanishi T (1969) 3*S*-(+)-3,7-Dimethyl-1,5,7-octatrien-3-ol in the essential oil of black tea. Agric Biol Chem 33:967–968

Nobumoto Y, Kubota K, Kobayashi A, Yamanishi T (1990) Structure of α-farnesene in the essential oil of oolong tea. Agric Biol Chem 54:247–248

Schultz TH, Flath RA, Mon TR, Eggling SB, Teranishi R (1977) Isolation of volatile compounds from a model system, J Agric Food Chem 25:446–449

Shigematsu Y, Shimoda M, Osajima Y (1994) Comparison of the odor concentrates of black tea. Nippon Shokuhin Kogyo Gakkaishi 41:768–777

Smith SL (1984) Coupled systems: capillary GC-MS and capillary GC-FTIR. J Chromatogr Sci 22:143–148

Straten SV, Maarse H (1989) Volatile components in food. TNO-CIVO Food analysis Institute, Zeist, The Netherlands, pp 680–710

Tsugita T, Imai T, Doi Y, Kurata (1979) GC and GC-MS analysis of head-space vplatiles by Tenax GC trapping techniques. Agric Biol Chem 43:1351–1354

Wang D, Ando K, Morita K, Kubota K, Kobayashi A (1994) Optical isomers of linalool and linalool oxides on tea aroma. Biosci Biotechnol Biochem 58:2050–2053

Williams PJ, Strauss CR, Wilson B (1980) Hydroxylated linalool derivertives as precursors of volatile monoterpenes of muscat grapes. J Agric Food Chem 28:766–771

Yamanishi T (1995) Flavor of tea. Food Rev Int 11:477–525

Yamanishi T, Nose M, Nakatani Y (1970) Futher investigation of flavor constituents in manufactured green tea. Agric Biol Chem 30:599–608

Analysis of Aroma Release from Scented Teas

N. WATANABE and I. PAL SINGH

1 Introduction

Tea, one of the most widely consumed beverages in the world, has diversified with time into non-traditional teas, such as instant teas, tea tablets, ready-to-drink tea mixes, liquid concentrates, canned teas, and scented teas. Increased cost of input and stiff competition have been cited as reasons for this diversification. Of these nontraditional teas, the scented teas have shown the greatest market potential.

A number of natural and synthetic flavors can be used to make scented teas. Fragrant flowers, such as jasmine or rose, spices like clove, cinnamon, or mint, or fruity flavors, can be used to make scented teas. Scenting is done by allowing these fresh flowers, fruity flavors, or spices to dry with freshly fired green or black tea. Jasmine tea, which is made by absorbing the fragrance of jasmine flowers on to the parched green tea leaves, is by far the most famous. The scenting process involves the mixing of moistened green tea leaves and just-opened jasmine flowers (Fig. 1). Luo et al. (1991) reported that the aroma of scented teas is derived mainly from the flower used and that the flowers used must reach full bloom gradually to release a higher amount of fragrance compounds during the scenting process.

The works summed up in this chapter embody recent experimental details related to the analysis of aroma of different scented teas, formation of fragrance in flowers used for scenting the teas, detection and isolation of aroma precursors, the enzymes concerned with aroma formation, their purification and identification, changes in flavor with storage time, and practical applications of the studies.

Flavors and teas are analyzed by GC and GC-MS. Sensory methods are usually relied on to govern the quality control for flavors and the teas. Scented teas, which are usually made by spray-drying methods, are not analyzed by GC or GC-MS, and only sensory methods are used for their evaluation. Abundant literature exists on the analysis of aroma from the teas and their flavors, but virtually no literature is available on the aroma of scented teas.

Aroma concentrate can be prepared by a modified Likens-Nickerson's simultaneous distillation and extraction (SDE) apparatus or by the column adsorption method followed by elution with ethyl ether/isopentane. Both methods have their own merits and demerits. Analysis of aroma concentrate is performed by GC and GC-MS. Shimoda et al. (1995) reported that the volatile compounds liberated from their nonvolatile glycosidic precursors by hydrolysis during SDE greatly influenced the chemical composition of the odor concentrate. Kinusaga and Takeo

Modern Methods of Plant Analysis, Vol. 19
Plant Volatile Analysis
Edited by H.-F. Linskens and J.F. Jackson

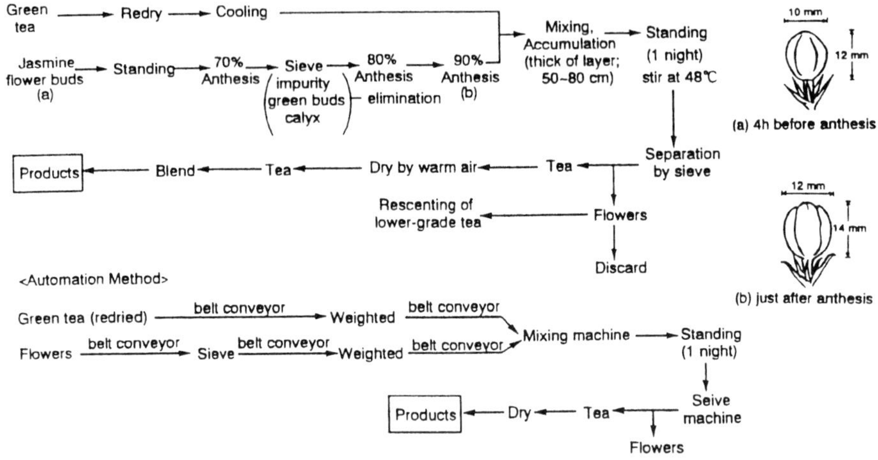

Fig. 1. Manufacturing procedure of jasmine tea

(1990) found that in green tea, the aroma compounds like linalool, geraniol, D-nerolidol, and indole were produced from their glycosides by the action of pectinase. Takeoka et al. (1992) also reported the presence of glycosides of volatile compounds like benzaldehyde, benzene alcohols, and β-ionone and its derivatives in nectarine fruit.

On the basis of studies on the formation of fragrance in flowers of *Jasminum sambac*, Watanabe et al. (1993) also hypothesized that most of the compounds constituting the alcoholic aroma are present as glycosides rather than glucosides. This was shown by treating the aroma precursor solution with crude enzyme prepared from flowers, naringinase, and β-glucosidase. Studies on direct isolation of aroma precursors resulted in the isolation of β-primeverosides of benzyl alcohol and 2-phenylethanol and 6′-*O*-malonyl β-D-glucopyranoside of linalool. Details of isolation of aroma precursors from leaves for oolong tea, as well as isolation and characterization of the enzymes concerned with liberating the aroma, will be presented.

2 Methods

2.1 Aroma of Chinese Scented Green Tea

2.1.1 Preparation of Aroma Concentrate

Yamanishi et al. (1988) and Luo et al. (1992) studied in detail the aroma of different types of Chinese green tea scented with flowers of *Jasminum sambac* (Moli tea), *Michelia alba* (Yulan tea), *Osmanthus fragrans* (Guihua tea), and *Citrus aurantium*. Aroma concentrate was prepared by a modified Likens-Nickerson's simultaneous distillation and extraction (SDE) method (Fig. 2).

2.1.2 Analysis of Aroma Concentrate

Aroma concentrates were analyzed by GC-MS and GC-FID. For GC-MS, a Hewlett-Packard 5790 A gas chromatograph combined with a JEOL JMS-DX 300 mass spectrometer and JEOL DA-5000 mass data analysis system was used for compound characterization. A Hitachi-M-80 B mass spectrometer was used for osmanthus scented tea. The GC column was 50 m × 0.25 mm i.d. fused silica WCOT coated with Carbowax 20M or DB-WAX (chemical bonded Carbowax 20M) 60 m × 0.25 mm column. Peak identification was performed by interpretation of mass spectra and by coincidence of retention time with authentic samples.

For GC-FID, a Shimadzu GC-9A gas chromatograph equipped with a flame ionization detector and a C-R2AX data processor was used. The GC column was 3 m × 3 mm i.d. glass column packed with PEG 20M 10% on chromosorb W 60/80 mesh. The column temparature was programmed from 60 to 190 °C at a rate of

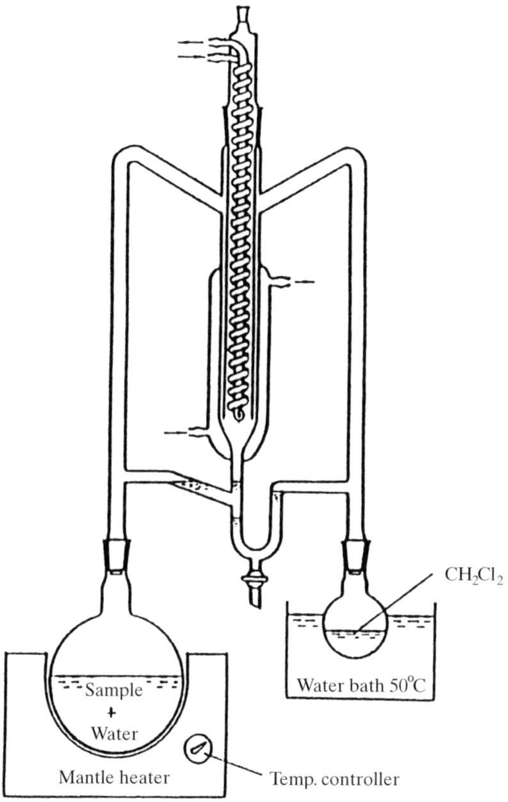

Fig. 2. Experimental setup for simultaneous distillation and extraction

2 °C min^{-1}. Nitrogen was used as a carrier gas at a flow rate of 20 ml min^{-1}. Peak identification was performed by coincidence of the retention time with the authentic compound, which was identified by GC-MS. Quantitative analysis of the aroma components in each scented tea was performed by calculations based on the data obtained by the recording data processor of GC-FID.

As revealed by the data shown in Table 1, and Fig. 3a–d, jasmine tea contains a high percentage of (*Z*)-3-hexenyl benzoate, benzyl acetate, linalool, and methyl anthranilate, while michelia scented tea contained twice as high a percentage of linalool as jasmine tea, and lacks the other three. High-quality jasmine tea contains a higher percentage of linalool and is made from high-quality green tea and fresh jasmine flowers. Low-grade jasmine tea is made by using low-quality green tea and replacing jasmine flowers by cheaper michelia flowers, thereby reducing the cost. At the same time, due to the higher percentage of linalool in michelia flowers, a higher concentration of linalool is maintained in low-grade jasmine tea. The aroma of osmanthus-scented tea is mainly due to ionone-related compounds. The main aroma constituents of *Citrus aurantium*-scented green tea are linalool, limonene, β-pinene, nerolidol, and geraniol, etc. The percentage of linalool, geraniol, and nerolidol increased from 18.3, 0.16, and 0.84 before scenting to 63.05, 1.62, and 3.11, respectively, after scenting. A high concentration of linalool, limonene, and nerolidol gives a very strong fragrance to *Citrus aurantium*-scented tea.

2.2 Comparison of Aroma Concentrate of Sen-cha Produced by Two Different Methods

In addition to the SDE method, aroma concentrate can also be prepared by the column adsorptive method. A very informative comparison of the odor concentrates obtained by two different methods, viz. SDE and column adsorption, has been presented (Shimoda et al. 1995). They found for green tea, Sen-cha, that the odor concentrate obtained by the column method was very close to the natural green tea flavor, whereas formation and decomposition of volatile components were observed by the SDE method. Here we will very briefly explain the differences between the experimental methods, as well as the quality of odor concentrates obtained.

2.2.1 Column Adsorption Method

Deionized hot water (80 °C; 1.0 l) was added to 50 g of Sen-cha and the leaves were filtered after 3 min through a coarse filter paper. The filtrate was immediately cooled to 40 °C by running tap water and passed through a column (28 × 250 mm) packed with 50 ml Porapak Q to adsorb the aromas. After washing with 80 ml of deionized water, the column was developed with 80 ml of a mixture of diethyl ether and isopentane (1 : 1), which eluted only the aromas. This solvent was found to be

Table 1. Comparison of main components in the aroma concentrates from four kinds of scented tea

Component	Percent in the aroma concentrate					
	Tea scented with					
	Jasminum sambac (moli tea)		*Michelia alba* (yulan tea)		*Osmanthus fragrans* (guihua tea)	*Citrus aurantium*
	(1) %	(2) %	(1) %	(2) %	%	%
Linalool	14.0	12.0	27.0	27.7	6.5	63.0
Linalool-3,6-oxide (*cis*)	0.2	0.3	1.1	1.3	1.1	1.4
Linalool-3,6-oxide (*trans*)	0.4	+	1.3	1.9	2.2	0.7
Linalool-3,7-oxide (*cis*)					2.4	1.1
Linalool-3,7-oxide (*trans*)					1.6	1.1
3,7-Dimethyl-1,5-octadien-3,7-Diol			2.9	2.9		0.7
3,7-Dimethyl-1,7-octadien-3,6-Diol						1.0
Geraniol	0.6	0.5	2.4	2.9	2.6	1.6
4-Terpineol						0.7
α-Terpineol			4.2	3.8		
γ-Terpineol						0.2
α-Farnesene	4.9	10.1	2.2	3.0		
β-Pinene						2.5
Limonene						5.8
t,t-Farnesol						2.4
Geranyl acetate						0.9
Linalyl acetate						2.2
Nerolidol	0.8	0.6	5.2	3.4	3.1	3.1
β-Ionone	+	+	3.3	3.1	14.0	0.9
β-Dihydroionone					8.2	
β-Dihydroionol					3.4	
β-Cyclocitral			1.9	2.0		
7,8-Dihydro-α-ionone			1.0	0.8		
(*Z*)-3-Hexen-1-ol	1.3	1.0	0.6	0.7	0.7	
Benzyl alcohol	6.4	4.4	+	+	+	0.7
2-Phenyl ethanol	0.2	+	2.4	2.2	+	0.6
(*E*)-2-Hexenal			1.1	1.5	1.2	
cis-Jasmone	0.7	+	1.6	1.5		
Methyl salicylate	3.3	2.3	1.8	1.8		
Methyl anthranilate	10.1	10.5				
(*Z*)-3-Hexenyl benzoate	24.9	22.6				
Phenethyl hexanoate			4.0	3.9		
4-Decanolide					5.2	
2-Methyl butyric acid			1.0	0.7		
Indole	3.8	7.8				

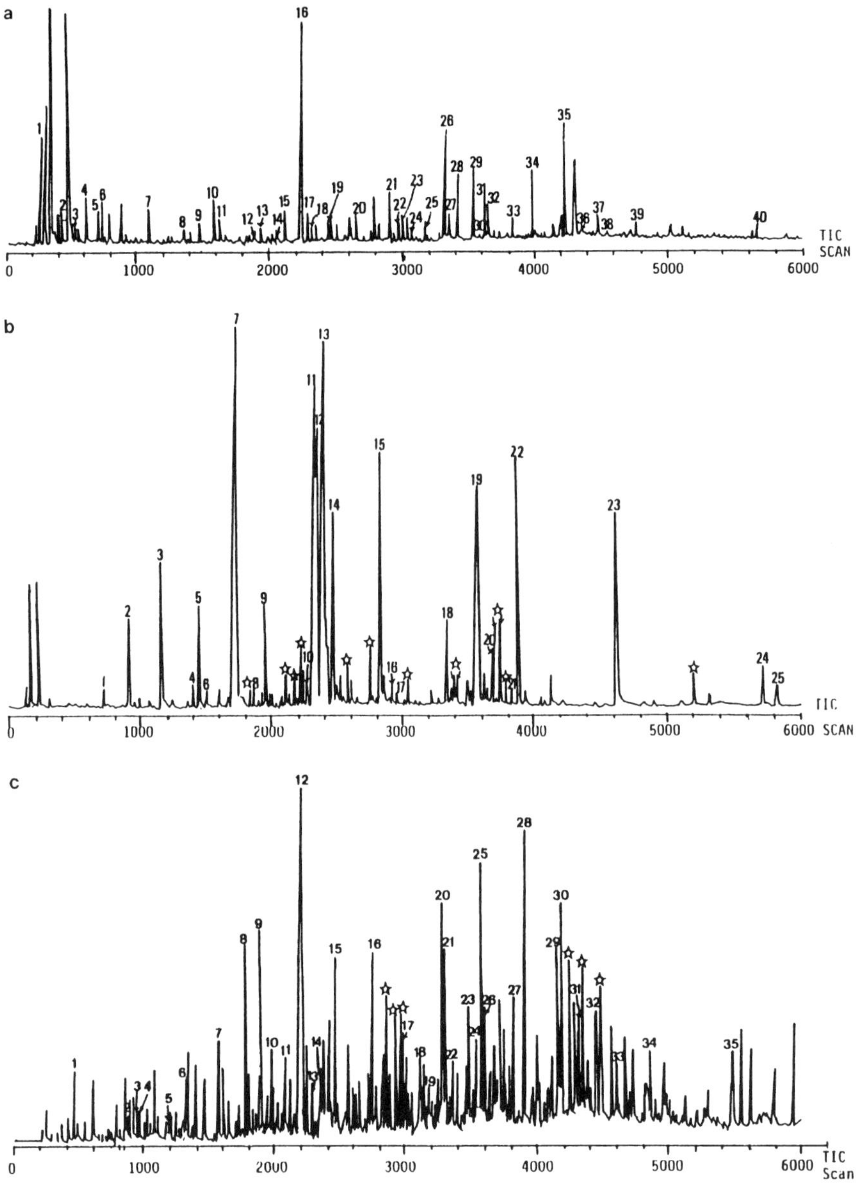
a
TIC
SCAN
0
1000
2000
3000
4000
5000
6000
b
TIC
SCAN
0
1000
2000
3000
4000
5000
6000
c
TIC
Scan
0
1000
2000
3000
4000
5000
6000

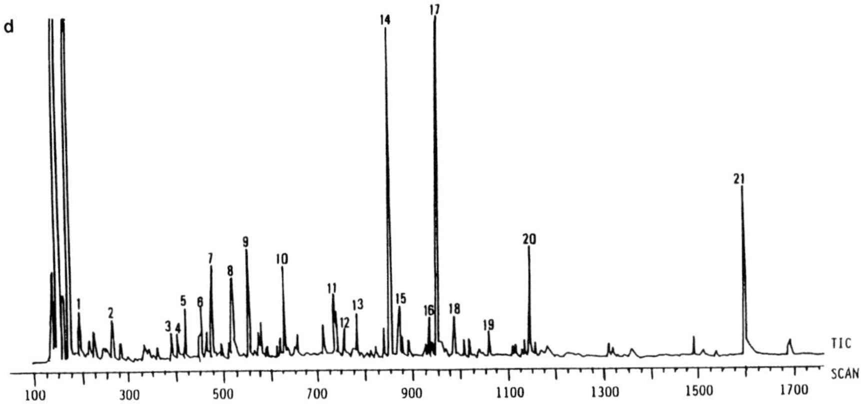

Fig. 3. **a** Gas chromatogram of the essential oil from green tea. Peak identification: *1* 2-methylpropanal; *2* pentanal; *3* ocimene; *4* hexanal; *5* 3-penten-2-one; *6* 4-methyl-3-penten-2-one; *7* pentanol; *8* (*Z*)-2-penten-1-ol; *9* hexanol; *10* (*Z*)-3-hexen-1-ol; *11* (E)-2-hexen-1-ol; *12* 1-octen-3-ol; *13* linalool-3,6-oxide; *14* (E-3, E-5)-heptadien-2-one; *15* benzaldehyde; *16* linalool; *17* octanol; *18* a sesquiterpene; *19* 1-ethylformylpyrrole; *20* (*Z*)-3-hexenyl hexanoate; *21* benzyl acetate; *22* α-farnesene; *23* γ-muurolene; *24* methyl salicylate; *25* nerol; *26* geraniol; *27* geranylacetone; *28* benzyl alcohol; *29* 2-phenylethanol; *30* benzyl cyanide; *31* β-ionone; *32* *cis*-jasmone; *33* phenol; *34* nerolidol; *35* (*Z*)-3-hexenyl benzoate; *36* decenoic acid; *37* methyl palmitate; *38* α-cadinol; *39* tricosane; *40* pentacosane. GC column temperature: 60 °C (4-min hold) $\xrightarrow{2\,^{\circ}\mathrm{C\,min}^{-1}}$ 200 °C (hold), MS ionization energy: 70 eV. **b** Gas chromatogram of the aroma concentrate from jasmine tea. Peak identification: *1* β-ocimene; *2* (*Z*)-3-hexenyl acetate; *3* (*Z*)-3-hexen-1-ol; *4* (*Z*)-3-hexenyl butyrate; *5* linalool-3,6-oxide; *6* α-copaene; *7* linalool; *8* caryophyllene; *9* methyl benzoate; *10* heptadecane; *11* β-sesquiphellandrene?; *12* benzyl acetate; *13* farnesene; *14* methyl salicylate; *15* benzyl alcohol; *16* nonadecane; *17* 2-phenylethanol; *18* nerolidol; *19* (*Z*)-3-hexenyl benzoate; *20* isoeugenol; *21* methyl palmitate; *22* methyl anthranilate; *23* indole; *24* benzyl benzoate; *25* neophytadiene; ☆ sesquiterpenoid (unidentified). GC column temperature: 60 °C (4-min hold) $\xrightarrow{2\,^{\circ}\mathrm{C\,min}^{-1}}$ 180 °C, MS ionization energy: 70 eV. **c** Gas chromatogram of the aroma concentrate from michelia-scented tea. Peak identification: *1* methyl 2-methylbutyrate; *2* 2-methylbutyric acid; *3* limonene; *4* 1,8-cineol; *5* β-ocimene; *6* 2,2,6-trimethylcyclohexanone; *7* (*Z*)-3-hexen-1-ol; *8* linalool-3,6-oxide(*cis*); *9* linalool-3,6-oxide(*trans*); *10* 2,4-heptadienal; *11* benzaldehyde; *12* linalool; *13* (E-3, E-5)-octadien-2-one; *14* β-elemene; *15* β-cyclocitral; *16* α-terpineol; *17* methyl salicylate; *18* nerol; *19* β-damascenone; *20* geraniol; *21* 7,8-dihydro-α-ionone; *22* benzyl alcohol; *23* 2-phenyl-ethanol; *24* 3,7-dimethyl-1,5-octadien-3,7-diol; *25* β-ionone; *26* *cis*-jasmone; *27* methyleugenol; *28* nerolidol; *29* phenethyl hexanoate; *30* 6,10,14-trimethylpentadecan-2-one; *31* 2,3-dimethyl-2-nonen-4-olide; *32* methyl palmitate; *33* decenoic acid; *34* dihydroactinidiolide; *35* dodecanoic acid; ☆ Sesquiterpenoid (unidentified). GC column temperature and MS ionization energy: as in Fig. 1. **d** Gas chromatogram of the aroma concentrate from osmanthus-scented tea. Peak identification: *1* hexanal; *2* (*E*)-2-hexenal; *3* (*Z*)-3-hexen-1-ol; *4* nonanal; *5* tetradecane; *6* linalool-3,6-oxide(*cis*); *7* linalool-3,6-oxide(*trans*); *8* pentadecane; *9* linalool; *10* hexadecane; *11* heptadecane; *12* linalool-3,7-oxide(*cis*); *13* linalool-3,7-oxide(*trans*); *14* β-dihydroionone; *15* geraniol; *16* BHT(artifact); *17* β-ionone; *18* β-dihydroionol; *19* nerolidol; *20* α-decalactone; *21* phytol. GC column temperature: 70 °C $\xrightarrow{2\,^{\circ}\mathrm{C\,min}^{-1}}$ 215 °C, MS ionization energy: 20 eV

superior because polyphenols were not eluted. Use of 100% ether eluted aromas as well as the polyphenols. After concentration, the polyphends formed a sticky, dark brown layer which makes sample injection for GC analysis difficult. If only isopentane was used, then the amount of aromas extracted was far less than that extacted with the ether/isopentane system. The eluate thus obtained was dried over anhydrous Na_2SO_4 for 12 h after the addition of an internal standard (5.0 µl of 1.0% cyclohexanol) and the solvent was evaporated to about 1 ml. The column can be regenerated by washing with diethyl ether (60 ml), purified methanol (80 ml), and deionized water (80 ml). Methanol was purified by passing through a column (35 × 500 mm) packed with an analytical grade active charcoal. The process was repeated for 15 batches, and the combined eluates were concentrated to about 200 µl.

2.2.2 Modified Likens-Nickerson Method (SDE Method)

The experimental setup was the same as in Section 2.1.1, except that separation was carried out under reduced pressure (150 mmHg, 70 °C) for 50 min. The condenser was cooled to −5 °C with a mixture of water and ethylene glycol.

The aroma concentrates obtained by both methods were analyzed by GC and GC-MS.

2.2.3 Preparative GC and Sniffing

For preparative GC, a Shimadzu GC-8A equipped with a 40 m × 1.2 mm i.d., chemically bonded PEG 20M (film thickness 1 µm) megabore open tubular column (Chemical Inspection and Testing Institute, Tokyo) and an FID were used. The oven temperature was programmed from 50 to 220 °C at rate of 2 °C min^{-1}. Flow rate of He was 18 ml min^{-1} with a splitless injection. Short capillaries (25 cm × 0.53 mm i.d.) with a chemically bonded phase of PEG 20M were used for trapping the components with a split ratio of 20 : 1. The separated compounds were eluted on a filter paper with a drop of ethyl ether, and the odor was sniffed.

2.2.4 Comparison of Odor Concentrates

The odor concentrate by the column method had a pleasant green tea odor with fresh, brisk, and sweet odor attributes, but that by the SDE method had a boiled green tea odor with woody, astringent, and stinking green odor attributes. A manifold increase in the amount of saturated alcohols such as *n*-hexanol, *n*-octanol, and *n*-nonanol by the SDE method may be the cause of the woody and fragrant odor of the concentrate, whereas an increase in the amount of unsaturated alcohols imparts a stinking green odor. All four types of linalool oxides increased by SDE, but *cis* pyranoid form was at the same level in both concentrates. These compounds may be produced by hydrolysis of their glycosidic aroma precursors during SDE. These terpene alcohols have floral, woody, and pleasant rose-like odors, but their amount is well below the sensory detection limit and hence their contribution to the note may be negligible. The large amount of nerolidol in SDE is responsible for the woody and floral odors of boiled green tea infusions (Table 2).

Table 2. Identification and comparative quantification of volatile compounds separated from green tea

Compound	Approximate concentration (μg/kg)	
	Column	SDE
n-Hexanol	tr[a]	161
(*E*)-3-Hexen-1-ol	27	42
(*Z*)-3-Hexen-1-ol	tr	5517
n-Octanol	16	456
(*E*)-2-Octenol	tr	258
n-Nonanol	tr	28
n-Pentanal	tr	99
n-Hexanal	17	306
(*Z*)-4-Heptenal	tr	45
(*E*)-2-Heptenal	tr	65
(*E*)-2-Octenal	tr	85
(*E*,*Z*)-2,4-Heptadienal	tr	249
2,2,6-Trimethylcyclohexanone	tr	56
3,5,5-Trimethyl-4-(3-oxo-1-butenyl)-2-cyclohexen-1-one	113	tr
Linalool oxide (*trans*-furanoid)	tr	119
Linalool oxide (*cis*-furanoid)	28	156
Linalool oxide (*trans*-pyranoid)	58	130
Linalool oxide (*cis*-pyranoid)	234	272
Linalool	40	933
D-nerolidol	275	593
(*E*,*E*)-2,4-heptadienal	tr	519
Benzaldehyde	37	184
Benzeneacetalaldehyde	38	56
Phenylethanone	tr	36
1-(3,4-Dimethylphenyl)ethanone	74	tr
1-(4-Ethylphenyl)ethanone	33	tr
β-lonone-5,6-epoxide (*trans*)	tr	45
β-lonone	tr	1157
β-lonone-5,6-epoxide (*cis*)	49	456
3-Ethyl-4-methyl-1 H-pyrrole-2,5-dione	285	tr
(*Z*)-3-hexenyl butanoate	tr	618
1-Ethyl-1 H-pyrrole-2-carboxaldehyde	35	490
3-Hexenyl hexanoate	295	tr
Coumaran	135	tr
lsothiocyanatocyclohexane	tr	187
N,*N*-Dibutylacetamide	tr	178

[a] Less than 5 μg/kg.

Aliphatic aldehydes such as pentenal, hexenal, heptenal, 2-pentenal, (*Z*)-3-hexenal, (*Z*)-4-heptenal, (*E*)-2-heptenal, (*E*)-2-octenal, (*E,Z*)-and (*E,E*)2,4-heptadienal, and 2,6-dimethyl-5-heptenal were increased markedly, about 26 times, with SDE. They could impart a grassy and oily aldehydic odor to the SDE concentrate. Benzaldehyde with an almond-like odor and benzene acetaldehyde with a floral odor also increased many times with SDE.

Alkenyl-2-ones, a characteristic group in volatile compounds of green tea could be liberated from free fatty acids by their oxidation to β-ketoacids followed by decarboxylation. Degradation products of carotenoids such as 2,2,6-trimethylcyclohexanone were formed in large quantities by SDE. These ketones, having a woody and hay-like odor, result in a great difference between the two odor concentrates. Phenylethanone was formed in SDE, while its derivatives disappeared. 1-(3,4-Dimethyl phenyl) ethanone and 1-(4-ethyl phenyl) ethanone have not been found in green tea but their authentic samples possessed green tea-like aroma with fragrant, sweet odors. The odor of green tea infusion was improved by adding the former at 100 ppb, but the contents of 1-(3,4-dimethyl phenyl) ethanone and 1-(4-ethyl phenyl) ethanone were lower in higher-grade Sen-cha, indicating that a quantitative balance of volatile flavor compounds was more important. β-Ionone and its epoxides, which increased with SDE, may be responsible for floral and fragrant notes. The derivatives of 2,5-pyrrolidine, which have a very intense sweet odor with preparative GC and sniffing, almost disappeared with SDE. Their disappearance definitely degrades the odor of green tea.

Among aliphatic esters, 3-hexenyl hexanoate, an important contributor to freshness in green tea, decreased with SDE. The increase in 1-ethyl-1*H*-pyrrole-2-carboxaldehyde with SDE imparts a pungent and bitter odor. Isothiocyanatocyclohexane was responsible for typical astringent odor while nitrogenous volatiles gave unpleasant odors to the SDE concentrate. The disappearance of coumaran and vanillin with SDE also degraded the green tea odor. It was concluded that decomposition and formation of volatile compounds degrade the flavor of the concentrate obtained by SDE, while the concentrate obtained by the column method is similar to the natural green tea flavor.

2.3 Changes in the Aroma of Jasmine Tea During Scenting

For making jasmine tea, green tea is repeatedly scented with jasmine flowers and dried. Its production cost, of which the flower represents about 25%, is the highest in teas. Luo et al. (1991) have shown that the utilizing efficiency of aroma is closely associated with processing procedures, number of times of scenting, and moisture content of the tea material.

2.3.1 Samples

Special grade green tea, jasmine tea samples from different processing stages of scenting, and first-grade tea samples with different moisture content.

Sample Treatment. Ten g of tea sample were extracted with 400 ml of hot distilled water and 25 ml CH_2Cl_2 by the SDE method. The solvent was removed by distillation to give a yellow aroma concentrate, which was analyzed by GC-FID under the same conditions as in Section 2.1.1.

2.3.2 Changes in Aroma

Effect of Drying. As shown in Table 3, the oil yield of tea was highest before drying and decreased by 15% after drying. The amount of linalool and α-farnesene decreased by 17 and 34%, respectively, while that of indole, benzyl alcohol, (*Z*)-3-hexenyl benzoate, and methyl anthranilate increased after drying.

Effect of Different Scenting Stages. As shown in Table 4, the oil yield increased with an increase in number of times of scenting, but the efficiency in absorbing aroma decreased with each scenting process. The amount of linalool and benzyl alcohol increased in the tea product as compared with their amount in green tea leaves. An incremental increase was observed in their amount with each stage of processing. A careful examination of gas chromatograms (Fig. 4) of jasmine flowers, green tea, and jasmine tea shows that those of fresh flowers and scented tea were similar, while both were quite different from green tea, suggesting that the aroma of jasmine tea is mainly derived from the aroma of the flower, and that the first stage of scenting is very important.

Effect of Moisture. As shown in Fig. 5, when the moisture content of green tea was less than 30%, the oil yields differed slightly with scenting but the quality indices of aroma increased gradually. Green tea with a lower moisture content before

Table 3. Changes of main aroma constituents of jasmine tea during drying

Constituent	Concentration	
	Before drying	After drying
Aroma oil yield (mg/20 g tea)	12.5	10.6
(*Z*)-3-Hexenyl acetate	2.03%[a]	1.76%[a]
(*Z*)-3-Hexenol	3.15	2.93
Linalool	18.88	15.7
Methyl benzoate	2.81	2.79
Benzyl acetate	32.54	30.68
α-Farnesene	12.95	8.58
Methyl salicylate	3.55	3.35
Benzyl alcohol	2.74	3.25
(*Z*)-3-Hexenyl benzoate	10.26	10.67
Methyl anthranilate	2.24	2.81
Indole	2.12	4.28

[a] Percent of total peak area.

Table 4. Changes in aroma contents of jasmine tea during the scenting process

	Jasminum sambac	Green tea	Scenting			Product
			1st	2nd	3rd	
Total (oil mg/10 g tea)		0.507	8.54[a]	14.74[a]	17.35[a]	18.68[a]
Linalool	8.71	0.093	1.95	2.94	2.95	3.30
(*Z*)-3-Hexenyl acetate	–	0.005	0.22	0.52	0.32	0.30
(*Z*)-3-Hexenol	1.15	0.009	0.24	0.65	0.36	0.36
(*Z*)-3-Hexenyl butyrate	–	0.021	0.05	0.01	0.08	0.11
Methyl benzoate	trace	0.008	0.18	0.44	0.35	0.36
Benzyl acetate	11.94	0.027	2.23	6.09	5.46	6.00
α-Farnesene	7.85	0.027	0.50	1.18	1.23	1.73
Methyl salicylate	–	0.004	0.24	0.51	0.36	0.38
Geraniol	0.26	0.001	0.04	0.06	0.05	0.05
Benzyl alcohol	3.38	0.007	0.56	0.86	0.74	0.81
cis-Jasmone, β-ionone	trace	0.031	0.04	0.06	0.05	0.04
Nerolidol	1.31	0.004	0.04	0.07	0.08	0.08
(*Z*)-3-Hexenyl benzoate	15.96	0.016	0.89	2.29	2.43	2.84
Methyl anthranilate	1.4	0.004	0.43	1.00	1.04	0.91
Indole	0.14	0.001	0.49	0.80	1.08	0.63

[a] Total aroma contents (mg/10 g of tea).

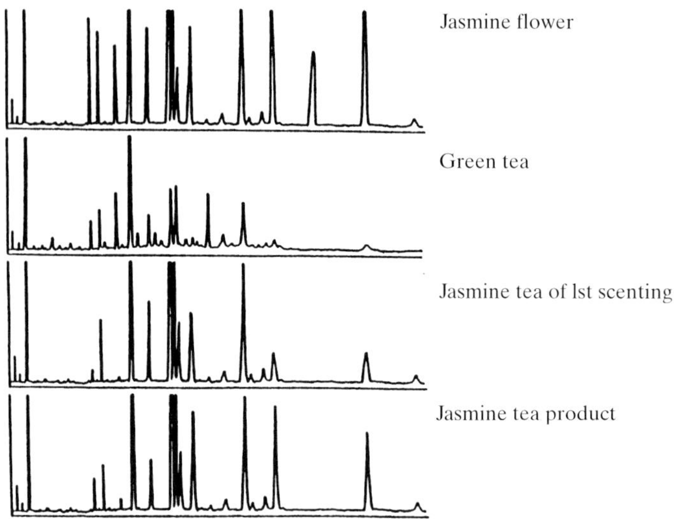

Fig. 4. Gas chromatograms of jasmine flower, green tea, and jasmine tea

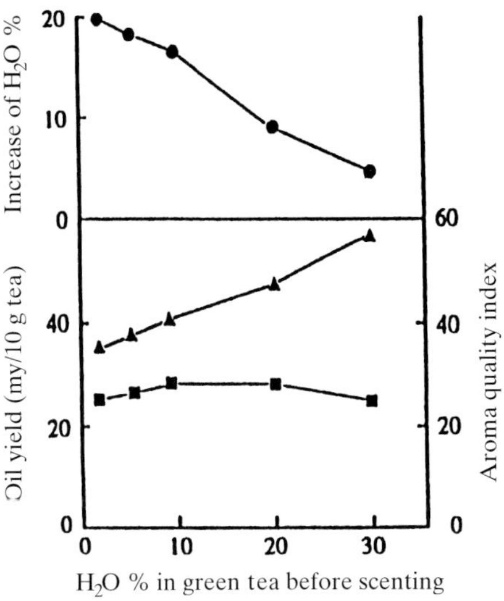

Fig. 5. Changes in moisture content (●), oil yield (■), and aroma quality index (▲) caused by the scenting process

scenting showed in a greater increase in moisture content and lower scenting efficiency in the scented product.

3 Flower Fragrance Formation

Since the aroma of scented teas is derived mainly from the flower used, the studies on flower fragrance formation are of the utmost importance in maintaining and improving the quality of scented teas. Here, we will explain the mechanism of fragrance formation in *Jasminum sambac* Ait., which is used for making jasmine tea. It is proposed that fragrance precursors in the flower buds are enzymatically transformed into volatile compounds during flower opening. Watanabe et al. (1993) designed a crude enzyme assay to clarify the mechanism of flower fragrance formation.

3.1 Crude Enzyme Assay

3.1.1 Flower Buds and Flowers for Crude Enzyme Assay

J. sambace Ait. was cultivated in China. The flower buds (average wt. 370 mg) with a size of 10 × 12 mm were harvested in the morning. The flower buds were left at 30 °C for 10 h to reach the full bloom stage. These flowers had a characteristic jasmine floral note, which the flower buds did not have.

3.1.2 Preparation of Crude Enzyme for Crude Enzyme Assay

Flower buds were homogenized in 10 × X (X = g fresh weight) ml of cold acetone chilled with solid CO_2 and filtered. The debris was washed with cold acetone until the flower fragrance almost disappeared. Drying in a dessicator under suction and ice cooling gave an acetone powder. Yield of acetone powder was 50 g from 200 g fresh wt. This acetone powder was directly used as a crude enzyme.

3.1.3 Preparation of Precursor Solution

Flower buds just before opening were extracted with 80% methanol three times under ice cooling. The combined extracts were concentrated in vacuo to yield an aqueous residue, which was chromatographed on an Amberlite XAD-2 column [column bed volume $Vt = 12 \times X$ (X = g dry wt.) ml]. The column was washed with water ($2 \times Vt$ ml) and pentane (Vt ml) to remove sugars and volatile materials that interfere with GC and GC-MS analysis of the aroma constituents liberated by the crude enzyme assay. Aroma precursors were eluted with methanol ($3 \times Vt$ ml) and concentrated. The resulting aqueous residue was diluted with water and used for the crude enzyme assay.

3.1.4 Crude Enzyme Assay

Precursor solution was treated in following different ways.

A. Ten ml of a precursor solution (20 g flower bud equivalent) was reacted with crude enzyme solution (3.75 ml, 250 mg flower equivalent).
B. Precursor solution was reacted with crude enzyme preparation inactivated by heating in a microwave oven for 1 min.
C. Precursor solution [precursor solution +0.1 M citrate buffer (pH 5)].
D. Crude enzyme +0.1 M citrate buffer (pH 5).
E. Inactivated crude enzyme + citrate buffer.
F. Precursor solution + β-glucosidase (1 mg).
G. Precursor solution + narnginase (1 mg).

Each reaction mixture was filled up to 20 ml with 0.1 M citrate buffer (pH 5) and was incubated at 30 °C in the presence of sodium azide (500 $\mu g\,ml^{-1}$). After 72 h the reaction mixture was filtered and 0.5 ml of ether solution of ethyl octanoate (20 $\mu g\,ml^{-1}$) was added as an internal standard. Duplicate sets of groups were assayed, one for a sensory test and the other for GC and GC-MS analyses. Aroma liberated from the precursor solution during incubation was extracted twice with ether. After drying over anhydrous Na_2SO_4, the combined extracts were concentrated by a stream of nitrogen and analyzed directly by GC and GC-MS.

3.2 GC and GC-MS Analyses After Crude Enzyme Assay

3.2.1 Conditions

A Hitachi model G-3000 gas chromatograph with an FID, equipped with a capillary column of TC-WAX or PEG-20M (GL Science Co. Ltd., Japan, 0.25 mm i.d. × 30 or 25 m) and a recorder (JASCO Model 807-IT, Japan) was used. Column temparature was held at 60 °C for 10 min, increased to 200 °C at a programme rate of 3 °C min^{-1}, and then held for 20 min at 200 °C. Injector and detector temparature were 250 °C each. N_2 with a flow rate of 1 ml min^{-1} was used as a carrier gas. An ethereal sample solution (0.8 μl) was injected through a splitter (1:90). For GC-MS, a JMS DX-302 GC-MS spectrometer (JEOL, Japan) with the same capillary column and data acquisition system DA-3000 was used. Analyzing parameters were: ionizing voltage: 70 eV; accelerating voltage: 3 kV; mass scan rate: 1 s for *m/z* 40–500. Results were verified by comparing relative retention time and MS data with those of authentic samples and a relative peak intensity to that of an internal standard (ethyloctanoate).

3.2.2 Results

As shown in Table 5, crude enzyme treatment of the precursor solution increased the amounts of 4-hydroxy-4-methyl-2-pentanone (from acetone), (*Z*)-3-hexenol, linalool, benzyl alcohol, and 2-phenyl ethanol. Naringinase treatment yielded higher amounts of linalool and 2-phenyl ethanol as compared with β-glucosidase treatment. β-Glucosidase yielded a higher amount of benzyl alcohol. These results

Table 5. Gas chromatographic analytical data of aroma compounds liberated from the precursor solution prepared from *Jasminum sambac*, by action of its crude enzyme preparation, β-glucosidase, and naringinase

Compound	Relative t_R	(A)[a,b]	(C)	(D)	(F)	(G)
4-OH-4-Me-2-pentanone	0.86	24.8[c]	0.1	0.4	0.1	–
(*Z*)-3-Hexenol	0.91	2.7	–	–	5.2	7.7
Ethyl octanoate	1.00	1.0	1.0	1.0	1.0	1.0
Acetic acid	1.09	59.4	0.3	0.7	0.7	0.6
Linalool	1.27	1.4	0.3	–	0.5	3.7
Benzyl alcohol	1.90	35.3	0.2	0.7	2.6	21.8
2-Phenyl ethanol	1.95	6.4	1.2	–	5.7	41.7
Methyl anthranilate	2.46	10.7	6.4	2.8	–	–

[a] Data for unidentified peaks not shown.
[b] See text, Section 3.1.4.
[c] Data are calibrated relative intensities so that relative peak intensity due to an internal standard (ethyl octanoate) is 1.

suggest that linalool and 2-phenyl ethanol are present as their glycosides rather than their glucosides. Benzyl alcohol and (*Z*)-3-hexenol are present as their glucosides. Later studies on the isolation of aroma precursors have shown that benzyl alcohol and 2-phenyl ethanol are present as their β-primeverosides (Inagaki et al. 1995) and linalool as its 6′-*O*-malonyl-β-D-glucopyranoside (Moon et al. 1994a).

3.3 Isolation of Aroma Precursors

Isolation of aroma precursors from *J. sambac* Ait. was carried out by enzymatic hydrolysis followed by GC and GC-MS analyses. Conditions for GC and GC-MS were the same as in Section 3.2.2.

3.3.1 Enzymatic Hydrolysis

A mixture of an aroma precursor sample equivalent to 2 g of fresh jasmine flower buds and acetone powder prepared from fresh jasmine flowers or naringinase in 10 ml of 0.1 M citrate buffer (pH 5) was kept at 30 °C for 12 h. The reaction mixture was subjected to workup as in Section 3.1.4 to give an aroma concentrate for GC and GC-MS analyses.

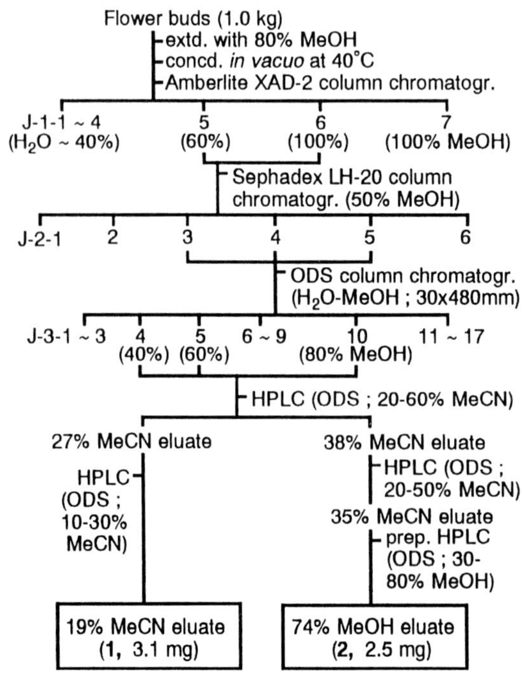

Fig. 6. Isolation procedure of **1** and **2** from *Jasminum sambac*

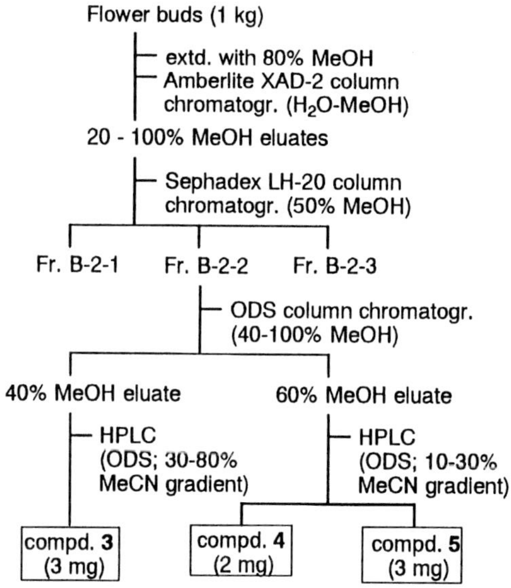

Fig. 7. Isolation procedure of **3**, **4**, and **5** from *Jasminum sambac*

3.3.2 Isolation

Flower buds of *J. sambac* Ait. Just before flower opening (600 g) were harvested in the morning in China and extracted with 80% methanol three times under ice cooling. Further isolation procedures for aroma precursors of linalool are given in Fig. 6, and for benzyl alcohol and 2-phenyl ethanol in Fig. 7. Linalyl β-D-glucopyranoside (**1**) and its 6′-*O*-malonate (**2**) were isolated as aroma precursors of linalool. However, linalyl β-D-glucopyranoside-6′-*O*-malonate (**2**) may be the true precursor, as malonyl moiety is easily eliminated during the isolation process. When a methanolic solution of **2** was passed through CM-Toyopearl 650 M (H^+) to obtain free acid and then treated with diazomethane, a facile elimination of malonyl moiety was observed. Benzyl 6-*O*-β-D-xylopyranosyl-β-D-glucopyranoside (**3**), a β-primeveroside, was isolated as precursor of benzyl alcohol. 2-phenylethyl β-primeveroside (**4**) and 2-phyenylethyl 6-*O*-α-rhamnopyranosyl-β-D-glucopyranoside (**5**), a β-rutinoside, were isolated as aroma precursors of 2-phenyl ethanol. Structures are shown in Fig. 8.

4 Studies on Oolong Tea

Oolong, a semifermented tea, has a reputation for its floral aroma mainly contributed by monoterpene alcohols like geraniol and linalool, and aromatic alcohols like 2-phenyl ethanol and benzyl alcohol. The features of aroma mainly depend on the fine differences in its manufacturing process. Oolong tea is manufactured by

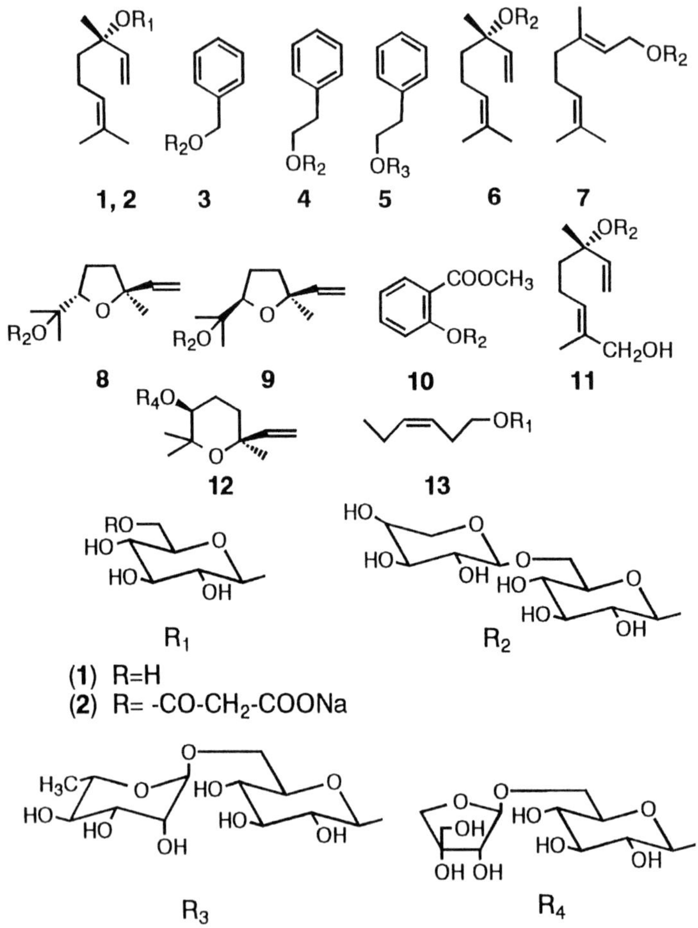

Fig. 8. Glycosidic aroma precursors

more complicated processes than other teas: solar withering, indoor withering, turning over, fixing, mass rolling, and drying. The floral alcoholic aroma, which is mainly produced during the withering process, is one of the most important factors to determine the quality of the tea.

4.1 Aroma Precursors

4.1.1 Sample Preparation

The samples of oolong tea were prepared in Fujian, China, as follows. The plucked fresh leaves (1st to 4th) were immediately pan-fired to inactivate the enzymes and dried as a leaf sample. Withered leaves were quickly pan-fired and dried as withered samples. The aroma concentrates of each tea sample (20 g) were prepared by the SDE method, as already described.

4.1.2 Detection of Aroma Precursors

Acid Hydrolysis by Two-Phase Treatment. A sample (10 g of dry leaves) was dissolved in 10 ml of water and extracted with ether to remove the volatiles. The resulting aqueous solution was concentrated and adjusted to pH 1 with H_2SO_4. Ethyl ether (10 ml) was added and the reaction mixture was heated at 40 °C with stirring for 2 h. The aqueous layer was neutralized with NaOH and extracted with ethyl ether to afford a sample for GC and GC-MS analyses.

Enzymatic Hydrolysis. As in Section 3.3.1

4.1.3 Isolation of Aroma Precursors

Guo et al. (1993) and Moon et al. (1994b) isolated several glycosidic aroma precursors from leaves of oolong tea. Isolation of alcoholic aroma precursors from tea leaves for oolong tea was carried out by acid or enzymatic hydrolysis of precursors followed by GC analysis. The leaf sample was extracted with boiling water for 10 min. The extract was concentrated, cooled, kept at 5 °C overnight, and centrifuged. The supernatant was mixed with active granular charcoal, packed in a column, and eluted with water, 20, 40, 60, and 100% MeOH. Fractions eluted with

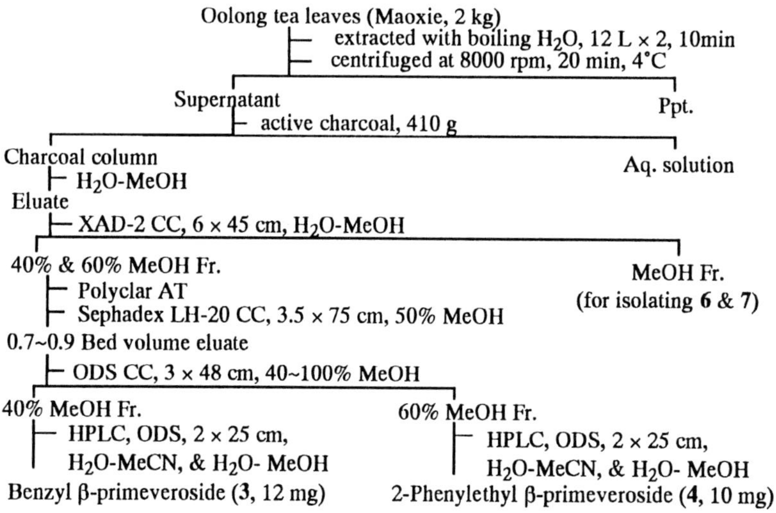

Fig. 9. Isolation procedure of **3** and **4** from oolong tea leaves (cv. Maoxie)

40–100% MeOH were concentrated and chromatographed on an Amberlite XAD-2 column (H_2O-MeOH). MeOH fraction was chromatographed on Sephadex LH-20 column (50% MeOH). Fractions of 0.8–1.0 × *Vt* were combined and chromatographed on ODS column (YMC-gel, H_2O-MeOH). The fraction eluted with 60% MeOH was subjected to preparative HPLC (ODS, YMC-Pack, 40–80% MeOH gradient elution) to give glycosides of geraniol and linalool identified as their β-primeverosides. Isolation procedures for glycosidic aroma precursors (β-primeverosides) of benzyl alcohol (**3**), 2-phenylethanol (**4**), linalool (**6**), geraniol (**7**), *trans*-and *cis*-linalool-3,6-oxides (**8** and **9**), and methyl salicylate (**10**) and 8-hydroxy geraniol (**11**) are given in Figs. 9–11. The isolation procedure for *cis*-linalool-3,7-oxide 6-*O*-β-D-apiofuranosyl-β-D-glucopyranoside (**12**) and (*Z*)-3-hexenyl β-D-glucopyranoside (**13**) is shown in Fig. 12. For these separations Polyclar AT was used to remove catechins. Spots were visualized on silica gel TLC (BuOH-AcOH-H_2O = 4 : 1 : 2) by spraying with vaniline-sulfuric acid reagent. Structures are shown in Fig. 8.

4.2 Isolation of Enzymes Concerned with Alcoholic Aroma Formation in Tea Leaves

To gain a clue to the mechanism of the alcoholic aroma formation, Guo et al. (1995) purified and characterized the enzyme/s responsible for the hydrolysis of

Oolong tea leaves (Shuixian, 2 kg)
├ extracted with boiling H_2O, 12 L × 2, 10 min
├ centrifuged at 8000 rpm, 20 min, 4°C

Supernatant | Ppt.
├ active charcoal, 380 g

Charcoal column | Aq. solution
├ H_2O-MeOH
Eluate
├ XAD-2 CC, 6 × 45 cm, H_2O-MeOH
MeOH Fr. (1.14 g)
├ Sephadex LH-20 CC, 3.5 × 75 cm, 50% MeOH
0.8~1.0 bed volume eluate (190 mg)
├ ODS CC, 3 × 48 cm, 60~100% MeOH
80% MeOH Fr. (104 mg)
├ ODS CC, 3 × 48 cm, 30~80% CH_3CN
35% CH_3CN Fr. (1.9 kg eq., 46 mg)
├ HPLC, ODS, 2 x 25 cm, 40~80% MeOH (80 min)

Linalool aroma precursor Frs.
├ repeating HPLC, 20~60% MeCN
(*S*)-Linalyl β-primeveroside (**6**, 6 mg)

Geraniol aroma precursor Frs.
├ repeating HPLC, 20~60% MeCN
Geranyl β-primeveroside (**7**, 12 mg)

Fig. 10. Isolation procedure of **6** and **7** from oolong tea leaves (cv. Shuixian)

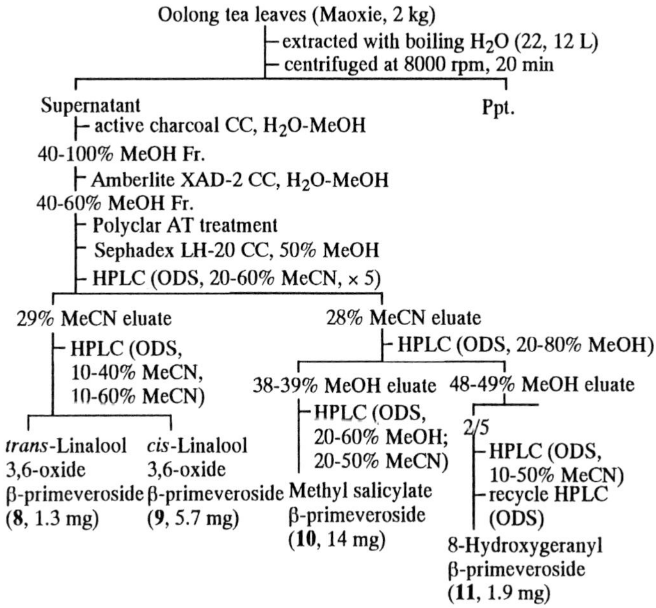

Fig. 11. Isolation procedure of **8**, **9**, **10**, and **11** from oolong tea

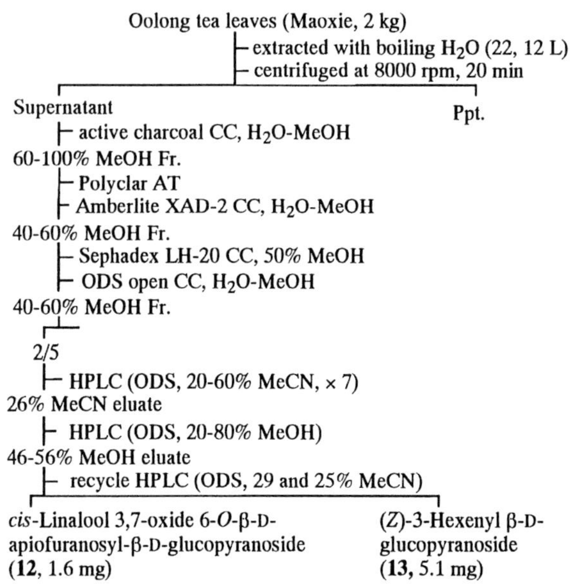

Fig. 12. Isolation procedure of **12** and **13** from oolong tea

glycosidic precursors from fresh leaves. β-Primeverosides of (S)-linalool (**6**) and geraniol (**7**) were treated with an acetone powder prepared from fresh tea leaves (cv. Yabukita), β-glucosidase (from almond), and naringinase (from *Penicillium decumbens*), respectively. The acetone powder had high activity to hydrolyze both **6** and **7**, whereas β-glucosidase and naringinase showed some activity toward **7**, but their activities toward **6** were quite low, strongly suggesting the presence of some endogenous glycosidase rather than a glucosidase.

Acetone powder was prepared from fresh tea shoots with a bud and two to three leaves. An enzymatic reaction of 0.1 g of acetone powder in 20 mM citrate buffer (pH 6) containing a crude aroma precursor (20 g fresh tea leaves equivalent) for 1 h liberated sufficient aroma precursors for GC analysis. The outline of purification procedure is given in Fig. 13. A crude enzyme sample extracted with buffer from acetone powder was fractionated by CM-Toyopearl chromatography. Each fraction was tested for hydrolysis activities toward *p*NP-β-D-galucopyranoside, *p*NP-β-D-xylopyranoside, *p*NP-β-D-galactopyranoside, and *p*NP-α-L-arabinopyranoside. Each *p*NP-glycoside [200 μl (10 mM *p*NP-Glc) or 400 μl (5 mM *p*NP-Xyl,-Gal, or -Ara)] was reacted with 30–100 μl of an enzymatic sample solution in 700 μl of 20 mM citrate buffer (pH 6) at 37 °C for 0, 3, 6, 9, 12, and 15 min. The reaction was stopped by adding 500 μl of 1-M sodium carbonate. The absorbance was measured in a 1-cm quartz cell at 405 nm. On the basis of the activity pattern, these fractions were combined into seven groups (A–G). Fractions E and F liberated alcoholic aroma from crude aroma precursor solution. These fractions showed *p*NP-Xyl and *p*NP-Glc hydrolyzing activities, respectively. Since the main activity to liberate aroma resided in fraction F, further purification was guided by activity toward *p*NP-Glc.

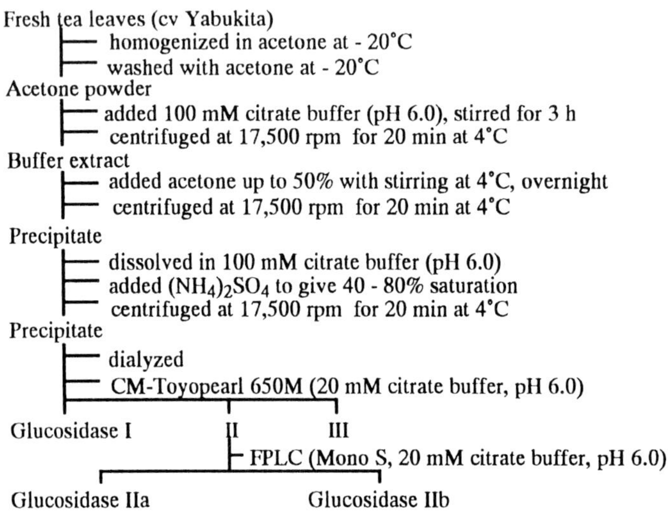

Fig. 13. Purification procedure of β-primeverosidase (glucosidase IIa) from tea leaves

The three active fractions, glucosidases (Gl) I, II, and III obtained by chromatography over CM-Toyopearl 650 M (TSK gel, 20 mM citrate buffer, pH 6, with NaCl gradient 0 to 0.5 M, 20 ml/fraction) were reacted with several β-primeverosides isolated from fresh tea leaves. Gl II had the main activity to hydrolyze β-primeverosides, and this activity was quite similar to that of the acetone powder, which showed that Gl II is the main enzyme concerned with alcoholic aroma liberation from tea leaves.

β-Primeverosidase activity of Gl II was confirmed by performing a hydrolysis experiment on authentic β-primeveroside and β-D-glucosides of benzyl alcohol and geraniol. As shown in Table 6, in each case the hydrolysis rate of geranyl β-primeveroside was higher than that of geranyl β-D-glucoside. In the case of benzyl glycosides, Gl II showed β-primeverosidase and Gl I showed β-glucosidase activity. This shows that Gl II acts as a β-primeverosidase.

To study the hydrolysis mechanism, Gl II was reacted with a mixture of β-primeverosides of geraniol, linalool, 2-phenylethanol, and benzyl alcohol. The reaction mixture was analyzed by TLC on a cellulose plate (BuOH-pyridine-H_2O-6:4:3), with aniline hydrogen phthalate as spray reagent. A spot near sucrose and maltose was observed, which was scraped off and extracted with water. The extract was analyzed by FABMS (negative) where a peak at *m/z* 311 ($[M-H]^-$) was observed, showing that the Gl II hydrolyzes the primeveroside at the sugar-aglycone linkage to yield the aroma component and primeverose. Günata et al. (1989) also proposed that in fruits as well as in tea leaves and flowers, rutinosides of terpenes were first hydrolyzed by α-rhamnosidase into glucosides which were converted to terpenoids by the enzymatic action of β-glucosidase.

Gl II from CM-Toyopearl chromatography was concentrated by ultrafiltration with Dia-flo ultrafiltration membrane (Amicon Corp.) at 4 °C. The concentrated solution was diluted with 20 mM citrate buffer (pH 6) and concentrated again. The same procedure was repeated twice to remove the salts in the sample. The sample was subjected to fast protein liquid chromatography on a Mono S HR (20 mM citrate buffer, pH 6, NaCl gradient, 0 to 0.5 M) to give two active fractions, Gl IIa and Gl IIb. FPLC of Gl IIa on Superdex 75 column (Pharmacia, 20 mM citrate buffer, pH 6) showed an active peak at about 60 kDa. SDS-PAGE analysis of Gl IIa at 30 mA for 3 h, under both reducing and nonreducing conditions, showed a single band at 61 kDa. Thus, Gl IIa was shown to be a monomeric enzyme.

Table 6. Hydrolysis rate (%) of primeverosides and glucosides by glucosidases I, II, and III

Alcoholic aroma glycoside	Glucosidase I	II	III
Geranyl-Pri	100.0	100.0	100.0
Geranyl-Glc	48.4	54.3	78.7
Benzyl-Pri	5.4	40.8	5.9
Benzyl-Glc	21.7	3.8	5.9

5 Effect of Flavoring Compounds on Stability

Although flavoring compounds bring strong flavor on the one hand, on the other hand, they cause a lower stability of the products; for example, in Earl Grey tea, aromatization with bergamot oil causes reduced stability of the products to oxidation processes due to the presence of many unsaturated terpenoids. As already noted above, there are not many analytical studies on tea infusions; in this case also sensory methods were used to evaluate the quality changes. The flavor was observed by sniffing and tasting 5–10-ml draughts. There is very little published information on flavor changes in scented tea on storage, but changes observed on storage of essential oils can be used for comparison.

A few hundred volatiles have been identified in tea and bergamot oil. Linalool and limonene are the major volatiles of bergamot oil-flavored tea. Their oxidation products affect the sensory quality of tea.

5.1 Oxidation of Bergamot Oil, Linalool, and Limonene Standards

A 100-ml portion of each substance was placed in 50-ml round-bottomed flasks, closed, and placed in a thermostat at 40 °C. Ten-ml portions were taken for GC and GC-MS analysis. Fifty-g portions of Earl Grey tea were stored in glass bottles at 60 °C to speed up the oxidation processes. The sample of bergamot tea was analyzed after 6 years storage at laboratory temperature in the original glass bottles covered by aluminum foil. Volatiles from tea were extracted with pentane, using a Likens-Nickerson's apparatus, and analyzed by GC and GC-MS.

Pure bergamo oil was stable at laboratory temparature (20 °C). The initial bergamot sample and that after 210 days of storage were analyzed. As is clear from Fig. 14, the concentration of different terpenoids such as α-pinene, β-pinene, and γ-terpinene decreased, while that of *cis*- and *trans*-linalool oxides, β-terpineol, and γ-terpineol increased. Many other low volatile compounds were produced. In addition to diols and oxides, many carbonyl compounds were also formed during oxidation of bergamot oil, linalool, and limonene standards. During the oxidation of linalool and limonene, the odor loses its fresh character. The intensity of the citrus note decreases, while that of the woody and bitter notes increases.

Scented tea is a very complicated mixture containing several hundred volatiles including odor substances present in tea leaves and those formed by oxidation and strecker reactions and also compounds from essential oils and their oxidation poducts. Hence, a complicated sensory profile, based on 28 descriptors (Table 7) was developed by Pokorny et al. (1995). As shown in Fig. 15, the odor and flavor acceptabilities were lower in a stored sample, and flavor became less balanced. The composition of flavor-active compounds was slightly different at ambient temparature, as they changed into their hydroperoxides, which were more stable. The stored sample possessed strong terpenic off-flavor and the oxidized, stored, dumpy, stale off-flavors were predominant.

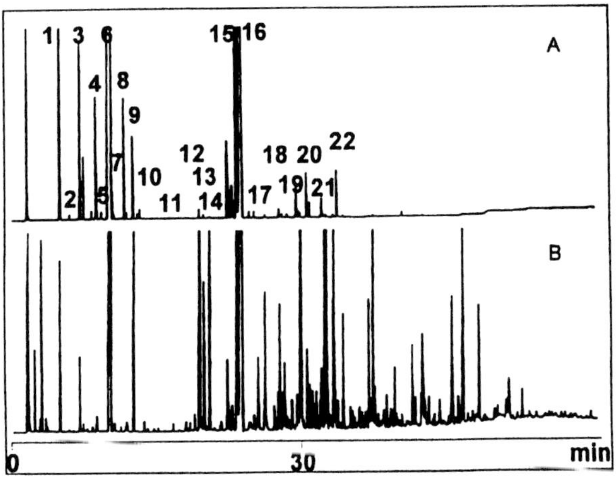

Fig. 14. Gas chromatograms of slightly oxidized (*A*) and highly oxidized (*B*) bergamot oil. Peak identification: *1* α-pinene; *2* hexanal; *3* β-pinene; *4* phelandrene; *5* heptanal; *6* limonene; *7* hexenal; *8* γ-terpinene; *9* *p*-cymene; *10* octenal; *11* nonanal; *12* *trans*-linalool oxide; *13* 2-furaldehyde; *14* *cis*-linalool oxide; *15* linalool; *16* linalyl acetate; *17* β-terpineol; *18* γ-terpineol; *19* neryl acetate; *20* geranyl acetate; *21* nerol; *22* geraniol

Table 7. List of descriptors for sensory evaluations of tea infusions

Label	Description	Label	Description
D1	Odor acceptability	D15	Terpenic flavor
D2	Odor intensity	D16	Sharp flavor
D3	Flavor acceptability	D17	Hot flavor
D4	Flavor intensity	D18	Minty flavor
D5	Rich flavor	D19	Woody flavor
D6	Full flavor	D20	Leaf-like flavor
D7	Balanced flavor	D21	Grassy flavor
D8	Bitter flavor	D22	Oxidized flavor
D9	Astringent flavor	D23	Stored flavor
D10	Biting flavor	D24	Old, damp, stale flavor
D11	Floral flavor	D25	Metallic flavor
D12	Fruity flavor	D26	Aromatic flavor
D13	Orange flavor	D27	Characteristic flavor
D14	Lemon peel flavor	D28	Foreign flavor

6 Improving the Quality of Stale Green Tea

Green tea, when stored for more than 1 year, loses its attractive greenish aroma, produces an undesirable smell, and is called stale tea. Guo et al. (1992) have successfully improved the commercial value of stale green tea by changing it to congou black tea. The method involves incubation of a dried sample of stale tea

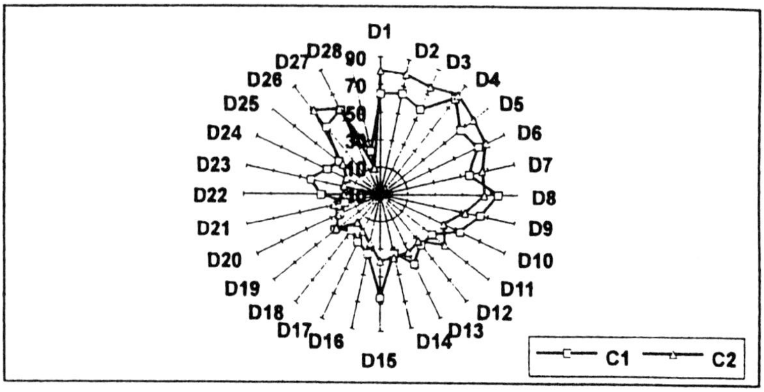

Fig. 15. Flavor sensory profiles of bergamot tea stored for 6 years at laboratory temparature (*C1*) and fresh Earl Grey/Pickwick tea (*C2*); the descriptors (*D1–D28*) are in Table 7

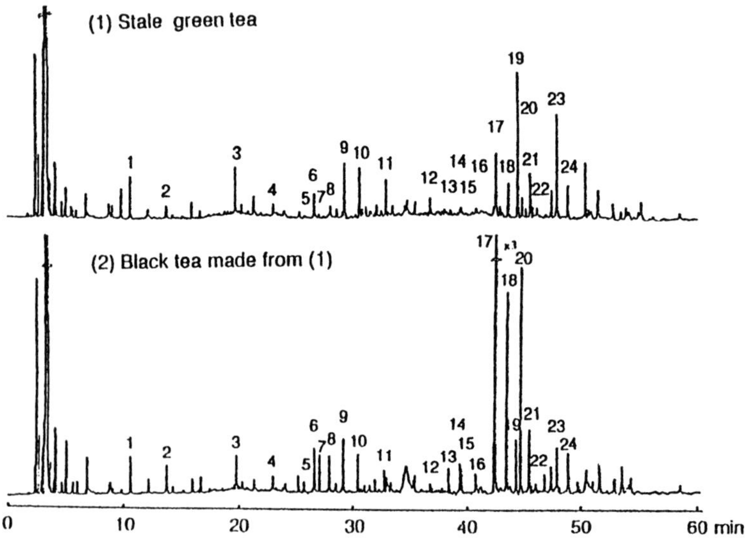

Fig. 16. Gas chromatograms of stale green tea and its black tea product. GC analysis: Hitachi 163 gas chromatograph; capillary column (PEG-20M, 25 m × 0.25 mm i.d.); temp. program: 60 °C for 10 min and then 3 °C min^{-1} to 180 °C; carrier gas, N_2 (1 ml min^{-1}). Peak identification: *1* 1-penten-3-ol; *2* (*E*)-2-hexenal; *3* (*Z*)-2-pentenol; *4* (*Z*)-3-hexenol; *5* linalool oxide I; *6* furfural; *7* linalool oxide II; *8* (*E*,*E*)-2,4-heptadienal; *9* benzaldehyde; *10* linalool; *11* 3,7-dimethyl-1,5,7-octatrien-3-ol; *12* α-terpineol; *13* linalool oxide III; *14* methyl salicylate; *15* linalool oxide IV; *16* nerol; *17* geraniol; *18* benzyl alcohol; *19* BHT; *20* 2-phenyl ethanol; *21* β-ionone; *22* *cis*-jasmone; *23* phenol + unknown; *24* nerolidol

together with new green tea by moistening with water and then mixing with homogenized fresh shoots (5% of the sample). The sample was incubated at 26 °C in a well-aerated place. After 10 h, leaves became red, and a sweet fragrant aroma of black tea appeared. The sample was then oven-dried at 110 °C. Aroma concentrates were prepared by the SDE method and analyzed by GC-MS. As shown in Fig.

16, the main constituents of black tea such as geraniol, linalool and its oxides, methyl salicylate, benzyl alcohol, 2-phenyl ethanol, and β-ionone increased markedly. Geraniol, which is very much less in green tea, became the most abundant peak in the black tea product.

7 Conclusions

In conclusion, the methods described here have helped in understanding the constituents of aroma of scented teas and the underlying mechanism of aroma release. The isolation of glycosides of alcoholic aroma compounds shows very clearly that they are present as their nonvolatile glycosides which are enzymatically transformed into aroma during scenting process. The assay designed for the study of flower fragrance formation can be used for detection and quantification of aroma precursors and the enzymes concerned, since the aroma of scented teas mainly depends upon alcoholic aroma compounds released by this mechanism.

References

Günata Z, Bitteur S, Baumes R, Brillouet JM, Tapiero C, Bayonove C, Cordonnier R (March 8, 1989) Process for obtaining aroma components and aroma from their precursors of a glycosidic nature, and aroma components and aromas thereby obtained. Eur Pat 0 332 381 A1

Guo W, Sakata K, Yagi A, Ina K, Luo SJ (1992) Preparation of congou black tea from stale green tea. Biosci Biotech Biochem 56:992–993

Guo W, Sakata K, Watanabe N, Nakajima R, Yagi A, Ina K, Luo SJ (1993) Geranyl 6-*O*-β-D-xylopyranosyl-β-D-glucopyranoside isolated as aroma precursor from tea leaves for oolong tea. Phytochemistry 33:1373–1375

Guo W, Yamauchi K, Watanabe N, Usui T, Luo SJ (1995) A primeveroside as a main glycoside concerned with alcocholic aroma formation in tea leaves. Biosci Biotech Biochem 59:962–964

Inagaki J, Watanabe N, Moon JH, Yagi K, Sakata K, Ina K, Luo SJ (1995) Glycosidic aroma precursors of 2-phenyl ethyl and benzyl alcohol from *Jasminum sambac* flowers. Biosci Biotech Biochem 59:738–739

Kinusaga H, Takeo K (1990) Deterioration mechanism for tea infusion aroma by retort pasteurization. Agric Biol Chem 54:2537–2542

Luo SJ, Guo W, Fu H (1991) Changes of aroma during process of Chinese jasmine tea. Proc Int Symp on Tea science. Aug 26–29 Shizuoka Japan, pp 57–61

Luo SJ, Guo W, Fu H (1992) Aroma of Chinese scented green tea with *Citrus aurantium* var. arama. Food Sci Hum Nutr •:347–350

Moon JH, Watanabe N, Sakata K, Inagaki J, Yagi A, AIna K, Luo SJ (1994a) Linalyl β-D-glucopyranoside and its 6′-*O*-malonate as aroma precursors from *Jasminum sambac*. Phytochemistry 36:1435–1437

Moon JH, Watanabe N, Sakata K, Yagi A, Ina K, Luo SJ (1994b) *trans* and *cis* Linalool-3,6-oxide 6-*O*-β-D-xylopyranosyl-β-D-glucopyranosides isolated as aroma precursors from tea leaves for oolong tea. Biosci Biotech Biochem 58:1742–1744

Pokorny J, Pudil F, Ulmannova K, Ficova E (1995) Effect of oxidation products of scented tea aroma compounds on flavor of tea infusion. Dev Food Sci 37A:815–825

Shimoda M, Shigematsu H, Shiratsuchi H, Osajima Y (1995) Comparison of odor concentrates by SDE and adsorptive column method from green tea infusion. J Agric Food Chem 43:1616–1620

Takeoka GR, Flath RA, Buttery RG, Winterhalter P, Gunter M, Ramming DW, Teranishi R (1992) Free and bound flavor constituents of white fleshed nectarines. In: Flavor precursors: thermal and enzymatic conversions. ACS Symp Ser 490:116–138

Watanabe N, Watanabe S, Nakajima R, Moon JH, Shimokihara K, Inagaki J, Etoh H, Asai T, Sakata K, Ina K (1993) Formation of flower fragrance compounds from their precursors by enzymatic action during flower opening. Biosci Biotech Biochem 57:1101–1106

Yamanishi T, Luo SJ, Guo W, Kobayashi A (1988) Aroma of Chinese scented green tea. Frontiers of flavor. Proc 5th Int Flavor Conf, 1–3 July 1987, Porto Karras, Chalkidiki, Greece, pp 181–190

Subject Index

Printing: Saladruck, Berlin
Binding: Buchbinderei Lüderitz & Bauer, Berlin